Chemistry and Biology of Ellagitannins

An Underestimated Class of Bioactive Plant Polyphenols

Chemistry and Biology of Ellagitannins

An Underestimated Class of Bioactive Plant Polyphenols

Editor

Stéphane Quideau

University of Bordeaux, France

World Scientific

NEW JERSEY • LONDON • SINGAPORE • BEIJING • SHANGHAI • HONG KONG • TAIPEI • CHENNAI

Published by

World Scientific Publishing Co. Pte. Ltd.
5 Toh Tuck Link, Singapore 596224
USA office: 27 Warren Street, Suite 401-402, Hackensack, NJ 07601
UK office: 57 Shelton Street, Covent Garden, London WC2H 9HE

British Library Cataloguing-in-Publication Data
A catalogue record for this book is available from the British Library.

CHEMISTRY AND BIOLOGY OF ELLAGITANNINS
An Underestimated Class of Bioactive Plant Polyphenols

ISBN-13 978-981-279-740-7
ISBN-10 981-279-740-8

Printed by Fulsland Offset Printing (S) Pte Ltd, Singapore

Preface

Emil Fischer, Karl Freudenberg and Paul Karrer were three iconic figures in the development of organic chemistry in the first half of the twentieth century; they were also pre-eminent amongst those attracted toward the study of vegetable tannins – extracts of plant materials whose use in the conversion of raw animal skins into leather was centuries old. Notwithstanding such efforts it was still possible for the chairman of a symposium dedicated to the subject of vegetable tannins held in Cambridge in 1956 to begin by stating "*After the early encouragement of Emil Fischer's outstanding contribution to the tannin problem, chemists realised that the problem was still one of great complexity, and tannin chemistry had gradually become one of the disorderly and untidy corners of organic chemistry*." It was generally recognised [Freudenberg (1920). Die Chemie der Natürliche Gerbstoffe, Springer: Berlin] that there was a broad division into condensed or non-hydrolysable and hydrolysable tannins but much else remained vague and uncertain. By contrast the progress made in our understanding of these previously intractable children of nature in the following fifty years has been remarkable. As is often the case in science much of this has been a result of the discovery of new techniques and methods – principally chromatography in its various guises and a panoply of new spectroscopic tools in structure determination. As is also often the case the contributions of particular individuals is frequently critical in charting the way forward. In the case of ellagitannins the efforts of two men in the 1950s were crucial in determining the nature and scope of the subject and its future development. They were the founding fathers of the subject as we know it today.

After the disruption following the second world war and at the age of fifty three, Otto Theodor Schmidt began what was to be the most productive and successful phase of his distinguished career namely the study of the structures of the ellagitannins. Where others before failed he isolated and studied ellagitannins from traditional sources such as Myrobalans and Algorabilla. It was a classical piece of work and inevitably employed classical techniques (often now long forgotten) such as crystallisation and chemical degradation! By today's standards the number of his publications is very modest but each has depth, the stamp of intellectual rigour and infinite attention to detail and each, without fail, has stood that most important test – the test of time. His work, along with that of his colleague Walter Mayer brought to light for the very first time some of the many characteristic structural groups found in ellagitannins – hexahydroxydiphenic acid, dehydrodigallic acid, dehydrohexahydroxydiphenic acid, valoneic acid, chebulic acid, brevifolin carboxylic acid – all derived from the fundamental building block gallic acid. It is one of the truisms of good scientific practice that one should always set aside time to think and as the work progressed Schmidt and Mayer began to elaborate an eminently satisfying framework and rationale, involving dehydrogenation and hydrolysis, for the biogenesis of the ellagitannins from their presumed galloyl ester precursors. This pioneering work of the Heidelberg school has securely underpinned all the subsequent chemical and biogenetic work in this field over the past half-century.

If Schmidt was a consummate practitioner of organic chemistry then E.C. Bate-Smith was an extraordinary talented polymath whose interests ranged from animal physiology to taste and astringency, to plant systematics and taxonomy. His investigations of astringency in foodstuffs led him to vegetable tannins and to their taxonomic distribution in the plant kingdom. After retirement this became his hobby and his passion, ('Bate-Smithery' as it was occasionally referred to). It constituted the first really serious excursion into the botanical aspects of vegetable tannins and demonstrated unequivocally their presence as normal phenolic metabolites in a much wider range of plants than had hitherto been appreciated. As such it was seminal and its results far-reaching. It opened up the field and has been thoroughly exploited in subsequent research not only in the search for new tannins but also from the point of view of plant systematics. In the specific case of ellagitannins Bate-Smith suggested that they were to be found in some 75 families of dicotyledonous plants. His methods were deceptively simple, those of the true amateur.

Thus Schmidt and Bate-Smith, in their differing ways, not only laid secure and lasting foundations for many of the developments in the chemistry and biochemistry of ellagitannins that have followed, but they also signposted many of the avenues which were now open to future research. Some measure of the explosion of activity in this field which ensued in the second half of the twentieth century may be gained by comparing Schmidt and Mayer's extensive review [Natürliche Gerbstoffe, *Angewandte Chemie*, (1956), 68, pp. 103–115] and Mayer's appreciation of Schmidt's work and career [*Liebig's Annalen der Chemie*, (1973), pp. 1759–1776] with the 1996 review of Ellagitannin Chemistry by Quideau and Feldman [*Chemical Reviews*, (1996), 96, pp. 475–503]. The present multi-author text now brings the subject right up to date and into the twenty-first century; its contents and the story of achievement it tells would have brought a smile of satisfaction to both Schmidt and Bate-Smith.

Almost inevitably the reader's attention is taken by the enormous range and number of ellagitannin metabolites which have now been isolated from plants – from simple 'monomeric' hexahydroxydiphenoyl esters, dehydroellagitannins, *C*-glycosidic ellagitannins, 'oligomeric' ellagitannins, to macrocyclic 'oligomeric' ellagitannins……The list seems endless and, in the context of the position fifty years earlier, it represents a monumental achievement, accomplished principally by researchers in Japan – Okuda, Yoshida, Nishioka, and Tanaka. Okuda and Yoshida and their colleagues discuss all of these developments in two chapters and bring order into what might have been chaos by analyzing and classifying the many structural themes within the ellagitannins class of phenolic metabolites. Gratifyingly they also show how each of the structural groups and sub-groups may be derived using the guidelines set out by Schmidt and Mayer in their biogenetic hypothesis, based upon the dehydrogenation of galloyl ester substrates, first elaborated in 1956. Similarly they also comment on the position of ellagitannins in relation to plant systematics – a topic very dear to the heart of Bate-Smith. Okuda indeed makes a very interesting suggestion linking the progressive oxidation of 'monomeric' ellagitannins to Cronquist's system of plant evolution. These ideas may well repay further attention.

First decribed by Mayer in the late 1960s, the *C*-glycosidic ellagitannins represent an intriguing and possibly unique, highly condensed sub-class. They are characterized by the C–C linkage between the C-1 of the open chain form of D-glucose and a carbon of a phenolic nucleus, derived originally from gallic acid. Jourdes, Lefeuvre and Quideau, in a fascinating and wide ranging account of these metabolites, discuss not only their possible biogenetic origins but give an in-depth review of their chemical reactivity. In particular they focus on the striking differences found between the diastereoisomers vescalagin and castalagin. Finally, and in this context, they reflect on the role of compounds such as vescalagin (found in oak) may have on the taste and colour of wines which are aged in oak barrels.

Beautifully complementing these descriptions of the chemistry and taxonomic distribution of ellagitannin metabolites are two chapters dealing with their biosynthesis and strategies aimed at their total chemical synthesis. In the first of these Gross charts the progress made, principally if not exclusively by his own group, towards an understanding of the mechanisms whereby first gallotannins and then ellagitannins are formed from gallic acid and D-glucose. Of particular interest is the recent discovery of an enzyme from *Tellima grandiflora* which is able to convert β-1,2,3,4,6-penta-*O*-galloyl-D-glucose into tellimagrandin II. In doing so the enzyme (probably of the laccase class) oxidatively couples the 4 and 6 galloyl ester groups in the substrate to the (*S*)-hexahydroxydiphenoyl group bridging these same two positions. It is an enormously encouraging first step towards the elucidation of further stages in this pathway. Parenthetically Professor Gross has thus provided welcome affirmation of the early speculations by Otto Schmidt, Mayer and others concerning the biogenesis of these metabolites.

Few would contradict the proposition that the direct chemical synthesis of ellagitannins presents tremendous challenges. Indeed there were no reports of the fruitful outcome of such attempts until the 1990s and in their 1996 review Quideau and Feldman give an elegant presentation of the possible strategies which may be adopted. These are further developed by Khanbabaee in a chapter on the synthesis of ellagitannins. Feldman was the first to successfully carry out biomimetic syntheses of ellagitannins (*e.g.*, tellimagrandin I and II, pedunculagin) by oxidative coupling of suitably protected preformed galloyl ester precursors and Khanbabaee develops in his essay the possibilities of a second synthetic strategy using as its key step the stereoselective esterification of a diol substrate by hexabenzyloxydiphenic acid (a compound first described by Schmidt in the 1950s). The example given of the synthesis of strictinin is a persuasive demonstration of this alternative rationale.

Plants have been widely used since antiquity as folk medicines particularly in Asia. The aerial parts of *Geranium thunbergii* are listed in Japan in official pharmacopoeias as an antidiarrhetic. Fruits of *Cornus officinalis* are likewise used as a tonic in traditional Chinese medicine. It has been suggested that the ellagitannins are the principal 'active' components of many of these medicines. Polyphenols such as these have attracted considerable interest in the past fifteen or so years because of possible benefits in human health care and the prevention of diseases such as carcinogenesis and arteriosclerosis. In this context anti-oxidant activity and radical scavenging, antiviral, antimicrobial, anti-inflammatory, and antitumour and anticancer properties have all been variously attributed to specific ellagitannins. These themes occur throughout the text and are also dealt with more specifically and from differing perspectives in chapters by Feldman, Törrönen, Tomás-Barberan and Tanaka. The weight of experimental evidence now available is impressive. For example several ellagitannins exhibit inhibitory action against anticancer targets such as DNA topoisomerases. Others display selective cytotoxicity against human solid tumour lines. The question whether some ellagitannins are not cytotoxic but rather act to enhance the human immune defences is taken up and developed by Feldman in his chapter. Antiviral and antimicrobial activities have also been widely reported and in this area the ability of various ellagitannins to inhibit replication of the *Herpes simplex* virus has attracted particular attention. Since some fruits and fruit juices contain significant amounts of ellagitannins the question of their effects as part of the diet on human health and their metabolism are also considered by Törrönen and Tomás-Barberan. Both address the vexed but critical question of the bioavailability of ellagitannins which is central to the whole basis of their possible therapeutic action. Tomás-Barberan diplomatically concludes that there are various unresolved issues within this process in humans that still need further research. Tanaka in his chapter makes some beautifully simple but critical observations on the solvation/desolvation processes which must underlie this whole phenomenon and that of complexation with other molecular species. Many questions remain. Thus the *C*-glycosidic ellagitannin vescalagin is formally derived from β-1,2,3,4,6-penta-*O*-galloyl-D-glucose by the loss of six hydrogen atoms. Yet it is crystalline and highly water soluble, quite different to its presumed biogenetic precursor. It is a subject which would reward further study. Notwithstanding the final outcome of such studies the reader should certainly follow Törrönen's advice that the consumption of the delicious ellagitannin-rich foods and beverages can be recommended!

The seeming cornucopia of possible medicinal/pharmacological properties associated with ellagitannins has now come right to the fore. It is a complex area and, for the uninitiated, the amount of information can be overwhelming. Perhaps now is the moment for the key underlying issues, such as bioavailability, to be addressed so as to develop a framework of ideas for the future, for as the poet T.S. Eliot wrote:

"*Where is the wisdom we have lost in knowledge*?
Where is the knowledge we have lost in information?"

Finally it is timely to return to the enigma of gallic acid itself, one to which Bate-Smith repeatedly drew attention. Whilst the occurrence of other hydroxybenzoic acids in the plant kingdom is at best sporadic and idiosyncratic, in certain plant families substantial amounts of polygalloyl and hexahydroxydiphenoyl esters are metabolized. The weight of experimental evidence favours the direct dehydrogenation of 3-dehydroshikimic acid – an intermediate in the shikimate pathway of aromatic amino-acid metabolism – as the route of biosynthesis to gallic acid. As such it is potentially unique for each phenolic group would then derive directly from the aliphatic oxygen functionalities of the substrate. Thereafter the ellagitannins are, *vide supra*, formed by further dehydrogenation of galloyl ester derivatives, usually of D-glucose. ***Dehydrogenation seems to be the key reaction throughout***. What is the rationale for this wholesale diversion of intermediate in the shikimate pathway by such means? It is an intriguing question waiting to be answered by investigations at an enzymic/genetic level, for as Eliot also remarked "*What we call the beginning is often the end.*"

E. Haslam, March 2008

Contents

Chapter 1

Ellagitannins Renewed the Concept of Tannins

Takuo Okuda,*[a] Takashi Yoshida,[b] Tsutomu Hatano[c] and Hideyuki Ito[c]

*[a]Emeritus Professor, Okayama University, Okayama 700-8530, Japan; [b]College of Pharmaceutical Sciences, Matsuyama University, Bunkyo-cho, Matsuyama, Ehime 790-8578, Japan; [c]Department of Pharmacognosy, Okayama University Graduate School of Medicine, Dentistry and Pharmaceutical Sciences, Tsushima, Okayama 700-8530, Japan; *corresponding author:* okudatak@white.megaegg.ne.jp

1.1 Old and New Concepts of Tannins

The pharmacological activities of tannins described in medicinal books before the recent achievements on ellagitannin chemistry were mostly those of gallotannins and condensed tannins of poor chemical uniformity. The gallotannins extracted from Chinese or Turkish gall, sometimes called tannic acid, are variable mixtures of polygallates of carbohydrates. They cause irritation on skin and mucous membranes, although they have been utilized in some traditional medicinal applications, and are technically defined on the basis of their general capacity to bind to proteins and nitrogen basic compounds such as alkaloids. The condensed tannins, mixtures of oligomeric and polymeric flavanols (*e.g.*, catechins), are chemically more unstable and heterogeneous, thus conforming to the old concept of tannins. They were mainly used for leathering and staining, although some plants containing them have been used as traditional medicines. Phlorotannins are highly unstable oligomers of phloroglucinol (*i.e.*, 1,3,5-trihydroxybenzene) produced by algae that have never been isolated without being first converted into their methyl or acetyl derivatives, and as such they constitute a third but rather peculiar group of tannins.

As for ellagitannins, although some members of this class of hydrolyzable tannins were obtained early on, it is the isolation and structural determination of over 500 pure compounds since 1975 from various plants, many of which used in traditional medicines, that brought remarkable changes in the definition and concept of "tannins" (Haslam, 1989, Okuda, 1995, 1999a, 2005, Okuda *et al.*, 1990, 1991, 1992a, 1993a, 1995, 2000, Quideau and Feldman, 1996).

1.1.1 About the chemical stability of ellagitannins

Geraniin, from *Geranium thunbergii*, which is one of the most popular medicinal plants used in Japan, was isolated as crystals, thus allowing its precise chemical analyses (Okuda *et al.*, 1982a) and X-ray crystallography (Luger *et al.*, 1998). The purified crystalline geraniin was surprisingly found to give almost no astringent taste, while its capacity to bind to hemoglobin and basic compounds, as evaluated by the relative astringency (RA) and relative affinity to methylene blue (RMB) index values (see Section 1.6.2), are comparable to those of other main tannins (Okuda *et al.*, 1985). The biological and pharmacological activities, successively found for geraniin and other purified ellagitannins, were remarkably different from those of the "tannins" vaguely imagined in the past. The biogenetic sequences of these newly found tannins allowed propositions about the chemical and biological correlations among hydrolyzable tannins produced in nature. The old concept of "tannins", which merely meant mixtures of hardly identifiable and unstable phenolics, has now been replaced by a new concept through which tannins, particularly ellagitannins, can be considered in a way similar to that of other types of natural organic products, such as terpenoids and alkaloids. Unlike the "tannins" of the old concept, these isolated ellagitannins generally remain intact when in contact with air. The various biological and pharmacological properties of these tannins can be determined for each individual compound.

Geraniin

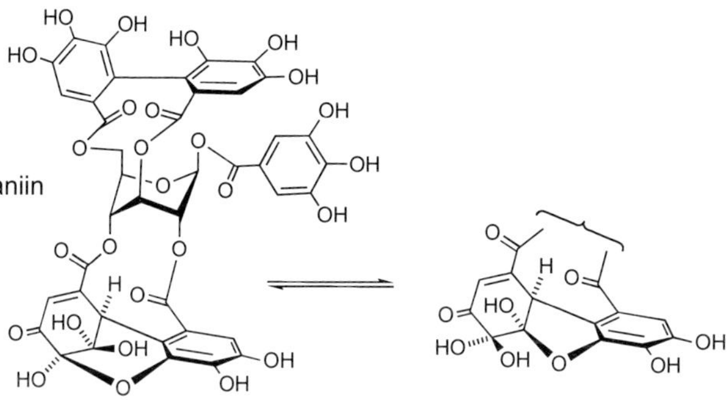

1.1.2 Definition of ellagitannins in the narrow and wider senses

Geraniin (Okuda *et al.*, 1976, 1982a) can be regarded as a keystone in the ellagitannin biooxidation process, since it is structurally classified as a dehydroellagitannin that is located at a junction in the biogenesis of the whole ellagitannin family. Geraniin has also been found in several oligomeric molecules as a composing monomer. Rapid developments were made in the field of ellagitannin chemistry after the discovery of geraniin, leading to the isolation of others dehydroellagitannins, as well as products of further biogenetic oxidation, and also oligomers, up to pentamers (Yoshida *et al.*, 1999, 2005).

While several ellagitannins had been isolated from the fruits of *Terminalia chebula* (*i.e.*, myrobalans) since 1947 (Schmidt and Mayer, 1956), and also from *Castanea* and *Quercus* species of the *Fagaceae* family (Mayer, 1971), new techniques of isolation, spectroscopy and biological screening (Okuda *et al.*, 1989a) enabled in more recent years rapid developments, which notably helped with the partial revision of some of their structures (Okuda *et al.*, 1980a, Yoshida *et al.*, 1980). Furthermore, these in-depth investigations also served to gather important insights on plant genealogy, for the structural diversity thus unveiled about ellagitannins was found to correlate to the evolution and classification of the plants that contain them (Okuda *et al.*, 2000).

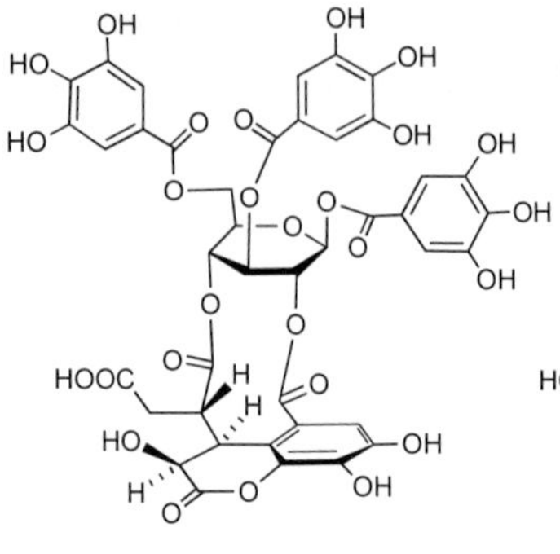

Chebulinic acid

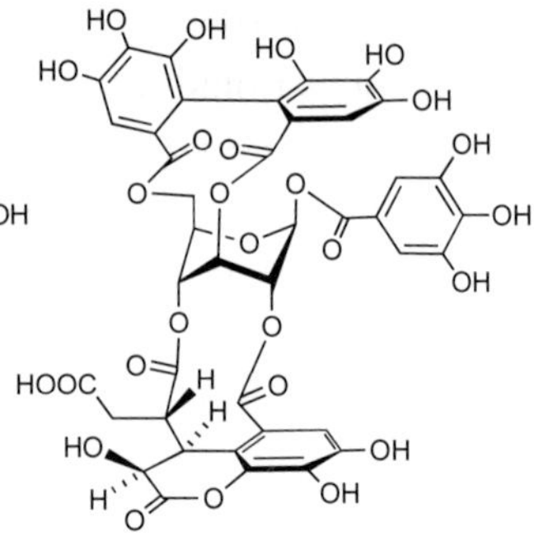

Chebulagic acid

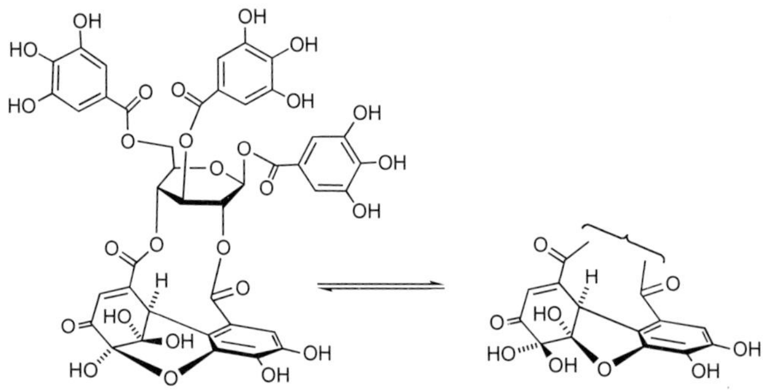

Terchebin

Ellagitannins can be defined in a narrow sense as hexahydroxydiphenoyl esters of carbohydrates or cyclitols, while the definition of ellagitannins in a wider sense also cover compounds derived from further oxidative transformations, including oligomerization processes (Okuda *et al.*, 1995). It is this latter and wider definition that will be taken into account throughout this book.

1.1.3 Stereochemistry of ellagitannins – Absolute configuration of HHDP, DHHDP and chebuloyl group

Chemical and circular dichroism (CD) spectral studies have shown that the absolute configuration of the atropisomeric biaryl HHDP groups at the O-2~O-3 and O-4~O-6 positions of the D-glucopyranose core of most ellagitannins is *S*, such as in the molecule of pedunculagin, whereas the configuration of the HHDP group at the O-3~O-6 positions, such as in geraniin, is *R*. The absolute configuration at the methine carbon of the DHHDP group at the O-2~O-4 positions in geraniin, terchebin and mallotusinic acid (Okuda and Seno, 1981) is *R*, whereas this configuration is *S* in isoterchebin (see structure **27** in Chapter 2) (Okuda *et al.*, 1982e/f). The methine carbon of the chebuloyl group in the molecules of chebulinic acid and chebulagic acid, which are biogenetically derived from geraniin, retains the stereochemical features of the DHHDP group at the O-2~O-4 positions of geraniin (Yoshida *et al.*, 1980, 1982).

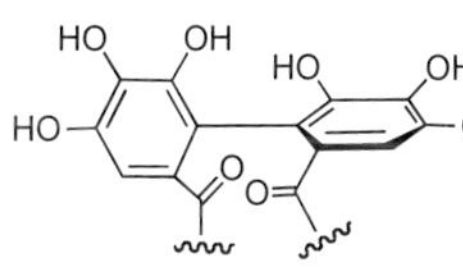

***R*-HHDP**

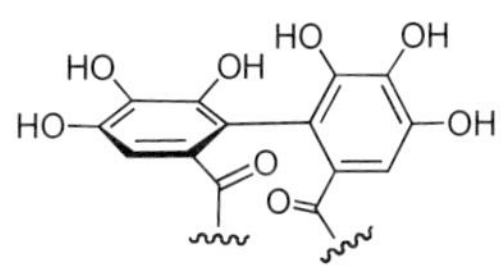

***S*-HHDP**

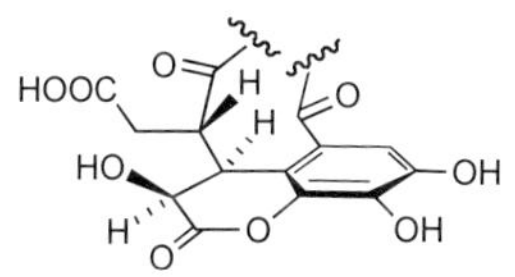

Chebuloyl

1.1.4 Condensation of dehydroellagitannins with other substances

Tannins are, in general, capable of interacting with co-existing substances, and are often bound to basic compounds, proteins and other high molecular mass compounds, as well as metallic ions. Besides the binding activities indexed by the aforementioned RA and RMB values, dehydroellagitannins also express structure-specific reactivity in condensation reactions with certain co-existing substances under mild conditions.

For example, a condensation product derived from geraniin and ascorbic acid, ascorgeraniin (or elaeocarpusin) was isolated from *Geranium thunbergii* and also from *Acer nikoense* and *Elaeocarpus sylvestris*. This compound also co-exists with geraniin in some other *Acer*, *Rhus* and *Cercidiphyllum* species. It has been prepared by condensation of geraniin with ascorbic acid in a moderately acidic aqueous or a methanolic aqueous solution at room temperature, thus demonstrating that it could be produced in the plant without any enzyme intervention (Okuda *et al.*, 1986a/b). An analog, putranjivain A, was isolated from several euphorbiaceous plants (Lin *et al.*, 1990).

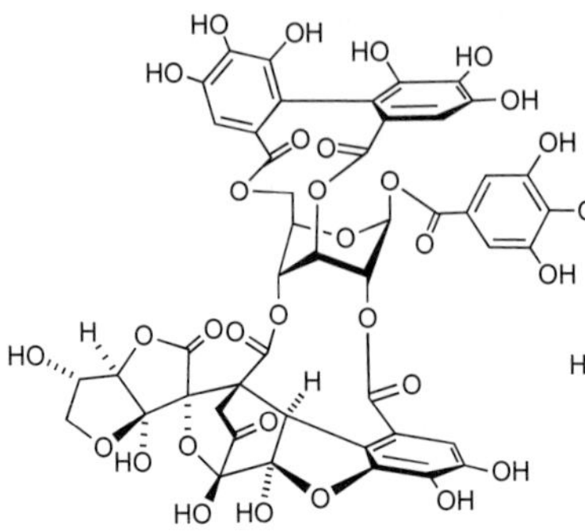

Ascorgeraniin (Elaeocarpusin)

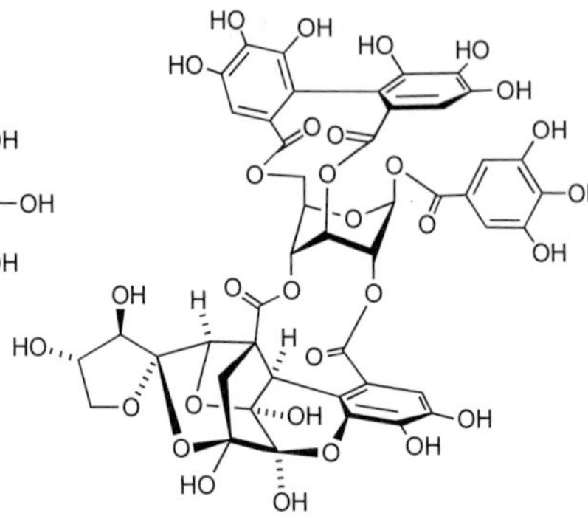

Putranjivain A

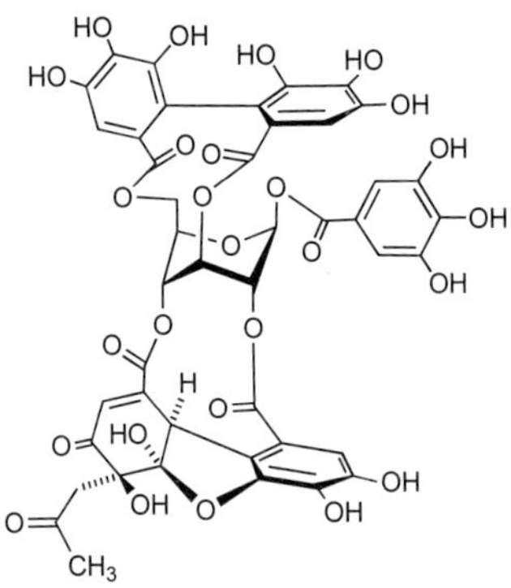

Phyllanthusiin D

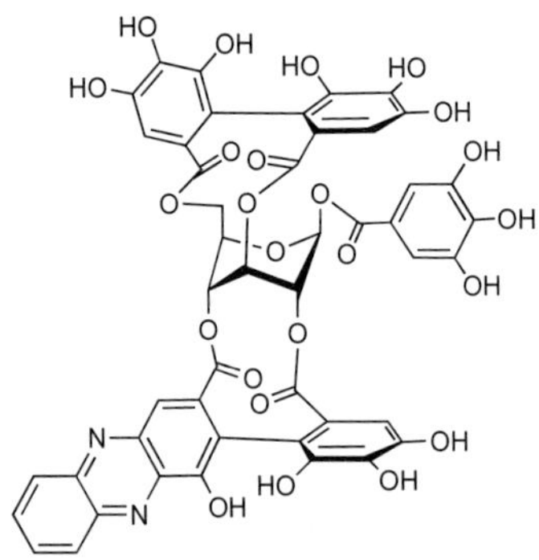

A phenazine derivative of geraniin

Phyllanthusiin D, a condensation product of geraniin with acetone, was isolated from acetone and aqueous acetone homogenates of *Phyllanthus flexuosus*, *Phyllanthus amarus*, and also from suspension cultures of *Geranium thunbergii* (Yazaki *et al.*, 1991). In this case, it is likely that phyllanthusiin D is simply an artefact formed during the extraction procedure, since it was produced when geraniin was refluxed in dry acetone containing a small amount of trifluoroacetic acid. One can here also mention the condensation reaction between geraniin and *ortho*-phenylenediamine in weakly acidic media that yields a phenazine derivative, a reaction commonly used to determine the presence of a DHHDP group in an ellagitannin molecule.

1.1.5 Accumulation of an ellagitannin of specific structure in a plant

Often a monomeric or an oligomeric ellagitannin is the main component of a plant species, and the pharmacological activity of that plant is sometimes attributable essentially to that component. Geraniin is the main component of *Geranium thunbergii* (*Geraniaceae*), making up over 10% by weight of the dry leaf. It is also the main component in other *Geranium* species (Okuda *et al.*, 1980b), usually accompanied by small amounts of analogs such as dehydrogeraniin, furosinin (Okuda *et al.*, 1982d), ascorgeraniin (Okuda *et al.*, 1986a) and geraniinic acids B and C (Ito *et al.*, 1999a).

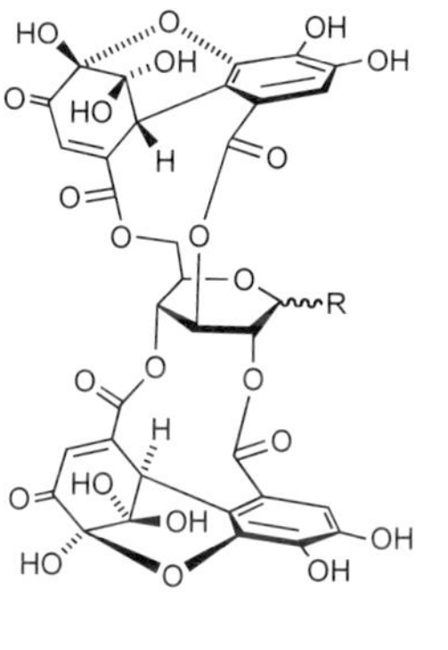

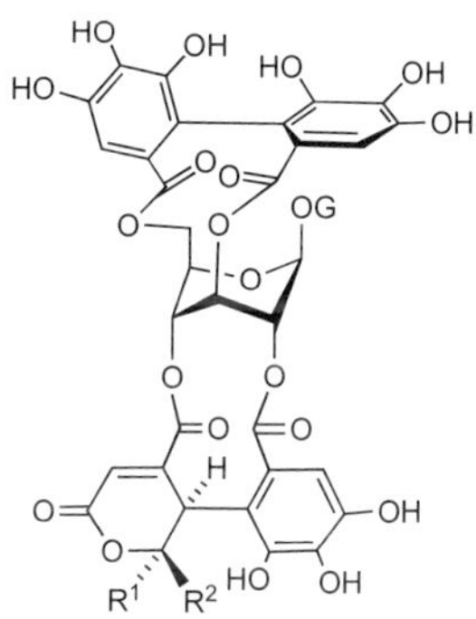

Dehydrogeraniin : R = (β)-OG
Furosinin : R = OH

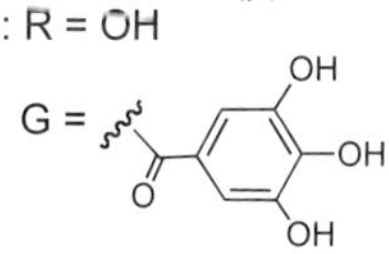

Geraniinic acid B : R^1 = COOH, R^2 = H
Geraniinic acid C : R^1 = H, R^2 = COOH

Dimeric agrimoniin, oenothein B (and its trimeric variant, oenothein A, see also Section 1.3.5) and coriariin A are also the main components in *Agrimonia pilosa* (Okuda *et al.*, 1982b), *Oenothera erythrosepala* (Hatano *et al.*, 1990a), and *Coriaria japonica* (Hatano *et al.*, 1986), respectively, and are usually accompanied by smaller amounts of the monomers composing these dimers and higher oligomers.

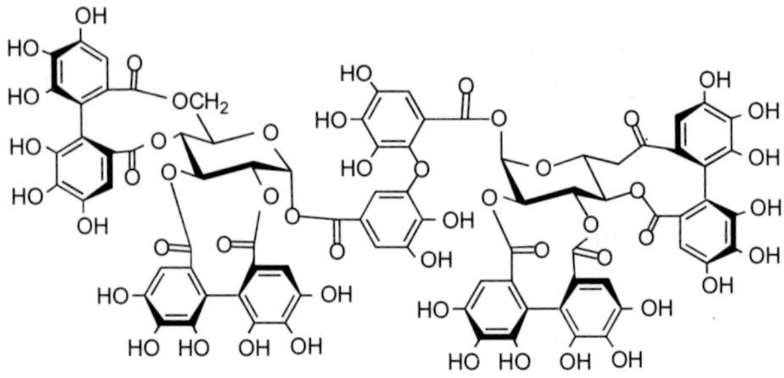

Agrimoniin

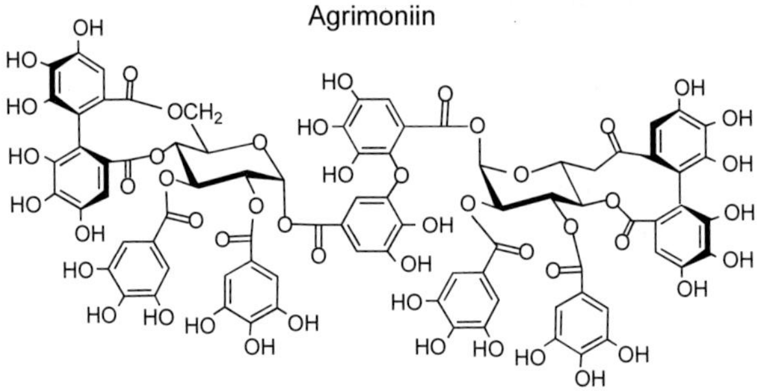

Coriariin A

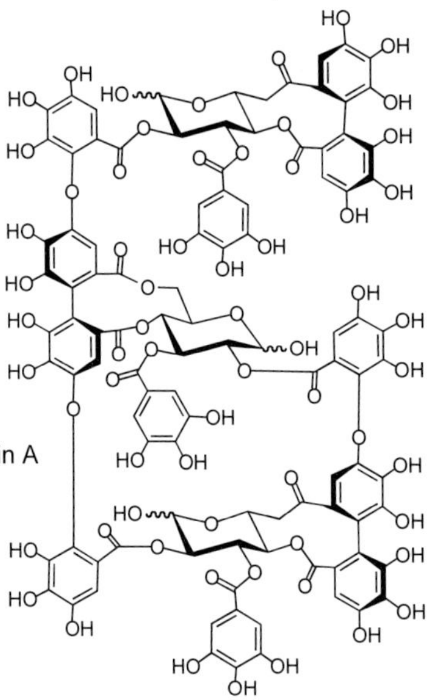

Oenothein A

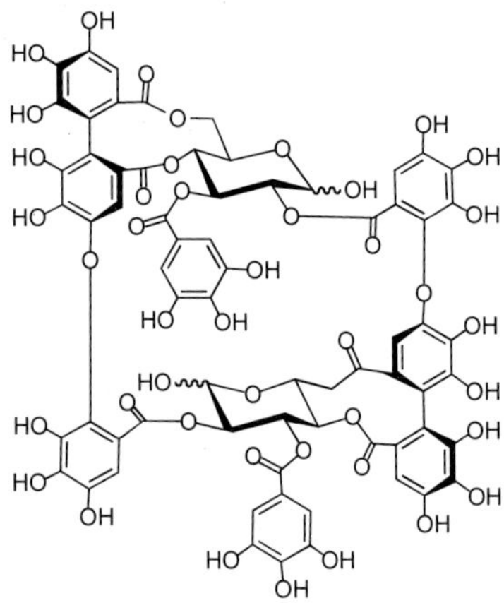

Oenothein B

1.2 Distribution of Ellagitannins in the Plant Kingdom

Ellagitannins of various structures, which are derived from biosynthetic stepwise oxidation of gallotannins (Okuda *et al.*, 2000) and subsequent oligomerization processes (Okuda *et al.*, 1993a), are generated by plant species of the *Dicotyledoneae* in the *Angiospermae*, mostly by plant species of the *Choripetalae*. This distribution in plants is similar to that of gallotannins, but ellagitannins are by far much richer in structural diversity and, unlike gallotannins, are isolable as pure and stable compounds. Ellagitannins are absent in most orders of *Sympetalae*, which rank higher in the *Dicotyledoneae* plant evolution system. This rather limited distribution of ellagitannins contrasts with the wider distribution observed for condensed tannins and caffetanins (caffeic acid esters), which are also found in *Monocotyledoneae* plant species in the *Gymnospermae*. It is interesting to note that ellagitannins are often the tannins identified as active principles in medicinal plants (Okuda *et al.*, 1989b), and that the condensed tannins expressing biological activities are often those that are galloylated, thus featuring structural motifs analogous to those of hydrolyzable tannins, such as in the active condensed tannins found in rhubarb, *Polygonum multiflorum*, *Saxifraga*, *stolonifera*, and *Diospyros kaki* (Okuda, 1999a).

1.3 Formation and Classification of Ellagitannins in Plants

Ellagitannins can be classified according to their biogenetic oxidation stages (Okuda *et al.*, 2000).

1.3.1 Oxidative biological transformations from gallotannins to ellagitannins and dehydroellagitannins

The characteristic unit of all ellagitannins, the hexahydroxydiphenoyl (HHDP) group, is the product of the first-stage biogenetic oxidation of galloyl groups. Linking one or two additional galloyl group(s) to the HHDP unit via C–O or C–C bond formation gives rise to several variations of the HHDP group, such as those shown below.

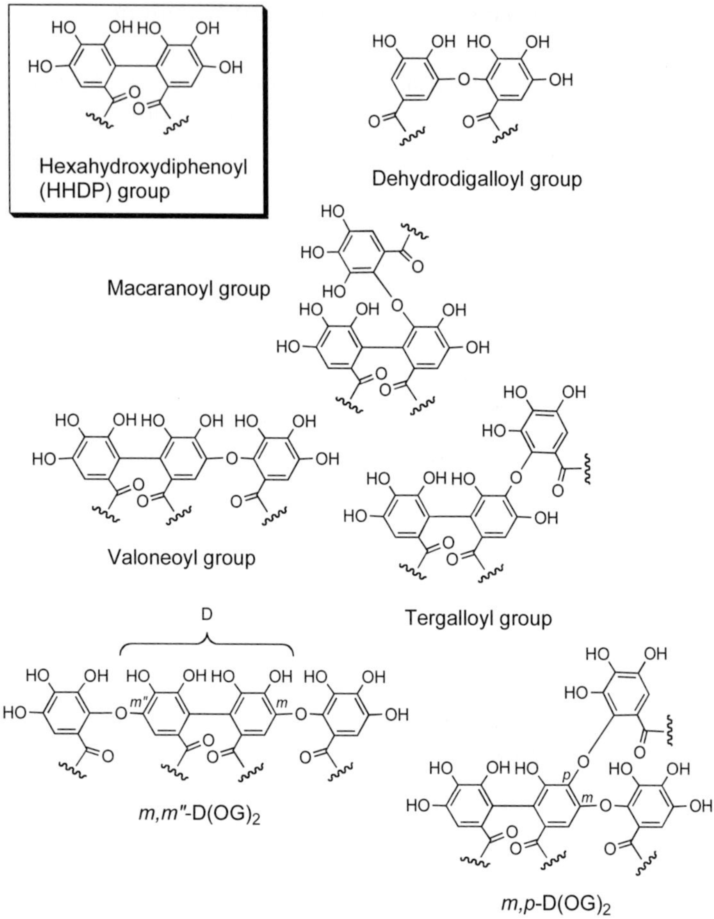

The HHDP group produced in the primary class of ellagitannins can then be oxidized to dehydrohexahydroxydiphenoyl (DHHDP) group (Fig. 1.1). The compounds that bear this DHHDP group are referred to as "dehydroellagitannins" and exemplified by *inter alia* geraniin, terchebin and furosinin. Among the special chemical reactivity features of the DHHDP group, a cyclohexenetrione linked to a pyrogallol, are (1) the aforementioned facile condensation with other compounds such as ascorbic acid and *ortho*-phenylenediamine (*vide supra*) that furnish ascorgeraniin and phenazine derivatives (Fig. 1.2) and (2) the equilibrium in aqueous or alcoholic solutions between five- and six-membered hemiacetal or acetal rings. Geraniin, in its crystalline form, adopts the six-membered hemiacetal ring form, but equilibrates back into a mixture of both cyclic hemiacetals or acetals when dissolved in an aqueous or an alcoholic solution. This behavior is reminiscent of that observed for D-fructose, which adopts a cyclic pyranose structure in the crystalline state, but equilibrates between fructopyranose and fructofuranose in aqueous solutions (Okuda *et al.*, 1982a).

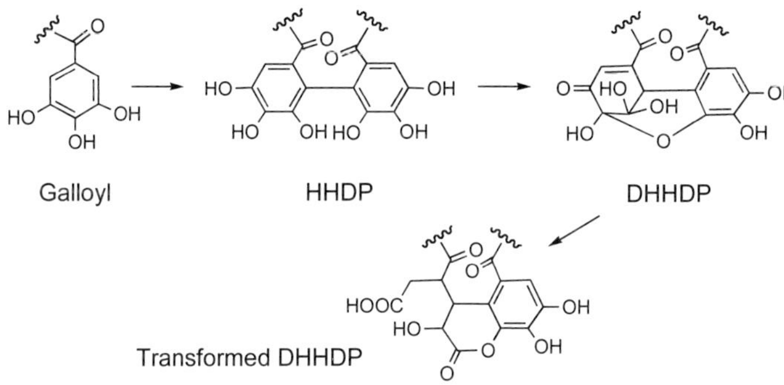

Fig. 1.1 Oxidative transformations from the galloyl and HHDP groups.

Further oxidative transformations of the DHHDP group yield several other subclasses of ellagitannins, some members of which are shown in the following part of this chapter and in Chapter 2.

1.3.2 Regiospecificity of the HHDP group on the glucose core, and its correlation to plant families

The positioning of the HHDP group or its oxidized variants on the glucose core is generally the same in ellagitannins produced by plant species of the same family. Thus, one type of ellagitannins bears the HHDP group at their glucose O-2~O-4 and/or O-3~O-6 positions, and another type bears it at their O-2~O-3 and/or O-4~O-6 positions. Ellagitannins of the former type such as geraniin, corilagin and granatin B are produced by plants of the *Geraniaceae*, *Combretaceae* and *Punicaceae* families, as well as in most species of euphorbiaceous plants. The latter type of ellagitannins exemplified by pedunculagin and casuarictin are found in plants of other families, *e.g.*, *Betulaceae*, *Coriariaceae*, *Cornaceae*, *Fagaceae*, *Hamamelidaceae*, *Lecythidiaceae*, *Lythraceae*, *Melastomataceae*, *Myrtaceae*, *Nyssaceae*, *Onagraceae*, *Rosaceae*, *Theaceae* and *Trapaceae*.

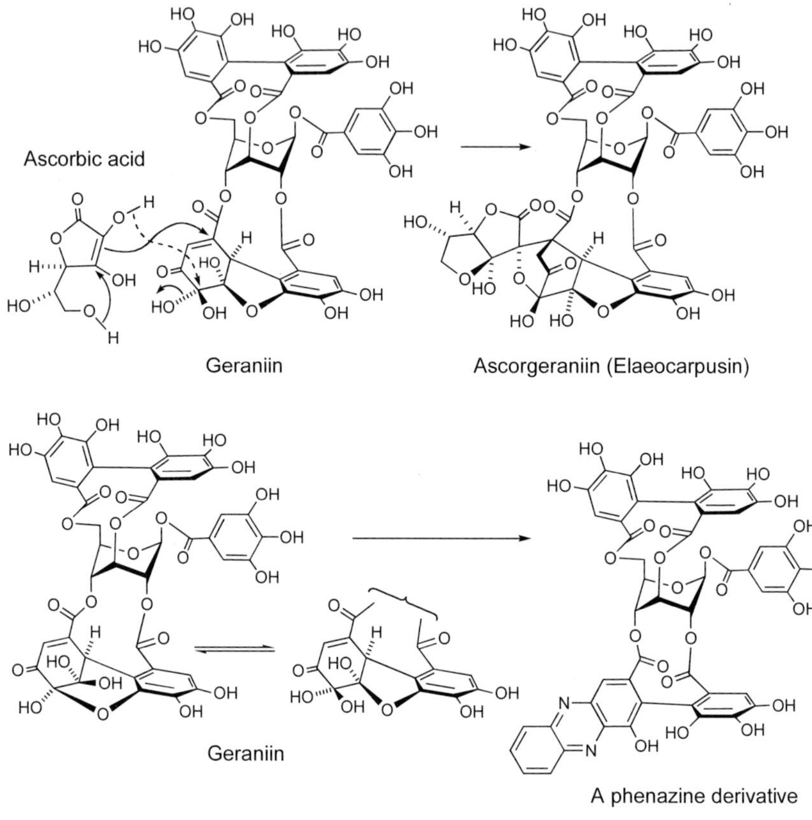

Fig. 1.2 Examples of condensation reactions of the geraniin DHHDP group yielding ascorgeraniin and a phenazine derivative.

1.3.3 *C*-glycosidic ellagitannins and complex tannins

1.3.3.1 *Occurrence of C-glycosidic tannins in plants*

The *C*-glycosidic tannins that are exemplified by casuarinin, stachyurin and casuariin, first isolated from *Casuarina stricta* (Okuda *et al*., 1982c,

1983a), are widely distributed in various plant species of *Casuarinaceae*, *Stachyuraceae*, *Myrtaceae*, *Betulaceae*, *Fagaceae*, *Hamamelidaceae*, *Lythraceae*, *Punicaceae*, *Melastomataceae*, *Rosaceae*, *Elaeagnaceae*, *Theaceae* and *Juglandaceae* families (Okuda *et al.*, 1982h). Castalagin and vescalagin were found in the woody *Castanea* and *Quercus* species (Mayer, 1971). The so-called complex tannins, which commonly referred to ellagitannins having a flavanol-based motif linked to C-1 of the glucose core through a C–C bond, occur in some species of *Fagaceae*, *Combretaceae*, *Myrtaceae*, *Theaceae* and *Melastomataceae*, and constitute a subclass of *C*-glycosidic ellagitannins (Yoshida *et al.*, 1992a).

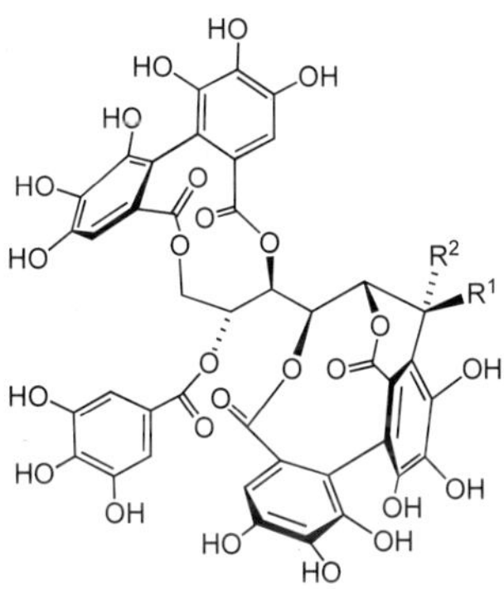

Casuarinin : R^1 = H, R^2 = OH
Stachyurin : R^1 = OH, R^2 = H

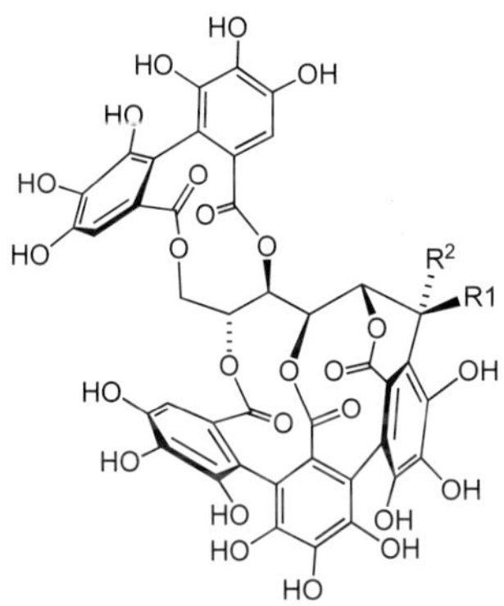

Castalagin : R^1 = H, R^2 = OH
Vescalagin : R^1 = OH, R^2 = H

Camelliatannin A

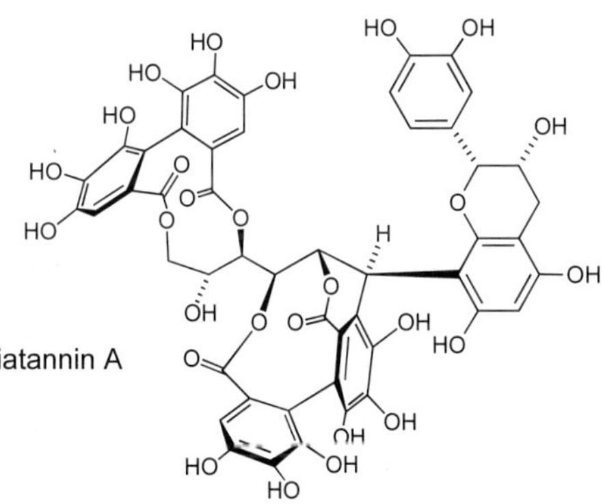

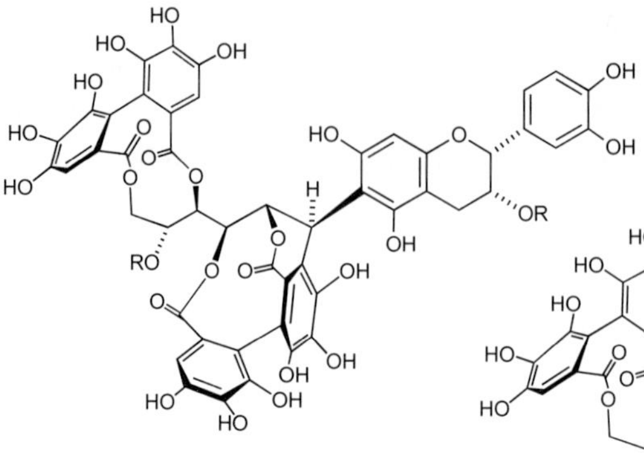

Camelliatannin B : R = H
Malabathrin A : R = G

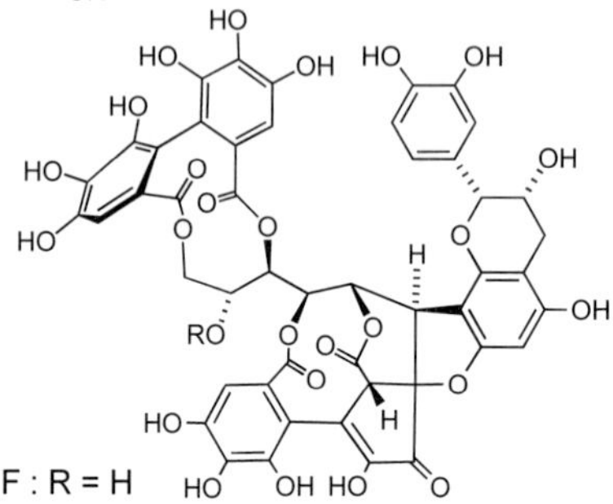

Camelliatannin F : R = H
Malabathrin E : R = G

1.3.3.2 *Biomimetic synthesis of C-glycosidic ellagitannins*

Casuarinin was biomimetically synthesized through an acid-catalyzed intramolecular phenol-aldehyde coupling reaction of liquidambin (Fig. 1.3, see also Fig. 9.6 in Chapter 9), an aldehydic ellagitannin presumed to be the key biosynthetic precursor of *C*-glycosidic ellagitannins (Okuda *et al.*, 1987). The complex tannins camelliatannins A and B were hemisynthesized by condensation of casuariin with (–)-epicatechin (Fig. 1.3), and also by conversion of camelliatannins E and C, respectively, via a treatment with polyphosphoric acid (Fig. 1.4). The transformation of camelliatannin A into cammelliatannin F, featuring a cyclopentenone ring, was achieved by heating camelliatannin A in a mixture of ethanol and acetic acid (Fig. 1.4, Hatano *et al.*, 1995).

1.3.4 Oligomerization of ellagitannins leading to pentamers

The first oligomeric hydrolyzable tannin isolated in 1982 was agrimoniin (*vide supra*), which remarkably displays α-glycosidic linkages on both of its constituting monomeric units (Okuda *et al.*, 1982b). Its isolation was followed by that of gemin A (*vide infra*), a dimer having both α and β linkages (Yoshida *et al.*, 1982), and gemins B-F (Yoshida *et al.*,

1985a/b), as well as various oligomers up to pentamers, including a dimer of geraniin, *i.e.*, acalyphidin D_1 (Yoshida *et al.*, 1992b, 1999, 2005). Such oligomers often express specific pharmacological activities (*vide infra*) that are not shared by monomeric ellagitannins and other tannins.

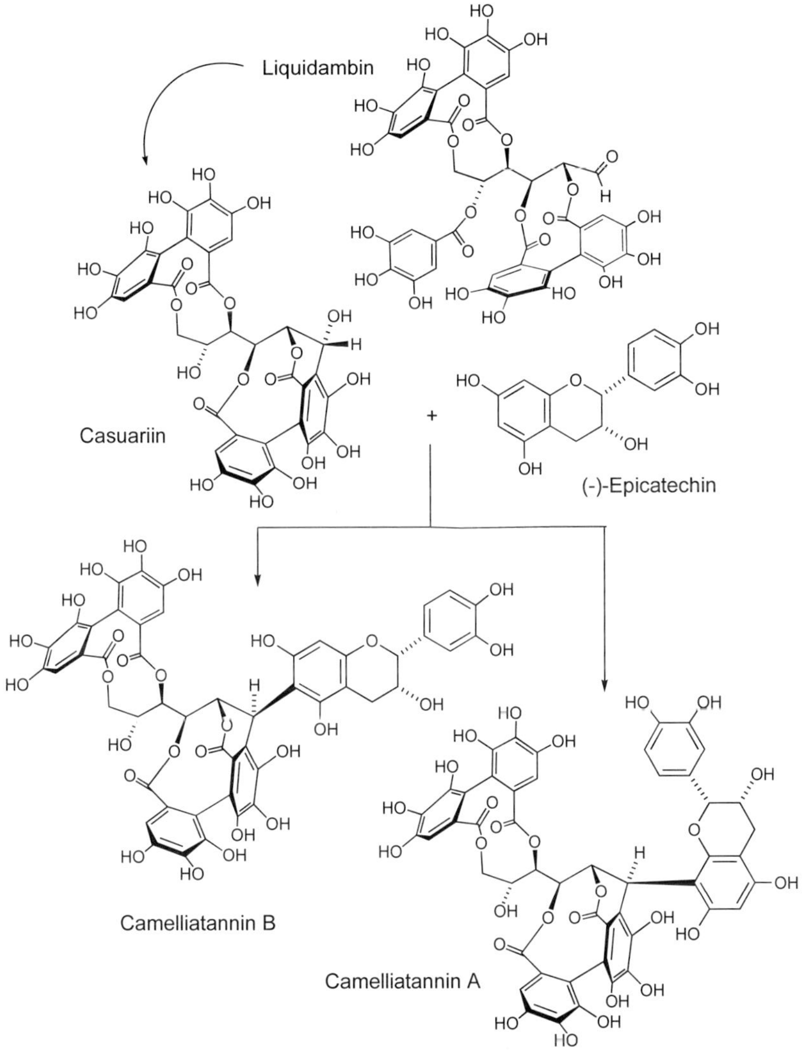

Fig. 1.3 Synthesis of camelliatannins A and B from casuariin and epicatechin. Casuariin derived from liquidambin via 5-*O*-desgalloylation of casuarinin (see Fig. 9.6 in Chap. 9).

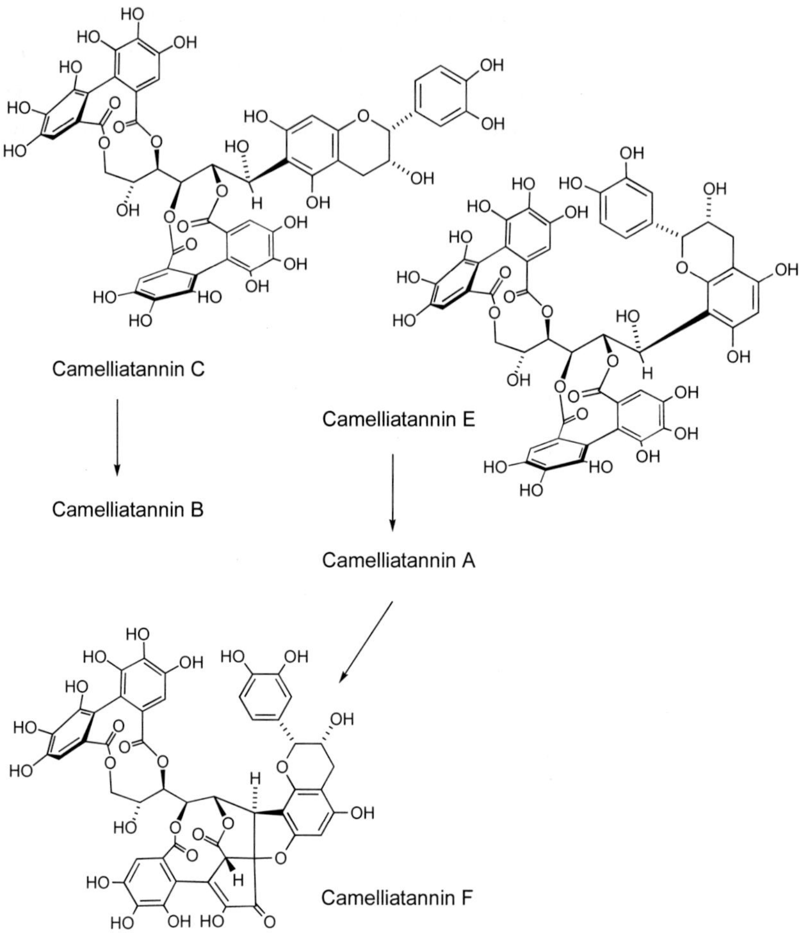

Fig. 1.4 Conversion of camelliatannins C and E into camelliatannins B and A, and conversion of camelliatannin A into camelliatannin F.

1.3.4.1 *Oligomers as main components in a plant species*

Agrimonia pilosa (*Rosaceae*), an antidiarrheic in Japan, produces agrimoniin as the main component accompanied by small amounts of potentillin, the monomer composing the agrimoniin molecule, and several other monomers. Agrimoniin is also the main component in *Agrimonia eupatria*. Oenothein B, a macrocyclic dimer, is the main component in *Oenothera erythrosepala* (*Onagraceae*) and is accompanied by the trimer oenothein A and tellimagrandin I, the monomer composing these oligomers (Hatano *et al.*, 1990a). Coriariin A is the main component in *Coriaria japonica* (*Coriariaceae*) (Hatano *et al.*, 1986). The monomers frequently found as constituents of these oligomeric molecules are tellimagrandins I and II, pedunculagin and casuarictin as further exemplified below by the structure of gemin A.

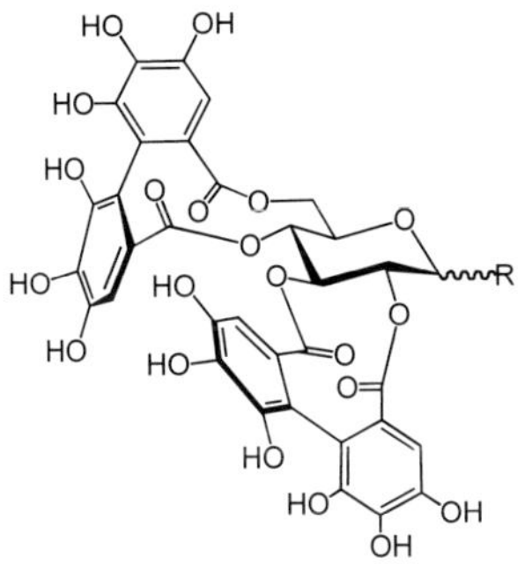

Pedunculagin : R = OH
Casuarictin : R = (β)-OG
Potentillin : R = (α)-OG

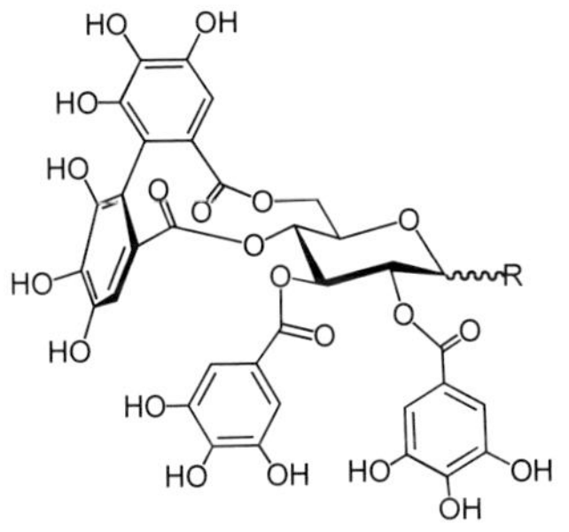

Tellimagrandin I : R = OH
Tellimagrandin II : R = (β)-OG

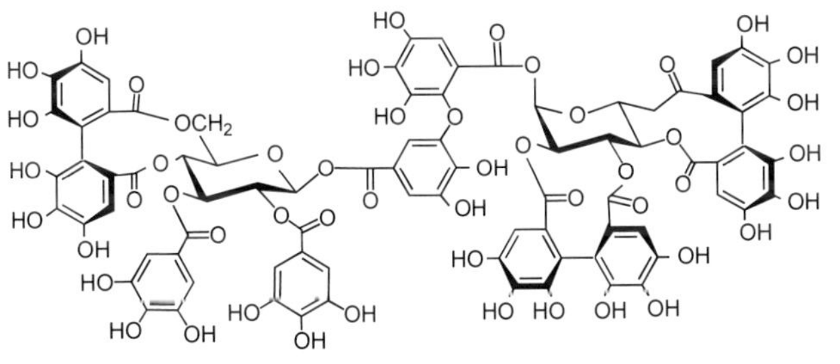

Gemin A

1.3.4.2 *Oligomerization via oxidative C–O and C–C coupling modes*

The oligomerization of ellagitannins mainly occurs via C–O oxidative coupling, but C–C oxidative coupling takes place in *C*-glycosidic ellagitannins, including complex tannins. The C–O coupling modes can be classified on the basis of the *O*-donating polyphenolic unit of a monomer and the *O*-accepting polyphenolic unit in another monomer composing the dimer, as shown below in Fig. 1.5. Generally, a given linking unit repeatedly participates in the construction of an oligomeric ellagitannin system.

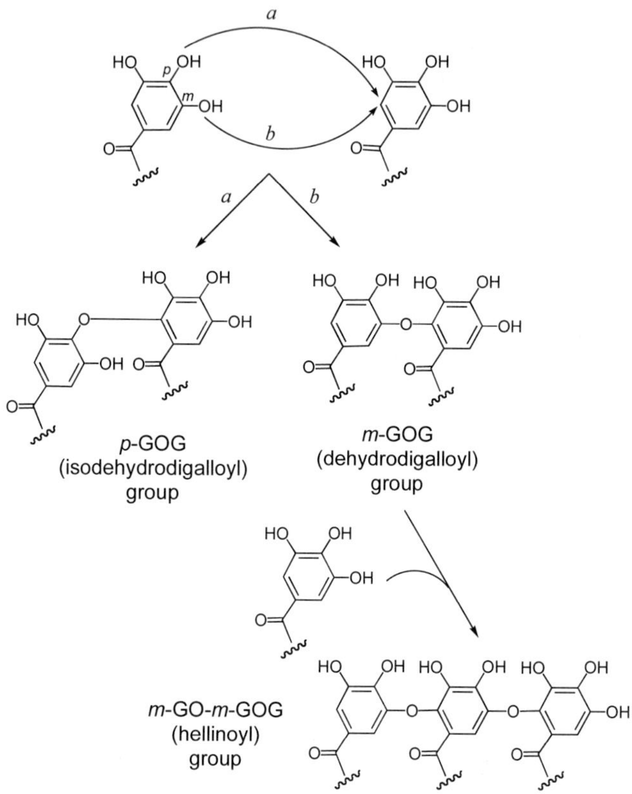

Fig. 1.5 Formation of the GOG and GOGOG oligomeric linking units.

1.3.4.2.1 The GOG- and GOGOG-type units

The GOG-type linking units are produced by C–O coupling between two galloyl groups (Fig. 1.5), as found in the *p*-GOG isodehydrodigalloyl group (*p-O* of a galloyl group C-linked to another galloyl group, mode *a*) and the *m*-GOG dehydrodigalloyl group (*m-O* of a galloyl group C-linked to another galloyl group, mode *b*). The GOGOG-type units are formed via an additional oxidative C–O coupling of a galloyl group with a GOG group. Agrimoniin and laevigatins B, C, D (Yoshida *et al.*, 1989a) and E, as well as gemin A, are examples of dimers featuring the GOG-type linkage, and tamarixinin A and hirtellin B (Yoshida *et al.*,

1991c) are examples of dimers featuring the *m*-GO-*m*-GOG-type unit, also referred to as the hellinoyl group as shown below.

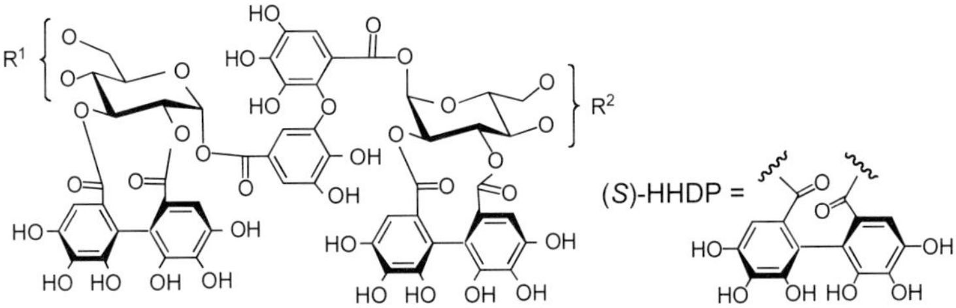

Laevigatin B : R^1 = H, H, R^2 = (*S*)-HHDP
Laevigatin C : R^1 = (*S*)-HHDP, R^2 = H, H
Laevigatin E : R^1 = R^2 = H, H

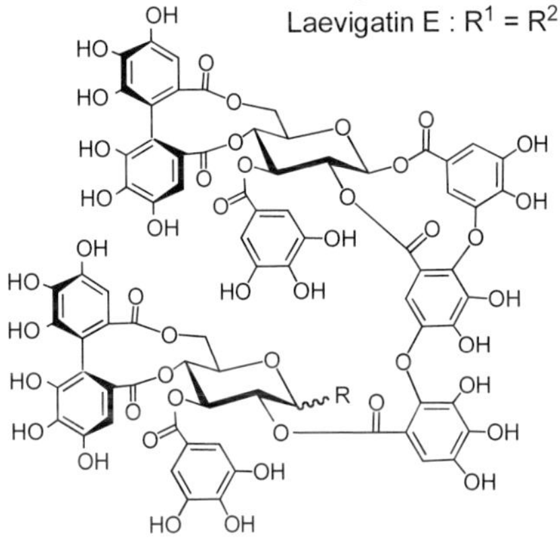

Tamarixinin A : R = OH
Hirtellin B : R = (β)-OG

1.3.4.2.2 The DOG and D(OG)$_2$-type units

In the DOG-type linking units, which are most frequently found in oligomers, the *O*-donating hydroxyl group is part of an HHDP group, and a galloyl group is the acceptor. The *m*-DOG and the *p*-DOG groups have been called valoneoyl and tergalloyl groups, respectively. The prefixes *m* and *p* referred to the position of the hydroxyl oxygen atom of the HHDP group being engaged in the diaryl ether bond. Rugosins D, E, F (dimers) and G (trimer) from several *Rosa* species (Okuda *et al.*, 1982g, 1990, Hatano *et al.*, 1990b), tetramers trapanin B from *Trapa japonica* (Hatano *et al.*, 1990c) and nobotanin K from *Heterocentrum roseum* (Yoshida *et al.*, 1989), and pentamers melastoflorins A-D from *Monochaetum multiflorum* (Yoshida *et al.*, 2005), are examples of the *m-DOG*-type linked oligomers. Eucalbanin C (*vide infra*) is a dimer having a *p*-DOG linking unit (Yoshida *et al.*, 1992c).

In a D(OG)$_2$-type linking unit, found in a smaller number of oligomers, two hydroxyl groups of an HHDP group engaged their oxygen atoms in diaryl ether bonds, as exemplified in oenothein A and woodfordin D from *Woodfordia fruticosa* (Yoshida *et al.*, 1991a) (see Section 1.3.5).

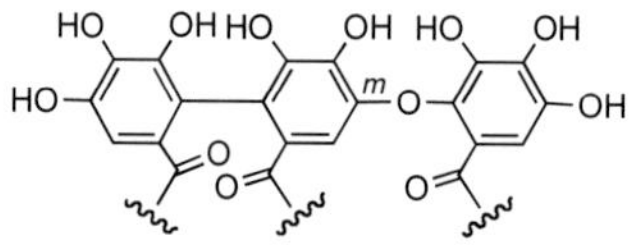

m-DOG (valoneoyl) group

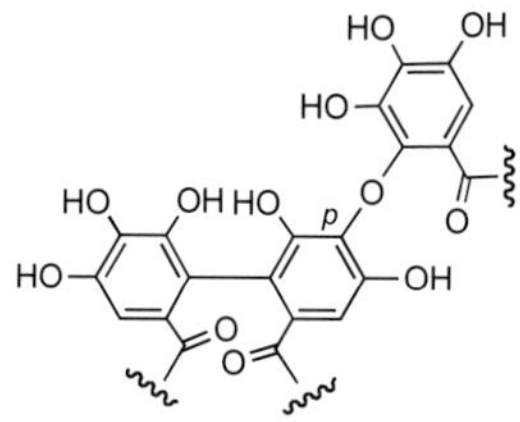

p-DOG (tergalloyl) group

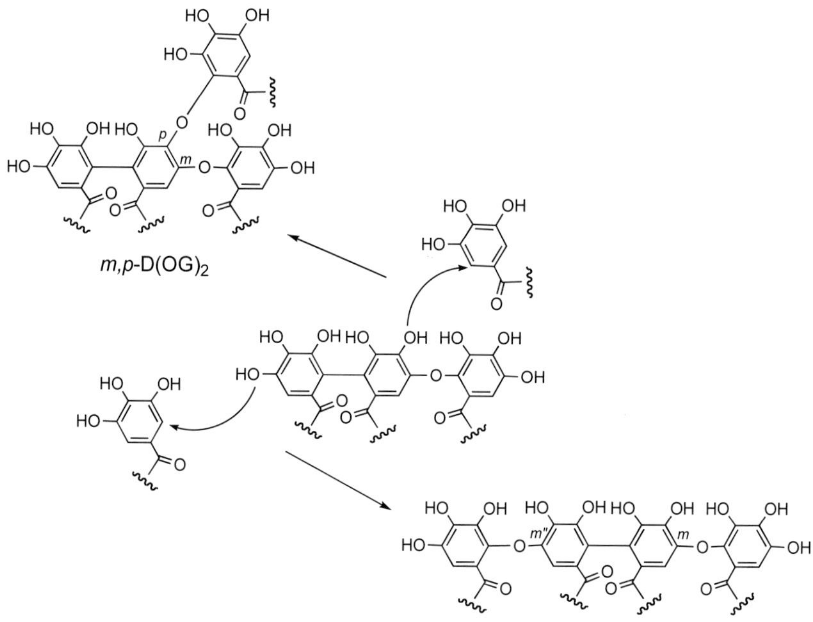

m,*m*″-D(OG)$_2$

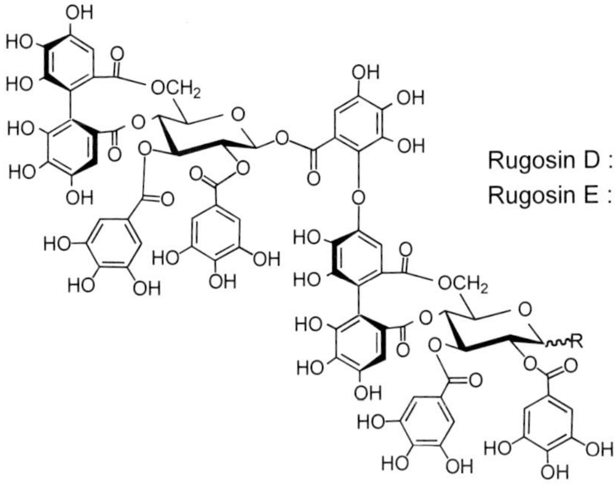

Rugosin D : R = (β)-OG
Rugosin E : R = OH

Rugosin F

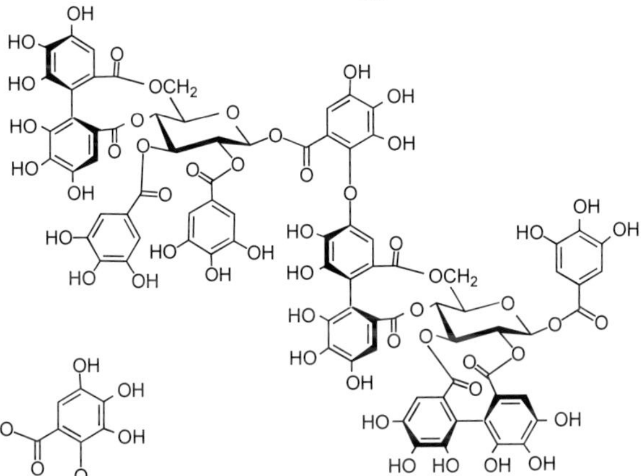

Rugosin G

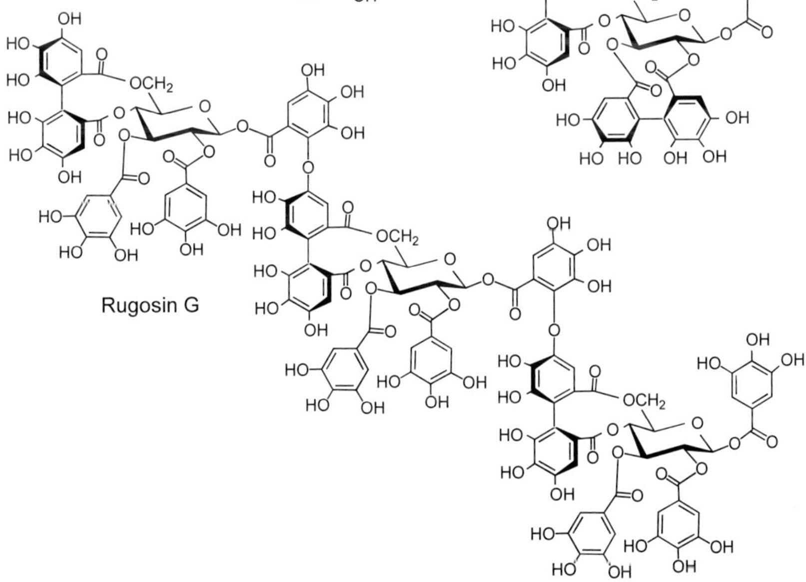

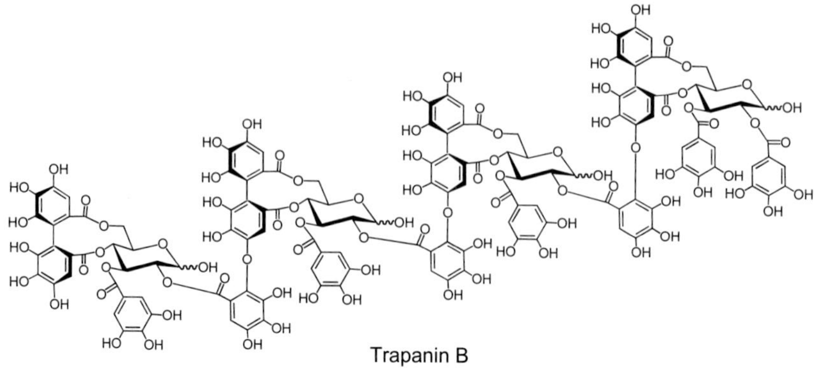

Trapanin B

Melastoflorin A

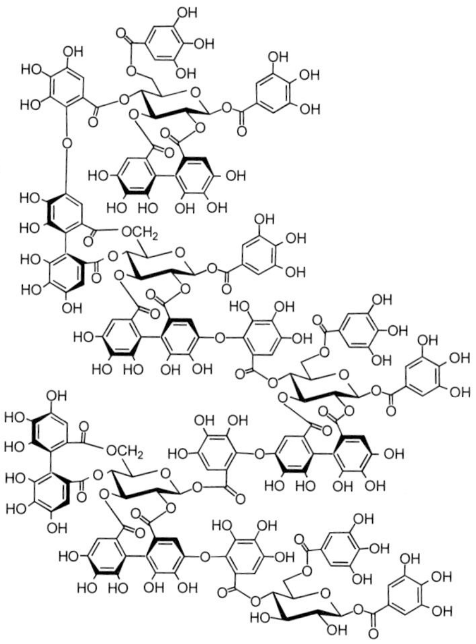

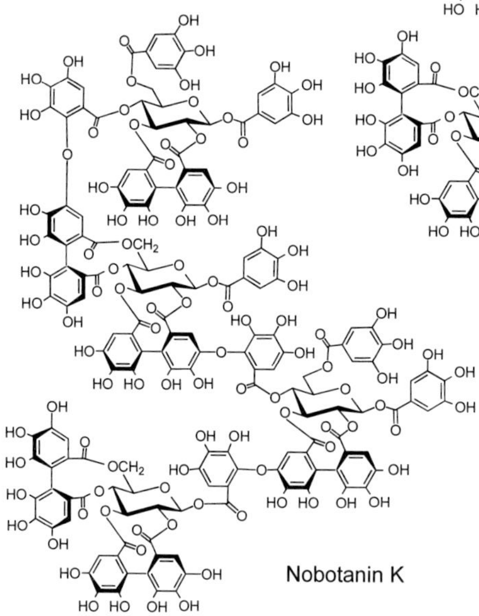

Nobotanin K

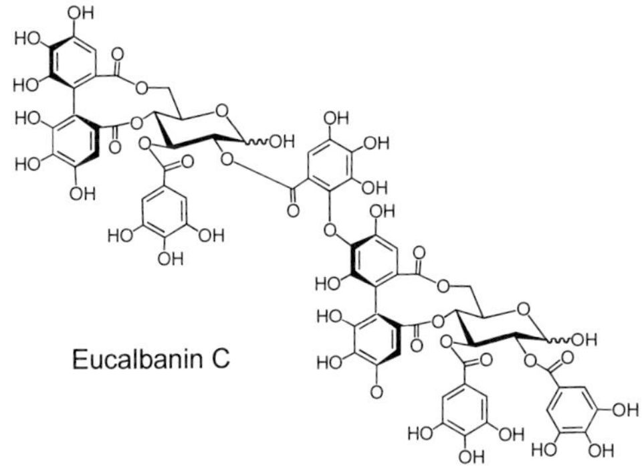

Eucalbanin C

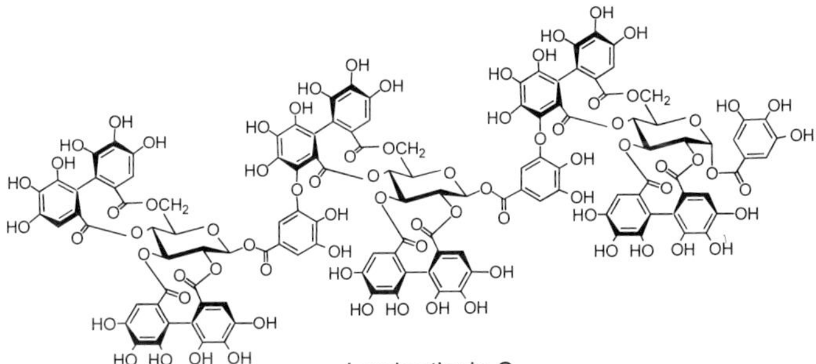

Lambertianin C

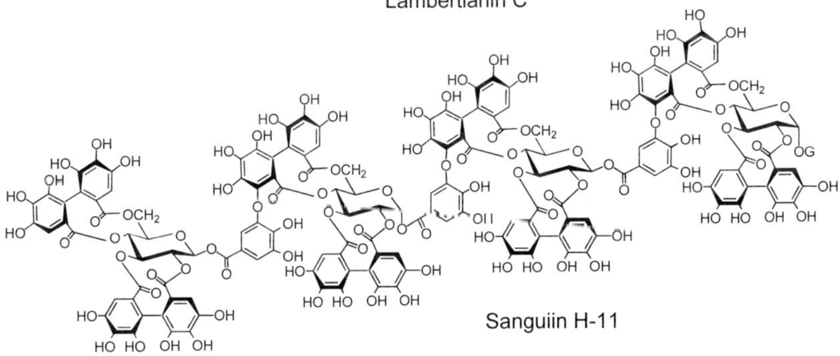

Sanguiin H-11

1.3.4.2.3 The GOD-type unit

The linking unit resulting from a donation of a galloyl hydroxyl oxygen to form an ether linkage to an HHDP group is classified as the GOD type. Roshenins A and B (Yoshida *et al.*, 1992d), lambertianin A and sanguiin H-6 are examples of the GOD-type dimers, and lambertianin C and sanguiin H-11 are trimeric and tetrameric examples (Tanaka *et al.*, 1985, 1993).

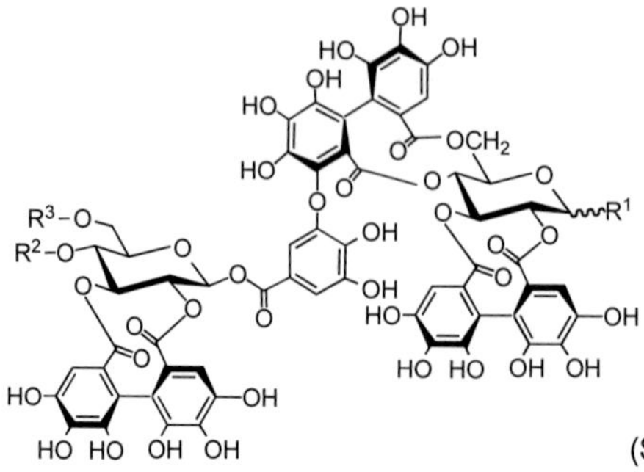

(S)-Sang =

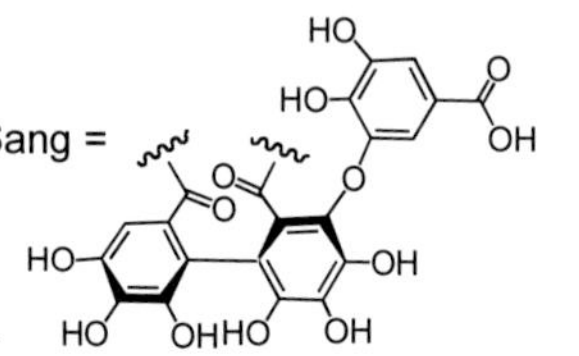

Roshenin A : R^1 = (α)-OG, R^2~R^3 = (*S*)-Sang
Roshenin B : R^1 = OH, R^2~R^3 = (*S*)-HHDP
Lambertianin A : R^1 = (β)-OG, R^2~R^3 = (*S*)-HHDP
Sanguiin H-6 : R^1 = (α)-OG, R^2~R^3 = (*S*)-HHDP

1.3.5 Macrocyclic oligomers

Macrocyclic dimers have been isolated from plants of several families. Oenothein B (Okuda *et al.*, 1982b), often accompanied by the trimer oenothein A (Yoshida *et al.*, 1991a), was isolated from *Oenothera* and *Epilobium* species in *Onagraceae* and *Lythrum anceps* in *Lythraceae*. Oenothein B was also isolated from *Woodfordia fruticosa* in *Lythraceae*, which also yields woodfordin C (Yoshida *et al.*, 1990a), a galloylated oenothein B, and woodfordin D, which is a galloylated oenothein A (Yoshida *et al.*, 1991). Camellin B, which displays an analogous macrocyclic structure, has been isolated from several *Thea* species and from *Schima wallichii* in *Camelliaceae* (Yoshida *et al.*, 1990b). Cuphiins D_1 (a galloylated woodfordin C) and D_2 (a regioisomer of woodfordin C) were isolated from *Cuphea hyssopifolia* (*Lythraceae*) (Chen *et al.*, 1999). Eugeniflorins D_1 (a galloylated oenothein B) and D_2, an analog having a hemiacetal-forming linking unit, were obtained from *Eugenia uniflora* (*Myrtaceae*) (Lee *et al.*, 1997). These oligomers feature the *m*-DOG linking unit. The *m*- and/or *p*-GOG-bearing dimers tamarixinin B, hirtellin C and isohirtellin C, the structures of which are evoked in the next section on oligomer transformations, also display macrocyclic motifs.

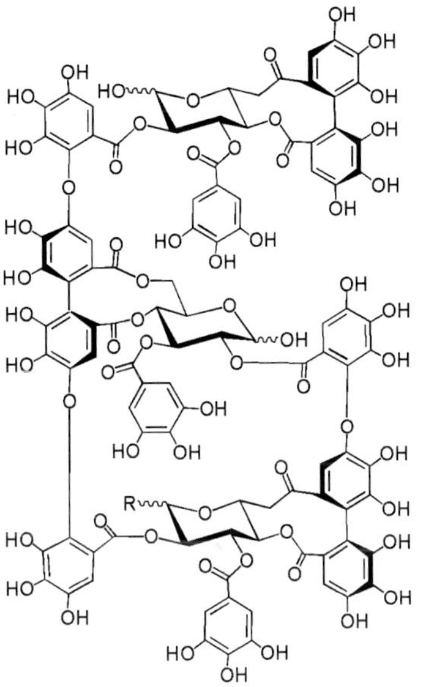

Oenothein A : R = OH
Woodfordin D : R = (α)-OG

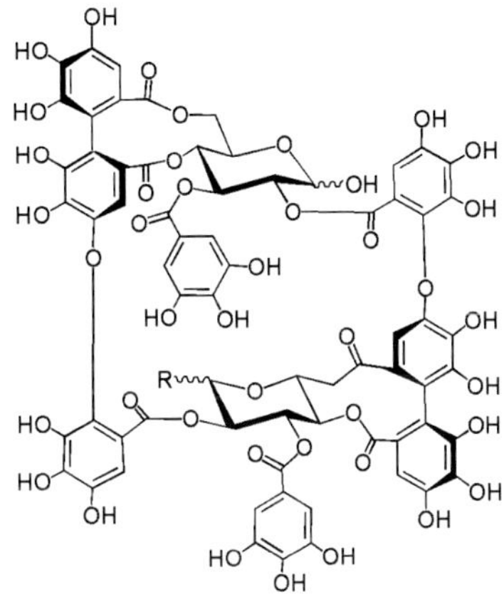

Oenothein B : R = OH
Woodfordin C : R = (α)-OG

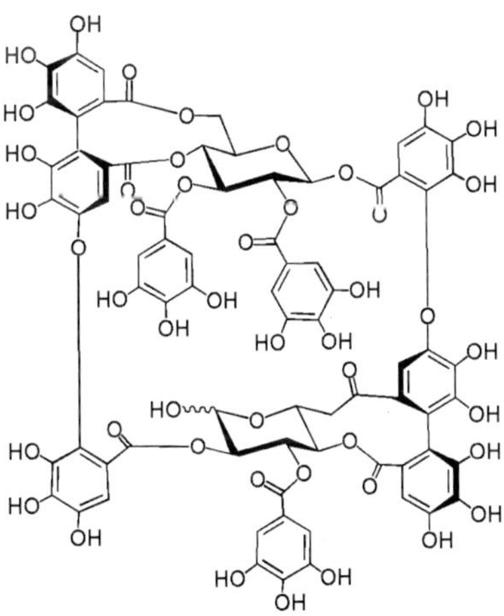

Camelliin B

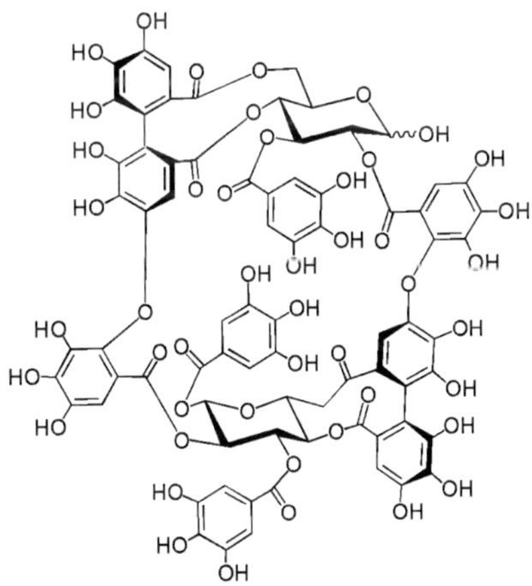

Eugeniflorin D_1

1.3.6 Structural transformations of ellagitannin oligomers

Although ellagitannins are generally stable compounds, some oligomers do undergo structural transformations under rather mild conditions.

1.3.6.1 *Isomerization of dimers via Smiles rearrangement*

Hirtellin C is a macrocyclic dimer possessing both the *m*- and *p*-GOG units that was isolated from *Reaumuria hirtella* (*Tamaricaceae*). It is readily isomerized into isohirtellin C in hot water at 95 °C via a Smiles rearrangement that enables the conversion of the *p*-GOG unit into a *m*-GOG unit. This rearrangement is presumably facilitated by the release of steric effects originating from hydroxyl groups in the *p*-GOG linking unit. Even at pH 7.4 and at only 40 °C, this isomerization reaction was almost complete within 30 min (Yoshida *et al.*, 1993a).

Application of this isomerization reaction was used to confirm the structure of tamarixinins B (a macrocyclic dimer) and C, both isolated from *Tamarix pakistanica* (*Tamaricaceae*). Indeed, their Smiles rearrangement led to analogs of established structures, tamarixinin B being thus converted into hirtellin C, and tamarixinin C into hirtellin A (Fig. 1.6, Yoshida, *et al.*, 1993b).

The tergalloyl group in eucalbanin C, a *p*-DOG-type dimer isolated from *Eucalyptus alba* (*Myrtaceae*), was quantitatively converted into eucalbanin B of the valoneoyl *m*-DOG-type in a phosphate buffer (pH 7.4) at room temperature (Fig. 1.7). The dilactonized tergalloyl group in eucalbanin A was also isomerized to the dilactonized valoneoyl group in cornusiin B (Yoshida *et al.*, 1992c).

1.3.6.2 *Hydrolysis of ellagitannin oligomers into monomers*

Partial hydrolysis of oligomers in boiling water or weak acids affords monomers. The linking unit usually remains attached onto one of the monomers thus released from the dimeric structure. This type of transformation, or partial degradation, is a useful tool to gather important informations in the course of the elucidation of an oligomeric structure.

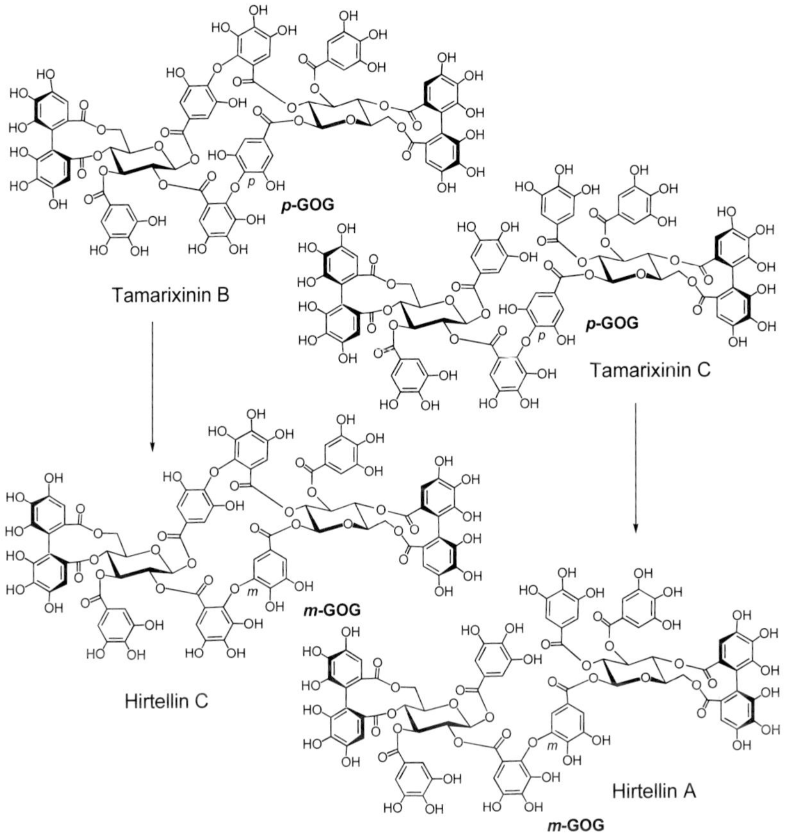

Fig. 1.6 Smiles rearrangement-mediated conversions of tamarixinin B into hirtellin C and tamarixinin C into hirtellin A.

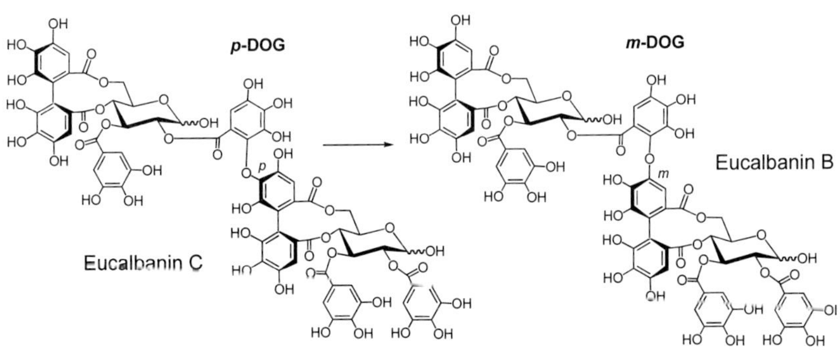

Fig. 1.7 Smiles rearrangement-mediated conversion of eucalbanin C into eucalbanin B.

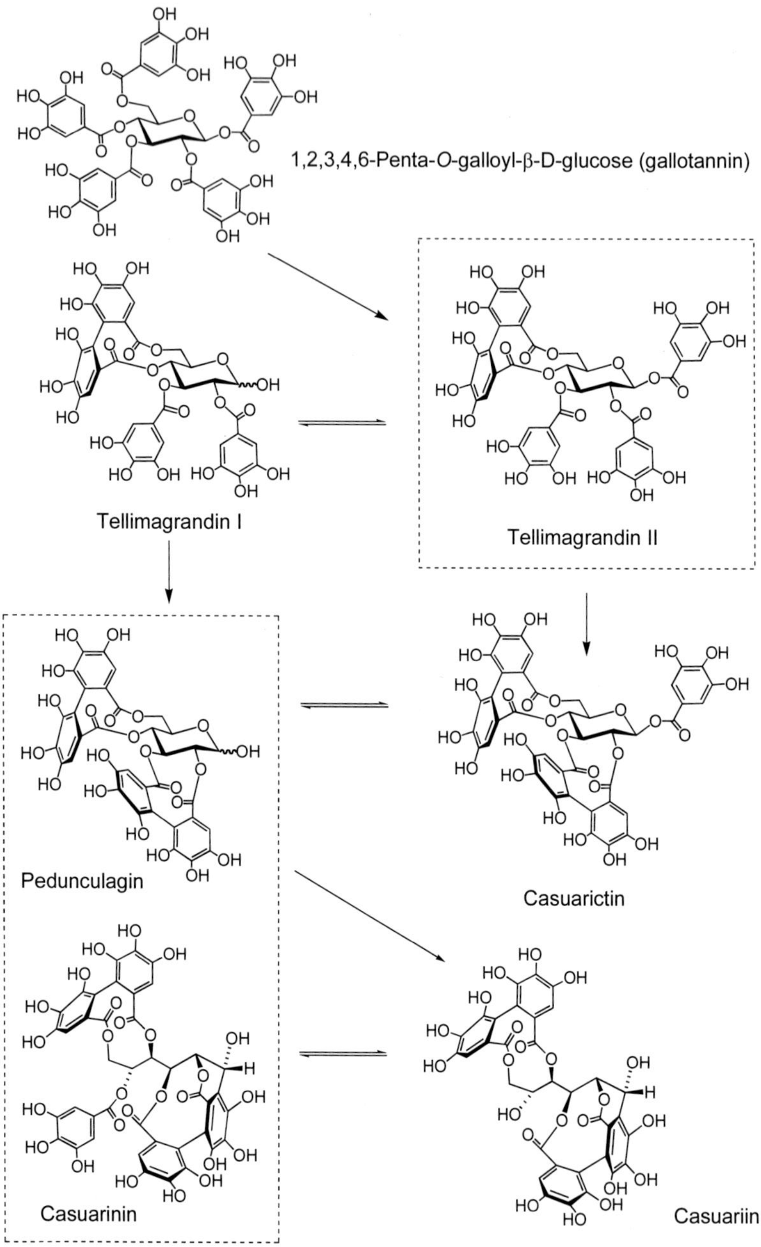

Fig. 1.8 Seasonal transformations of ellagitannins in *Liquidambar formosana*.

1.3.7 Seasonal transformations of ellagitannins in a plant

While the ellagitannin structures occurring in a plant species are generally invariable throughout a year, as observed for geraniin in *Geranium* species, seasonal structural change of main tannins occurs in some woody plants. An example is the seasonal transformations of gallo- and ellagitannins in *Liquidambar formosana*. Tellimagradins, casuarictin and gallotannins, which are abundant in young leaves in April, are replaced by casuarinin by July, and the latter, together with pedunculagin, are the main tannins in the leaves from summer to autumn until the leaves fall down. This seasonal transformation, which interestingly parallels the oxidative biogenetic route followed by ellagitannin structures, is depicted above (Fig. 1.8, Okuda *et al.*, 1987).

1.3.8 Production of ellagitannins by tissue cultures

The callus and shoot cultures of *Hetrocentron roseum*, under illumination with fluorescent lamps, produce large amounts of casuarictin (a *C*-glycosidic monomer) and nobotanin M (a dimer) (Yazaki and Okuda, 1990). Oenothein B and other macrocyclic dimers were produced by callus culture of *Oenothera laciniata* and shoot tissue culture of *O. tetraptera* (Taniguchi *et al.*, 1998, 2002). Geraniin and other ellagitannins were accumulated by *Aleurites fordii* callus culture (Taniguchi *et al.*, 2002).

1.4 Correlation of Ellagitannins of Various Oxidation Stages with Plant Evolution Systems

1.4.1 Classification of hydrolyzable tannins based on the oxidation stages of their polyphenolic functions

Hydrolyzable tannins of various biogenetic oxidative stages can be classified into types I to IV according to the degree of oxidation of their polyphenolic groups. The different polyphenolic groups and examples of compounds of each type are as follows (Okuda *et al.*, 2000):

Type I: gallotannins (galloyl group, *e.g.*, 1,2,3,4,6-penta-*O*-galloyl-β-D-glucose, *i.e.*, PGG)
Type II: ellagitannins (HHDP group, *e.g.*, pedunculagin)
Type III: dehydroellagitannins (DHHDP group, *e.g.*, geraniin)
Type IV: oxidized ellagitannins [chebuloyl group, *e.g.*, chebulagic acid, geraniinic acids A and B, phyllanthusiins A-C, repandusinic acid, heterophylliin E]

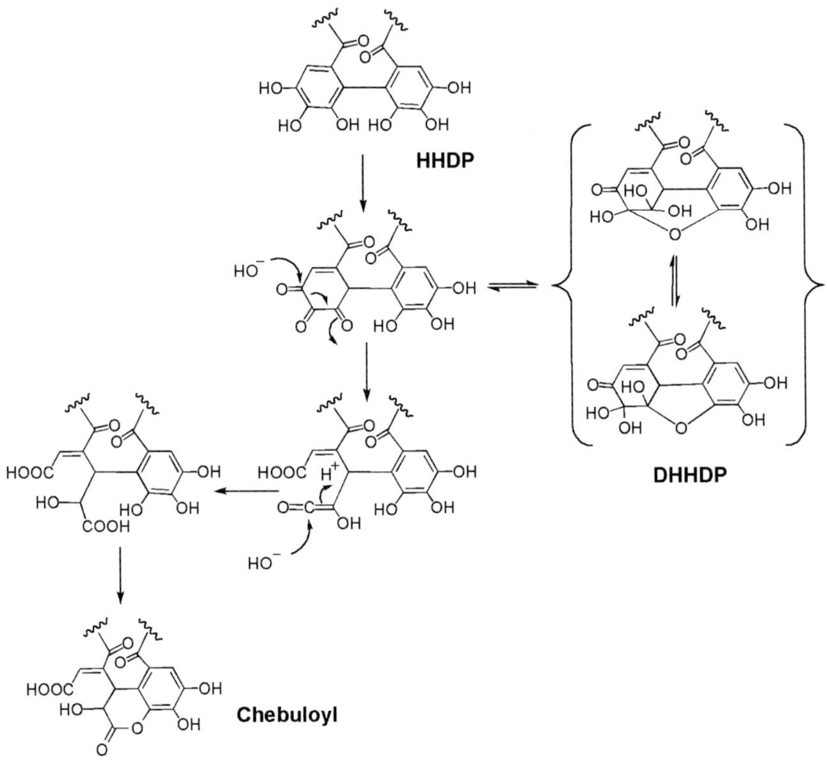

In addition to these oxidative transformations, various additional transformations occur in plants, yielding monomers that can be classified into types I+ to IV+:

Type I+: *C*-glycosidic gallotannins (3,4,11-tri-*O*-galloylbergenin)
Type II+: *C*-glycosidic ellagitannins, *e.g.*, casuarinin; complex tannins,

e.g., camelliatannin A; ellagitannins linked to a phenolic or polyphenolic moiety through an ether linkage, *e.g.*, coriariin B; gluconic acid version of ellagitannins, *e.g.*, shephagenin A (Yoshida *et al.*, 1996)

Type III+: dehydroellagitannins linked to a phenolic or polyphenolic moiety through an ether linkage, *e.g.*, mallotusinic acid (Okuda *et al.*, 1981)

Type IV+: oxidized dehydroellagitannins linked to a phenolic or polyphenolic through a C–C bond, *e.g.*, camelliatannin F.

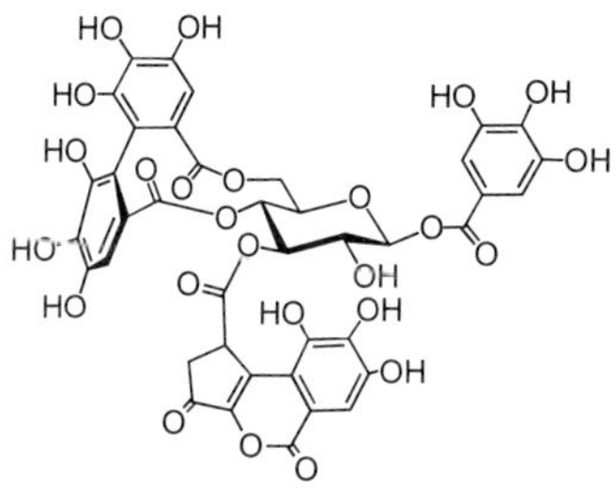

Heterophylliin E

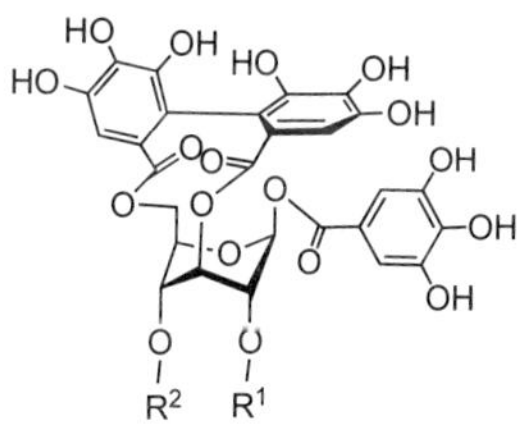

Phyllanthusiin A : R^1~R^2 = **A**
Phyllanthusiin B : R^1~R^2 = **B**
Phyllanthusiin C : R^1~R^2 = **C**
Repandusinic acid A : R^1 = H, R^2 = **D**
Geraniinic acid B : R^1~R^2 = **E**
Geraniinic acid C : R^1~R^2 = **F**

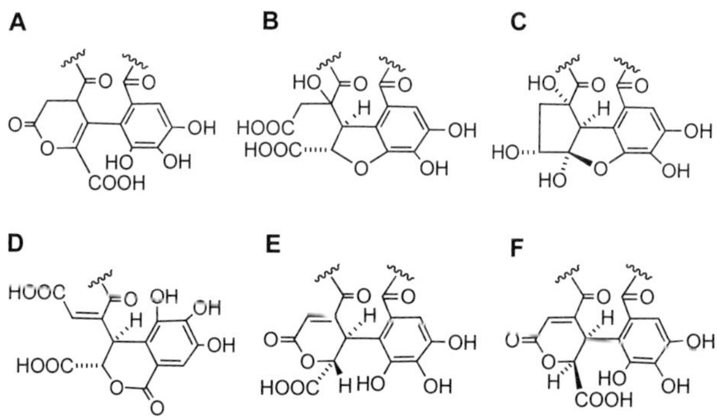

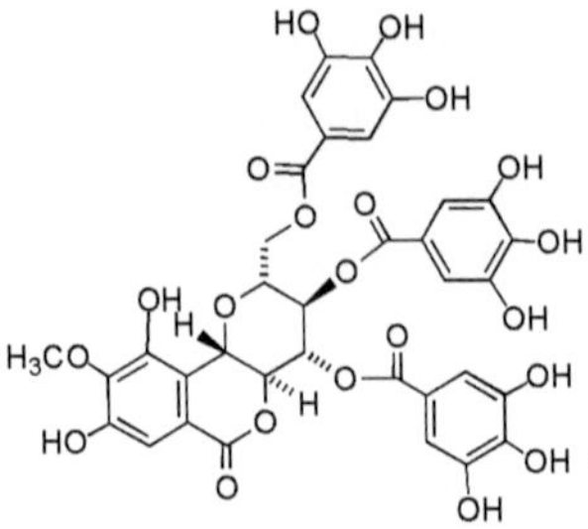

Trigalloylbergenin

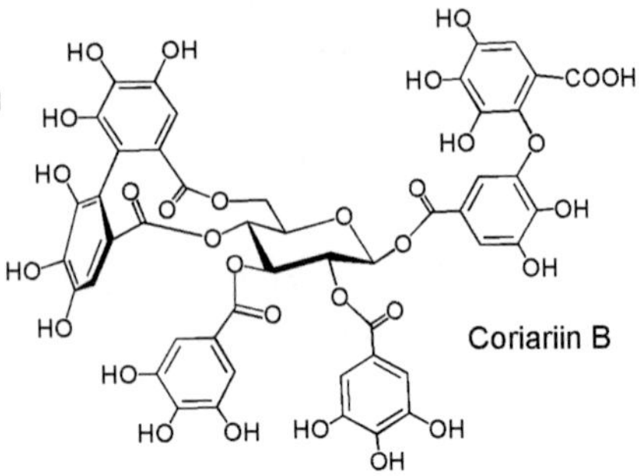

Coriariin B

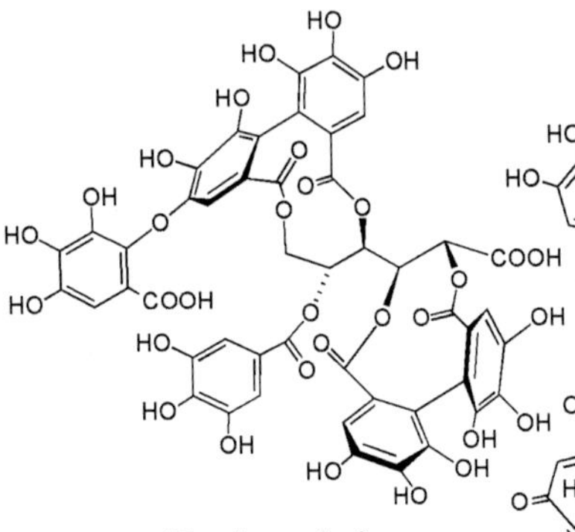

Shephagenin A

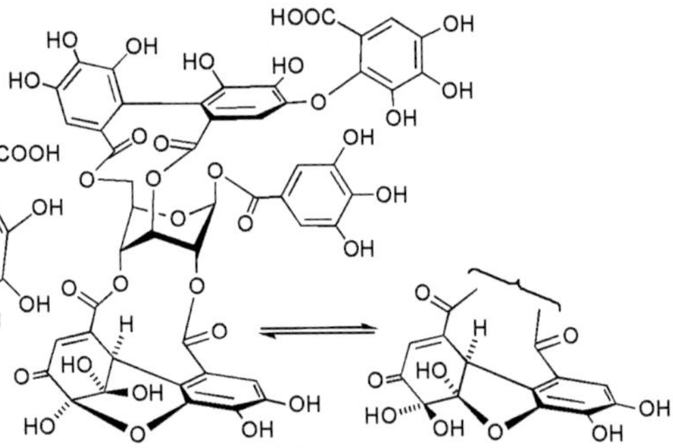

Mallotusinic acid

1.4.2 Correlation of the oxidation stages with Cronquist's system of plant evolution

The progressive oxidative transformations of monomeric ellagitannins can be correlated with morphological plant evolution system. Since hydrolyzable tannins generally express strong antioxidation properties, the potency of which being correlated to their biogenetic oxidation stage, a correlation of their structures with the plant evolution system may be more significant than that of other plant metabolites. Thus, a search of such a correlation within subclasses of the Cronquist's system of plant evolution for *Dicotyledonae* (Fig. 1.9), *i.e.*, *Rosidae*, *Dilleniidae*, *Hamamelidae*, *Caryophyllidae* and *Magnolidae* (NB: ellagitannins are not found in *Asteridae*) unveiled the following aspects and several other features:

i) Oxidized tannins of the types III and IV are frequently found in the *Rosidae*, while they are found only in a small number of plant species belonging to the *Dileniidae* and *Hamamelidae*. These oxidized tannins are not found in the *Caryophyllidae* and *Magnolidae*, the earliest subclasses of the *Dicotyledonae*.

ii) Correlations within a subclass are as follows. Rosales, the earliest order in the *Rosidae*, mostly produce type-I and type-II ellagitannins. The oxidative transformations to types III and IV progress according to the evolution of the orders, *i.e.*, Rosales → Sapindales → Geraniales, without being accompanied by any production of oligomers, and Rosales → Euphorbiales, being accompanied by production of oligomers. The oxidative transformations in the *Dilleniidae* seem to progress from Dilleniales (type I) to Theales (types I, II and IV+) (Okuda *et al.*, 2000).

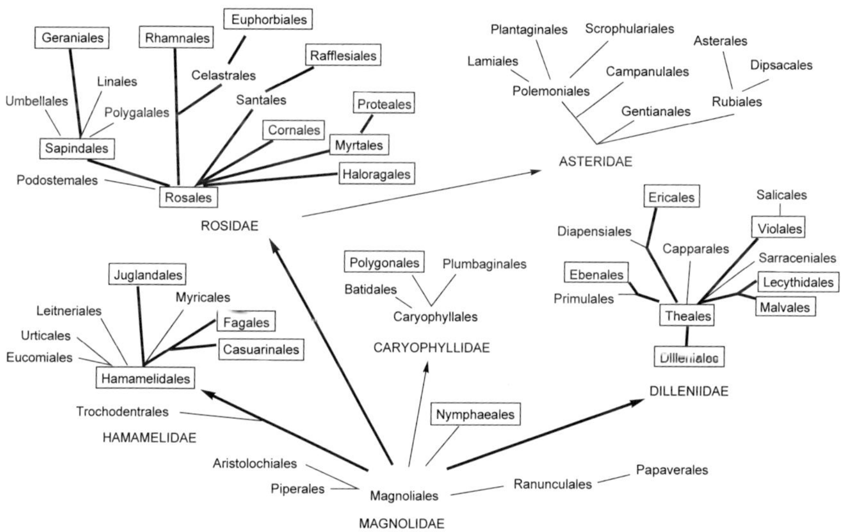

Fig. 1.9 Cronquist's plant evolution system of *Dicotyledonae* —: evolutionary route of ellagitannin-producing plants. □: orders to which ellagitannin-producing plants belong.

Analogous considerations may be made for the formation of C–O and C–C bonds (see Chapter 2) during oligomerization of ellagitannins, as well as during their macrocyclization, which requires formation of an additional bond:

i) Macrocyclic ellagitannin dimers, *e.g.*, oenothein B and woodfordin C.
ii) Macrocyclic ellagitannin trimers, *e.g.*, oenothein A and woodfordin D.
iii) Macrocyclic dehydroellagitannin dimers, *e.g.*, eugeniflorin D_2 from *Eugenia uniflora*.

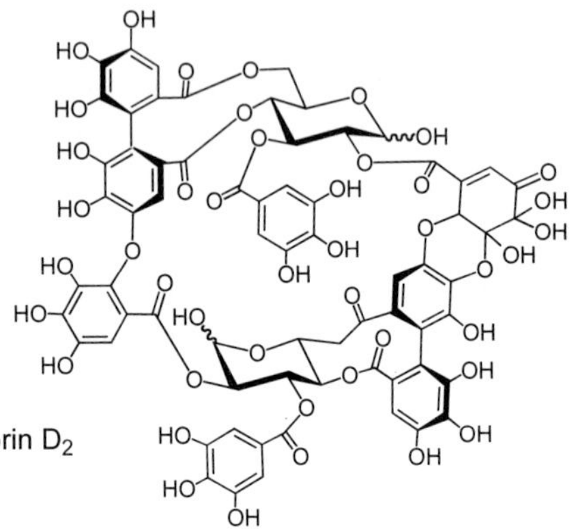

Eugeniflorin D_2

Geraniales
Geraniaceae
Geranium: I, III, IV

Euphorbiales
Euphorbiaceae
Alchornea: III, III+
Aleurites: III, III+, IV
Antidesma: III, III+
Euphorbia: I, II, II+, III, III+, IV
Excoecaria: I, II, II+, III, III+
Macaranga: I, II, III, IV
Mallotus: I, I+, II+, III, III+, IV
Phyllanthus: I, II, III, IV

Fig. 1.10 Ellagitannins types in Geraniales and Euphorbiales plant orders.

1.4.3 Isolation of oxidized ellagitannin oligomers in specific plant orders

Highly oxidized ellagitannin dimers of types III, III+ and IV have been isolated from euphorbiaceous plants in Euphorbiales, besides from *Geraniaceae* in Geraniales (Fig. 1.10). Euphorbin E from *Euphorbia hirta*, composed of geraniin and oxidized isoterchebin molecules (Yoshida, *et al.*, 1990c), acalyphidin D_1 (Amakura, *et al.*, 1999) composed of two geraniin molecules with an oxidized linking unit, and eurobustin C (Hatano *et al.*, unpublished data) from *E. robusta*, a macrocyclic dimer with brevifolin carboxylic acid as one of the linking units, are such examples. Many oligomers having geraniin as the composing unit have also been isolated from Euphorbiaceous plants (Yoshida *et al.*, 1999).

Euphorbin E

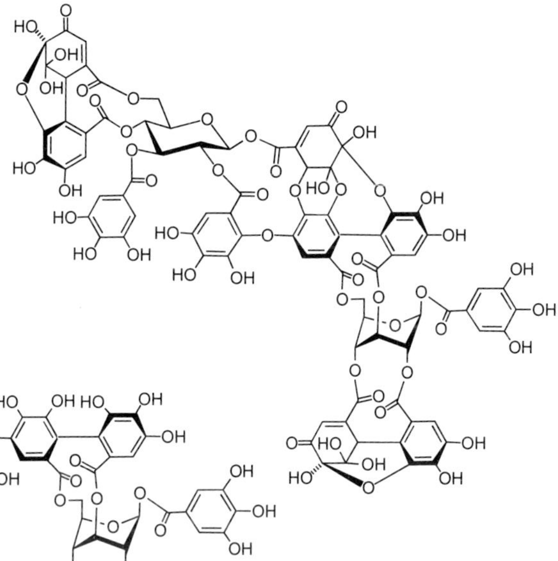

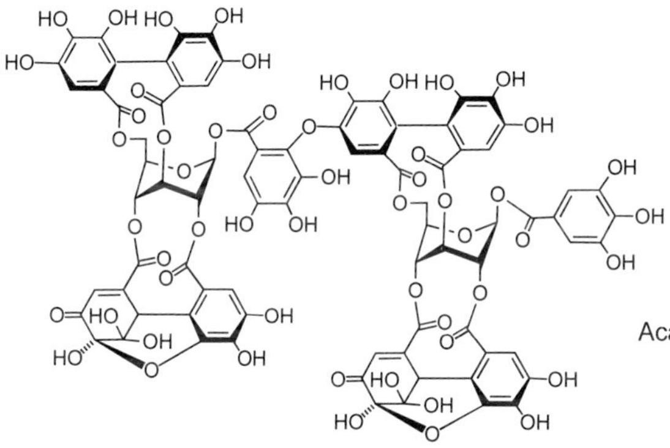

Acalyphidin D_1

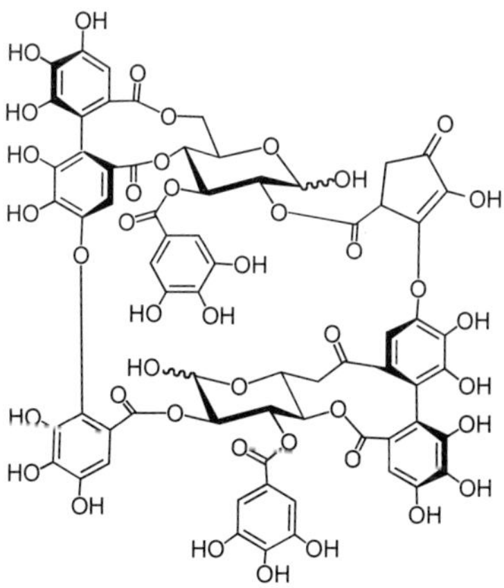

Eurobustin C

1.5 Main Ellagitannin-Rich Medicinal Plants

Many plant species containing ellagitannins have been used for the treatment of diseases, especially in Asia (Okuda *et al.*, 1992c). It is notable that the comparison of the amounts and pharmacological properties of ellagitannins and other components in plants shows that some of these ellagitannins could be playing the main role in the medicinal application of these plants (Fig. 1.11).

Fig. 1.11 Selection of ellagitannin-rich medicinal plants (main ellagitannin).

The aerial part of *Geranium thunbergii* (*Geraniaceae*), producing geraniin, is one of the most popular medicinal plants in Japan, and is one of the official medicines in the pharmacopoeia, often used as an antidiarrheic. This medicinal plant has also been widely used for controlling intestinal function, mainly for preventing constipation. These medicinal effects will be principally due to protection of mucous membrane in intestine mainly by geraniin, and not retardation or acceleration of peristalsis of intestine, as observed in a pharmacological experiment with extracted intestine. The potent antioxidant activity and related activities of geraniin could also be participating in these effects. These two ways of application against constipation and diarrhea, opposite to each other at first sight, would be attributable to these activities helped by the mild property of geraniin.

Mallotus japonicus (*Euphorbiaceae*) is also a folk medicine used in Japan. Besides its bark that contains bergenin and oligogalloylated bergenins, utilized as an anti-ulcer medicine, its leaves that yield geraniin and mallotusinic acid (a type-III+ ellagitannin) have also been used for their stomachic effects (Okuda and Seno, 1981). Fruits of *Trapa japonica* (*Oenotheraceae*), which contain trapanin B (a tetramer), were also used as a stomachic and tonic medicine (Hatano *et al.*, 1990c).

The herb of *Agrimonia pilosa* (*Rosaceae*), yielding agrimoniin, the first isolated dimeric ellagitannin, is used as an antidiarrheic and a hemostatic medicine in Japan, although it is not as popular as *G. thunbergii.* Moreover, it has been used clinically as an anti-cancer medicine in China. *Agrimonia eupatria*, a vulnerary, cholagogic and anti-aphonic plant that grows in Europe and other parts of the world, also produces agrimoniin. Particularly worthnoting is the host-mediated antitumor activity observed for agrimoniin and several other analogous ellagitannins (see Section 1.7.2.3 and Chapter 6, Section 6.2).

Myrobalans, the fruits of *Terminalia chebula* (*Combretaceae*) that grows in India and Southeast Asia, yield chebulinic acid, chebulagic acid and terchebin. It is one of the most frequently used plant parts in Ayurveda, the traditional Indian medicine.

The herb of *Oenothera biennis* (*Onagraceae*), trivially called evening primrose, and other *Oenothera* species were used by Native Americans to quiet nervous sensibility. The herb of *Oenothera erythrosepala*, from which were isolated the two macrocyclic oligomers oenothein A and B, is presumed to be a descendant from an American wild species, once cultivated.

The root peel of *Punica granatum* (*Punicaceae*) is an anthelmintic widely used against tapeworm. The ellagitannins granatins A and B and punicalin promote the anthelmintic effect of isopelletierine, which is not very effective when used alone. The fruit peel of this tree has also been used in Central and Western Asia as a gargling liquid for throat diseases.

The fruit of *Cornus officinalis* (*Cornaceae*), rich in cornusiins A, D and E (dimers), and C and F1 (trimers) (Hatano *et al.*, 1989a/b), is a tonic in several prescriptions of traditional Chinese medicine. The flower of *Camellia japonica* (*Theaceae*) was used as a hemostatic in China.

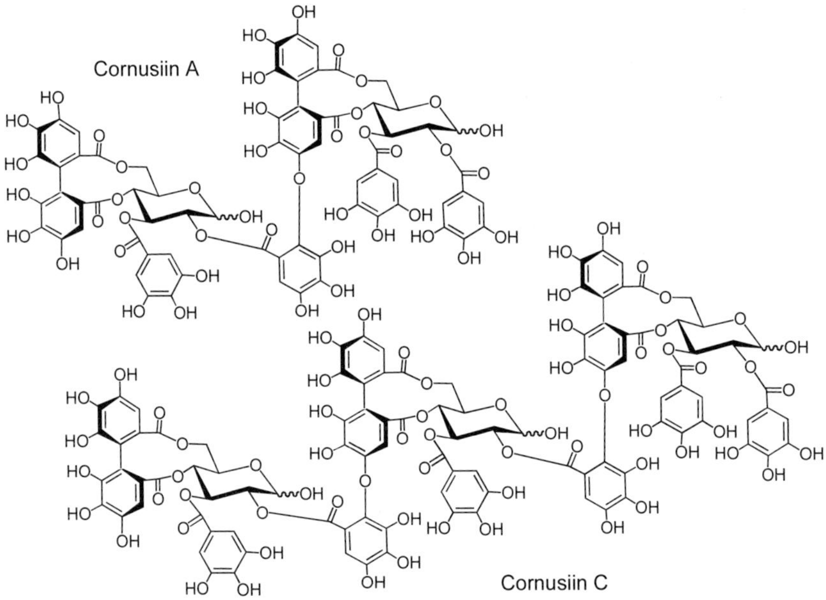

Cornusiin A

Cornusiin C

The leaves of *Castanea crenata* (*Fagaceae*), which produces acutissimins A and B (Ishimaru *et al.*, 1987), has been used as a tonic and an antitussive medicine, and also for treating a rash produced by lacquer poisoning. The herb of *Sanguisorba officinalis* (*Rosaceae*), which contains sanguiins H-6 and H-11, has been used as a hemostatic and antidiarrheic in China. The herb of *Geum japonicum* (*Rosaceae*), which contains *inter alia* dimeric gemins A, B and C, was used in Japan as a diuretic. The herb of *Rubus triphyllus* (*Rosaceae*), which contains sanguiins H-6 and H-11, was used as an antidiarrheic and a tonic medicine in Japan (Tanaka *et al.*, 1985).

1.6 Properties and Primary Activities of Ellagitannins

The remarkable chemical stability of most ellagitannins allows accurate evaluation of their biological and pharmacological activities, in contrast to other types of tannins for which such evaluations are often difficult to perform.

1.6.1 Reduction, stabilization and precipitation of other substances by tannins, and solubilization of precipitates by excess tannin

At room temperature, ellagitannins, like other tannins, reduce metallic ions such as Cu^{2+}, Fe^{3+} and Cr^{6+} into Cu^{+}, Fe^{2+} and Cr^{3+}, respectively (Okuda *et al.*, 1982i). Such reductions are presumably accompanied by oxidation of the tannin molecules into quinonoid species. The discoloration of natural pigments, *e.g.*, shikonin and β-carotene, that occurrs during storage of their respective solution in ethanol in the presence of oxygen under light exposure was remarkably suppressed by addition of geraniin. This suppressing effect was further enhanced in the presence of metallic ions such as Ca^{2+} and Mg^{2+}. The precipitates produced by mixing punicalin with Fe^{3+} or Cu^{2+} at a concentration of punicalin of 5.0×10^{-3} M were solubilized at higher concentrations of punicalin, and the precipitates produced by mixing chebulinic acid with quinine, cinchonine, berberine or papaverine were also solubilized by increasing the concentration of chebulinic acid over 1.0 mg/10ml (Okuda *et al.*, 1982j). Such solubilizations of precipitates are attributable to the higher solubility of the complexes formed when using higher concentrations of these tannins.

1.6.2 Indexes of tannin binding activity and reversal of tannin biological activities

The leather making activity of tannins is attributed to their aptitude to form multiple hydrogen bonds to collagen in hide. The binding of tannins with alkaloids has been exploited for preparing some medicines such as complexes of tannic acid with *inter alia* berberine and diphenhydramine in order to suppress the offensive taste of these compounds. Gallotannin mixtures have been mainly used for this purpose because of their ready availability. The efficacy with which a given tannin molecule binds to hemoglobin or methylene blue relatively to that of tannic acid JP (*i.e.*, Japanese Pharmacopoeia) (RA or RMB) or to that of geraniin (RAG or RMBG) offers convenient indexes that serve to rapidly evaluate the binding activities of various tannins (Okuda *et al.*, 1985). The latter indexes RAG and RMBG are more reliable than the former ones, because of the structural uniformity of the standard compound geraniin.

The effects of tannins on enzymes can drastically vary and even be reversed depending upon the concentration at which the tannin molecule is used, as well as upon the structural class to which it belongs. For example, the inhibitory effects of the ellagitannins chebulinic acid and granatin B on *Streptococcus mutans*, a carcinogenic bacterium, at 10^{-5} M are less potent than those observed at 10^{-6} M, but they are reinforced by further increasing the concentration of these ellagitannins (Kakiuchi *et al.*, 1986). Geraniin, mallotusinic acid, chebulinic acid and chebulagic acid enhanced adrenocorticotropic hormone (ACTH)-induced lipolysis in fat cells at a concentration of 20 μg/ml or 5 μg/ml, but all of these ellagitannins have no effect on the insulin-induced lipogenesis from glucose. These activities of ellagitannins are contrary to those observed with condensed tannins, which weakly inhibited ACTH-induced lipolysis, whereas they enhanced insulin-stimulated lipogenesis from glucose (Kimura *et al.*, 1983).

1.6.3 Antioxidant activities

One of the most notable activities of tannins and related polyphenols is their potent antioxidant activity (Okuda *et al.*, 1992b, 1993b, Okuda,

1997a/b, 1999b). One of the roles of tannins in plant tissues, particularly in those around the vascular bundle where their concentration is generally high, might have to do with the prevention or at least the retardation of oxidation in the plant body.

The antioxidant activity of tannins was initially demonstrated by their suppression of the autoxidation of ascorbic acid (Yoshida, *et al.*, 1981). The inhibitory effect of tannins on Cu(II)-catalyzed autoxidation of ascorbic acid was examined by kinetic studies and ESR measurements showing that the inhibitory effects by several ellagitannins (*e.g.*, geraniin, mallotusinic acid and corilagin), and ellagic acid, which is produced by hydrolysis of ellagitannins, are markedly higher than that by polyphenols of low molecular masses, such as gallic acid, and also significantly higher than that by pentagalloylglucose (PGG). These effects are attributable to the potent radical scavenging activity of ellagitannins as substantiated by signals of stable free radicals in their ESR spectra. Unlike ellagitannins, polyphenols of low molecular masses usually gave unstable or no ESR signals. However, the antioxidant effect of ellagic acid, in spite of its rather small size, is quite high in accordance with the high stability of its free radical (Fujita *et al.*, 1987).

The radical scavenging capacity of ellagitannins of various chemical structures has also been evaluated on the basis of their effects on the 1,1-diphenyl-2-picrylhydrazyl (DPPH) radical (Fig. 1.12). Ellagitannins generally showed more potent effects than α-tocopherol and ascorbic acid, as measured by the discoloration of the purple DPPH radical solution. The generation of stable free radical from an alkyl gallate upon scavenging the DPPH radical was demonstrated by ESR measurements and by high-yielding isolation of the dialkyl ester of hexahydroxydiphenoic acid (HHDP) produced by mutual coupling of transient C-centered galloyl radicals (Yoshida *et al.*, 1989b). The antioxidant activity of ellagitannins, thus attributable to their radical scavenging effect, was also evidenced by their significant inhibitory action on the carbon tetrachloride- and galactosamine-induced cytotoxicities in primary cultured rat hepatocytes (Hikino *et al.*, 1985).

The antioxidant effect of tannins on lipids in biological systems was shown by inhibition of lipid peroxidation induced by adenine 5'-diphosphate (ADP) and ascorbic acid in rat liver mitochondria, and by inhibition of lipid peroxidation induced by ADP and NADPH in rat liver microsomes. All tannins, except some small polyphenols and methylated polyphenols among twenty-five compounds, showed significant inhibitory effects in these two systems at a concentration of 1 μg/ml, ellagitannins being generally much more potent than condensed tannins. The peroxidation was almost completely inhibited by pedunculagin and isoterchebin at a dose of 5 μg/ml (Okuda *et al.*, 1983b).

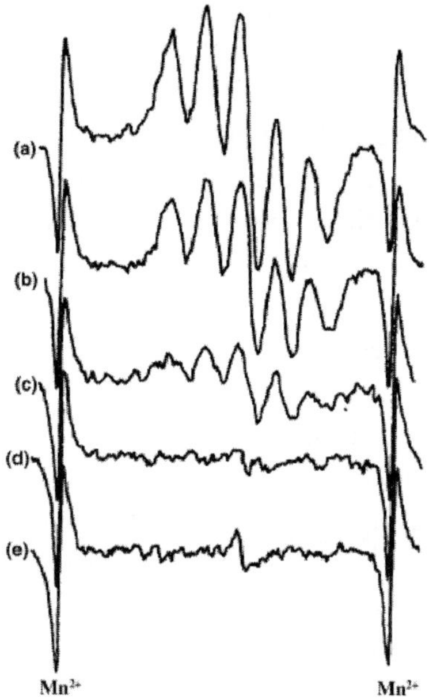

Fig. 1.12 Evidence of the radical scavenging effects of geraniin on the ESR spectrum of DPPH. The ESR spectra were recorded in the absence [(a)] and in the presence [(b) 5.0 × 10^{-7} M, (c) 1.0 × 10^{-6} M, (d) 2.5 × 10^{-6} M, (e) 5.0 × 10^{-6} M] of geraniin.

In a study of the effects of tannins on arachidonic acid metabolism, geraniin and corilagin inhibited the formation of the lipoxygenase product 5-HETE (*i.e.*, 5-hydroxyeicosatetraenoic acid) in rat peritoneal polymorphonuclear leukocytes, dose dependently at concentrations ranging from 10^{-3} to 10^{-6} M, whereas the formation of the cyclooxygenase products HHT, thromboxane B_2 and 6-keto-$PGF_{1\alpha}$ was not inhibited at these concentrations (Kimura *et al.*, 1986).

In an investigation of the protective effects of tannins against oxidative damage induced in mouse ocular lenses by incubating them with xanthin-xanthine oxidase, ADP and Fe^{3+} (*i.e.*, X-XOD system), lipid peroxide concentration in the lens was markedly lowered by geraniin, but the effect was low when using polyphenolic small molecules (Iwata *et al.*, 1987).

The effects of twenty-five tannins, including ellagitannins and small polyphenols, on the concentration of superoxide anion radical generated in the hypoxanthine-xanthine oxidase system, were evaluated by ESR measurements of the adduct of 5,5-dimethyl-1-pyrroline-*N*-oxide (DMPO) and the radical. Increasing the ellagitannin concentration in the solution of superoxide generating mixture inhibited the appearance of signals of the DMPO adduct in a dose-dependent manner. The scavenging effect of ellagitannins on the superoxide anion radical increased generally with increase of the number of phenolic hydroxyl groups in the molecule. The radical scavenging mechanism of ellagitannins and related polyphenols on the superoxide anion radical was thus substantiated (Hatano *et al.*, 1989c).

1.7. Biological and Pharmacological Activities

1.7.1 Antiviral, antimicrobial and immunomodulatory activities

Monomeric and dimeric ellagitannins, as well as some gallotannins, potently inhibited *Herpes simplex* infection, as reported by Fukuchi and co-workers (Fukuchi *et al.*, 1989). Dimeric ellagitannins efficaciously inhibited reverse transcriptase from RNA tumor virus (Kakiuchi *et al.*, 1985). Several ellagitannins, including gemin D (a monomer), gemin A, camelliin B and nobotanins A, B and F (dimers), and trapanin B and nobotanin K (tetramers), among 87 tannins and related polyphenols, were shown to inhibit HIV-induced cytopathic effects and HIV-specific antigen expression, but condensed tannins expressed no similar activity (Nakashima *et al.*, 1992).

Ellagitannins and other tannins exhibited interesting antimicrobial activities on some drug-resistant bacteria in the presence of some other antimicrobial agents (see Chapter 2). *Helicobacter pylori*, a Gram-negative spirillum that may cause chronic gastritis, gastric ulcer, duodenal ulcer and also stomach cancer, was potently inhibited by tellimagrandin I and corilagin at a minimum inhibitory concentration (MIC) of 6.25 μg/ml.

The effect of β-lactam against MRSA (methicillin-resistant *Staphylococcus aureus*) aquiring multi-drug resistance was restored by corilagin and tellimagrandin I, as well as by several other polyphenols (see Chapter 2 for further details). Potent effects against leishmanises, a group of diseases with extensive morbidity and mortality in developing countries, were observed using several dehydroellagitannins, the ellagitannin hippophaenin A and also several gallotannins, while the effect of proanthocyanidins was generally less pronounced. Differences were found between *Leishmania* promastigotes and *L.* amastigotes in the anti-leishmanial activity of each polyphenol. These intriguing differences may be indicative of an activation of leishmanicidal macrophage function, which led researchers to rely on several functional bioassays, including a biochemical assay for nitric oxide (NO), a fibroblast-lysis assay for release of tumor necrosis factor (TNF-α), and a cytopathic effect inhibition assay for interferon (IFN)-like properties, for carrying out in-depth investigations of the activity of tannins on leishmanises (see Chapter 2).

1.7.2 Antitumor activities

A large amount of hard data has been gathered during the last two decades on the inhibitory activities of polyphenols, including ellagitannins and analogs, on tumor incidence and propagation. This is in sharp contrast to what was commonly thought earlier on, when precise chemical evidence of tannins was not available. Indeed, induction of cancers by some plant species was believed to be due to their high content in "tannins". Today, the cytotoxic activity of several tannins of defined structure has been reported. Recent evidences of the antitumor activities of ellagitannins are reviewed thereafter.

1.7.2.1 *Inhibition of mutagenicity of carcinogens*

The mutagenicity of Trp-P-1 (*i.e.*, 3-amino-1,4-dimethyl-5*H*-pyrido[4,3-*b*]indole) and MNNG (*i.e.*, N-methyl-N'-nitro-N-nitrosoguanidine), and also that of *N*-OH-Trp-P-2 (*i.e.*, 3-hydroxyamino-1-methyl-5H-pyrido[4, 3-b]indole), a directly-acting mutagen, were strongly inhibited by ellagitannins from medicinal plants, such as geraniin, mallotusinic acid, pedunculagin and agrimoniin, and also by (–)-epigallocatechin gallate (EGCG). Since ellagic acid was found to inhibit the mutagenicity of 7β,8α-dihydroxy-9α,10α-epoxy-7,8,9,10-tetrahydrobenzo[a]pyrene (B[a]p diol epoxide) and since it is produced by hydrolysis of ellagitannins, variation of the antimutagenic activity of *Geranium thunbergii* was investigated along the extraction of geraniin during which its hydrolysis occurs. Interestingly, the results showed that after an initial and rapid increase of the inhibitory effect on Trp-P-1 due to an increase of the concentration of geraniin as a result of its extraction from the plant, a marked downward modulation of that effect was observed due to the hydrolysis of geraniin. An upward modulation of the inhibitory effect on B[a]p diol epoxide occurred simultaneously due to the hydrolytic release of ellagic acid from geraniin (Okuda *et al.*, 1984a).

1.7.2.2 *Inhibition of tumor promotion*

Tumor promotion is a much longer process than its initiation during the two-stage chemical carcinogenesis, and its inhibition is therefore regarded as an important objective in cancer prevention. The tumor promotion on mouse skin by 12-*O*-tetradecanoylphorbol-13-acetate (TPA), after initiation with dimethylbenz[a]anthracene (DMBA), was significantly inhibited by ellagic acid and several ellagitannins isolated from *Cowania mexicana* and *Coleogyne ramosissima*. These compounds were also found to inhibit Epstein-Barr virus early antigen (EBV-EA) activation induced by TPA. The TPA-induced ornithine decarboxylase (ODC) activity and the TPA-stimulated hydroperoxide production were inhibited by several ellagitannins and other polyphenols (Ito *et al.*, 1999b). Inhibitors of TNF-α release nowadays constitute attractive potential candidates for the development of cancer preventing strategies.

In this vein of investigation, geraniin and corilagin were identified as potent inhibitors (Okabe *et al.*, 2001).

1.7.2.3 *Host-mediated antitumor activity*

Several oligomeric ellagitannins specifically inhibited tumor (Sarcoma-180 and MM2) growth after having been administrated either before or after intraperitoneal inoculation of tumor cells into mice abdomen. This effect was found only for these oligomers among over a hundred tannins and related polyphenols thus screened. Among these active oligomers were macrocyclic oligomers, such as oenothein B and woodfordin C (dimers), oenothein A and woodfordin D (trimers), and woodfordin F (tetramer) (Miyamoto *et al.*, 1997). This effect was attributed to an enhancement of the immune response of host animals, which was supported by their stimulation of IL-1 production from human peripheral macrophages (Miyamoto *et al.*, 1993, see Chapter 6).

1.7.3 Induction of apoptosis

Ellagitannins induced apoptotic cell death, which was characterized by internucleosomal DNA cleavage and apoptotic body in human promyelocytic leukemic HL-60 cells and evaluated by agarose gel electrophoresis and fluorescence activated cell sorter, at levels of potency higher than those determined for condensed tannins. However, the most active compound thus screened was the simple phenol gallic acid (Inoue *et al.*, 1994, Sakagami *et al.*, 1995, 1999).

1.7.4 Effects on liver functions and others

Intramuscular administration of geraniin siginificantly lowered levels of glutamyl oxaloacetic transaminase (GOT), glutamyl pyruvic transaminase (GPT) and lipid peroxides in serum (Nakanishi *et al.*, 1999). Intramuscular injection of geraniin and ellagic acid significantly suppressed experimental hepatic injuries induced by carbon tetrachloride, D-galactosamine, and thioacetamide in rats, and a protective effect against liver damages was confirmed by histological observation (Hikino *et al.*, 1985). Geraniin was also found to suppress the increase of lipid peroxide level in the serum caused by inhalation of carbon tetrachloride.

The effects of tannins as evaluated via oral administration are particularly worthy of further investigation. The oral administration of geraniin-rich extract of *Geranium thunbergii* significantly lowered the lipid peroxide level in the serum and the liver of rats in which liver injury was induced by feeding them with peroxidized oil. The levels of serum cholesterol, GOT and GPT in the rats treated with peroxidized oil were also reduced in the presence of geraniin (Kimura *et al.*, 1984).

1.7.5 Absorption and metabolism of ellagitannins in animals

The metabolic conversion of ellagic acid in animals into 3,8-dihydroxybenzo-[b,d]pyran-6-one and related compounds was reported in 1980 (Doyle and Griffiths, 1980). This compound and 3-hydroxy-6*H*-dibenzo[b,d]pyran-6-one were detected in the urine and serum of a sheep that was fed with *Terminalia oblongata* leaves containing chebulagic acid, punicalagin and teroblongin (*i.e.*, 1-α-*O*-galloylpunicalagin). 3-*O*-Glucuronide of 3-hydroxy-6*H*-dibenzo[b,d]pyran-6-one was also isolated from the urine and serum of the sheep (Okuda *et al.*, 1995). These aspects and other related to the bioavailability of ellagitannins are discussed in greater details in Chapters 7 and 8.

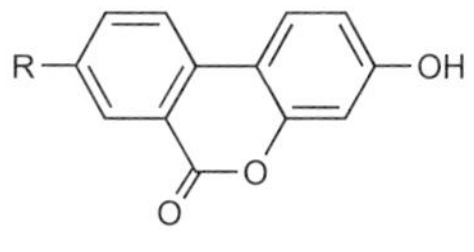

3,8-Dihydroxy-6*H*-dibenzo[b,d]pyran-6-one : R = OH
3-Hydroxy-6*H*-dibenzo[b,d]pyran-6-one : R = H

1.8 Bibliography

Amakura, Y., Miyake, M., Ito, H., Murakaku, S., Araki, S., Itoh, Y., Lu, C.-F., Yang, L.-L., Yen, K.-Y., Okuda, T. and Yoshida, T. (1999). Acalyphidins M_1, M_2, and D_1, ellagitannins from *Acalypha hispida*, *Phytochemistry*, 50, pp. 667–675.

Chen, L.-G., Yen, K.-Y., Yang, L. L., Hatano, T., Okuda, T. and Yoshida, T. (1999). Macrocyclic ellagitannin dimers, cuphiins D_1 and D_2, and accompanying tannins from *Cuphea hyssopifolia*, *Phytochemistry*, 50, pp. 307–312.

Doyle, B. and Griffiths, L. A. (1980). The metabolism of ellagic acid in the rat, *Xenobiotica*, 10, p. 247–256.

Fujita, T., Komagoe, K., Sasaki, Y., Uehara, I. Okuda, T. and Yoshida, T. (1987). Inhibition mechanism of tannins on Cu(II)-catalyzed autoxidation of ascorbic acid, *Yakugaku Zasshi*, 107, pp. 17–22.

Fukuchi, K., Sakagami, H., Okuda, T., Hatano, T., Tanuma, S., Kitajima, K., Inoue, Y., Inoue, S., Ichikawa, S., Nonoyama, M. and Konno, K. (1989). Inhibition of *Herpes simplex* virus infection by tannins and related compounds, *Antiviral Res.*, 11, pp. 285–298.

Haslam, E. (1989). *Plant Polyphenols*, Cambridge University Press: Cambridge.

Hatano, T., Hattori, S and Okuda, T. (1986). Coriariin A and B, new dimeric and monomeric hydrolysable tannins, *Chem. Pharm. Bull.*, 34, pp. 4092–4097.

Hatano, T., Ogawa, N., Kira, R., Yasuhara, T. and Okuda, T. (1989a). Cornusiins A, B and C, dimeric, monomeric and trimeric hydrolysable tannins from *Cornus officinalis*, and orientation of valoneoyl group in related tannins, *Chem. Pharm. Bull.*, 37, pp. 2083–2090.

Hatano, T., Yasuhara, T. and Okuda, T. (1989b). Cornusiins D, E and F, new dimeric and trimeric hydrolysable tannins from *Cornus officinalis*, *Chem. Pharm. Bull.*, 37, pp. 2665–2669.

Hatano, T., Edamatsu, R., Hiramatsu, M., Mori, A., Fujita, Y., Yasuhara, T., Yoshida, T. and Okuda, T. (1989c). Effects of tannins and related polyphenols on superoxide anion radical, and on 1,1-diphenyl-2-picrylhydrazyl radical, *Chem. Pharm. Bull.*, 37, pp. 2016–2021.

Hatano, T., Yasuhara,T., Matsuda, M., Yazaki, K., Yoshida, T. and Okuda, T. (1990a) Oenothein B, a dimeric hydrolysable tannin with macrocyclic structure, and accompanying tannins from *Oenothera erythrosepala*, *J. Chem. Soc. Perkin Trans. 1*, 2735–2743.

Hatano, T., Ogawa, N., Shingu, T. and Okuda, T. (1990b). Rugosins D, E, F and G, dimeric and trimeric hydrolysable tannins with valoneoyl group(s), from flower petals of *Rosa rugosa*, *Chem. Pharm. Bull.*, 38, pp. 3341–3346.

Hatano, T., Okonogi, A., Yazaki, K. and Okuda, T. (1990c). Trapanins A and B, oligomeric hydrolyzable tannins from *Trapa japonica* Flerov., *Chem. Pharm. Bull.*, 38, pp. 2707–2711.

Hatano, T., Han, L., Taniguchi, S., Singu, T., Okuda, T. and Yoshida, T. (1995). Camelliatannins C and E, new complex tannins from *Camellia japonica* leaves. *Chem.Pharm. Bull.*, 43, pp. 1629–1633.

Hikino, H., Kiso, Y., Hatano, T., Yoshida, T. and Okuda, T. (1985). Antihepatotoxic actions of tannins, *J. Ethnopharmacol.*, 14, pp. 19–26.

Inoue, M., Suzuki, R., Koide, T., Sakaguchi, N., Ogihara, Y. and Yabu, Y. (1994). Antioxidant, gallic acid, induces apoptosis in HL-60 RG cells, *Biochem. Biophys. Commun.*, 204, pp. 898–904.

Ishimaru, K., Nonaka, G. and Nishioka I. (1987). Isolation and characterization of acutissimins A and B, novel tannins from *Quercus* and *Castanea* species, *Chem. Pharm. Bull.*, 38, pp. 602–610.

Ito, H., Hatano, T., Namba, O., Shirono, T., Okuda, T. and Yoshida, T. (1999a). Modified dehydroellagitannins. Geraniinic acids B and C, and phyllanthusiin F, *Chem. Pharm. Bull.*, 47, pp. 1148–1151.

Ito, H., Miyake, M., Nishitani, E., Mori, K., Hatano, T., Okuda, T., Konoshima, T., Takasaki, N., Kozuka, M., Mukainaka, T., Tokuda, H., Nishino, H. and Yoshida, T. (1999b). Antitumor promoting activity of polyphenols from *Cowania mexicana* and *Coleogyne ramosissima*, *Cancer Letters*, 143, pp. 5–13.

Iwata, S., Fukaya, Y., Nakazawa, K. and Okuda, T. (1987). Effects of tannins on the oxidative damage of mouse ocular lens, *J. Ocular Pharmacol.*, 3, pp. 227–238.

Kakiuchi, N., Hattori, M., Namba, T., Nishizawa, M., Yamagishi, T. and Okuda, T. (1985). Inhibitory effects of tannins on reverse transcriptase from RNA tumor virus, *J. Nat. Prod.*, 48, pp. 614–621.

Kakiuchi, N., Hattori, M., Nishizawa, M., Yamagishi, T., Okuda, T. and Namba, T. (1986). Inhibitory effect of various tannins on glucan synthesis by glucosyltransferase from *Streptococcus mutans*, *Chem. Pharm. Bull.*, 34, pp. 720–725.

Kimura, Y., Okuda, H., Okuda, T., Yoshida, T., Hatano, T. and Arichi, S. (1983). Effects of various tannins and related compounds on adrenocorticotropic hormone-induced lipolysis and insulin-induced lipogenesis from glucose in fat cells, *Chem. Pharm. Bull.*, 31, pp. 2501–2506.

Kimura, Y., Okuda, H., Mori, K., Okuda, T. and Arichi, S. (1984). Effects of various extracts of *Geranii herba* and geraniin on liver injury and lipid metabolism in rats fed peroxidized oil, *Chem. Pharm. Bull.*, 32, pp. 1866–1871.

Kimura, Y., Okuda, H., Okuda, T. and Arichi, S. (1986). Effects of geraniin, corilagin and ellagic acid isolated from *Geranii Herba* on arachidonate metabolism in leukocytes, *Planta Med.*, 52, pp. 337–338.

Lee, M.-H., Nishimoto, S., Yang, L.-L., Yen, K.-Y., Hatano, T., Yoshida, T. and Okuda, T. (1997). Two macrocyclic hydrolysable tannin dimers from *Eugenia uniflora*, *Phytochemistry*, 44, pp. 1343–1349.

Lin. J.-H., Ishimaru, M., Tanaka, T., Nonaka, G.. and Nishioka I. (1990). Structures of macaranins and macarinins, new hydrolysable tannins possessing macaranoyl and tergalloyl ester groups, from the leaves of *Macaranga sinensis*, *Chem. Pharm. Bull.*, 38, pp. 1844–1851

Luger, P., Weber, M., Kashino, S., Amakura, Y., Yoshida, T., Okuda, T., Beurskens, G., and Dauter, Z. (1998). Structure of the tannin geraniin based on conventional X-ray data at 295K and on synchrotron data at 293 and 120K, *Acta Cryst.*, *B* 54, pp. 687–694.

Miyamoto, K., Murayama, T., Nomura, M., Hatano, T., Yoshida, T. and Okuda, T. (1993). Antitumor activity and interleukin-1 induction by tannins, *Anticancer Res.*, 13, pp. 37–42.

Miyamoto, K., Murayama, T., Yoshida, T., Hatano, T. and Okuda, T. (1997). Anticarcinogenic activities of polyphenols in foods and herbs. In *Antinutrients and Phytochemicals in Food*, ed. by Shahidi, F., American Chemical Society: Washington, DC., pp. 245-259.

Mayer, W. (1971). Über die Gerbstoffe aus dem Holz der Edelkastanie und Eiche, *Das Leder*, 22, pp 277–283.

Niemetz, R. and Gross, G. G. (2005). Enzymology of gallotannin and ellagitannin biosynthesis, *Phytochemistry*, 66, pp. 2001–2011.

Nakanishi, Y., Okuda, T. and Abe, H. (1999). Effects of geraniin on the liver in rats. III. *Natural Medicines*, 53, pp. 22–26.

Nakashima, H., Murakami, T., Yamamoto, N., Sakagami, S., Tanuma, T., Hatano, T., Yoshida, T. and Okuda, T. (1992). Inhibition of human immunodeficiency viral replication by tannins and related compounds, *Antiviral Res.*, 18, pp. 91–103.

Okabe, S., Suganuma, M., Imayoshi, Y., Taniguchi, S., Yoshida, T. and Fujiki, H. (2001). New TNF-α releasing inhibitors, geraniin and corilagin, in leaves of *Acer nikoense*, Megusurino-ki, *Biol. Pharm. Bull.*, 24, pp. 1145–1148.

Okuda, T., Yoshida, T. and Nayeshiro, H. (1976). Geraniin, a new ellagitannin from *Geranium thunbergii*, *Tetrahedron Lett.*, pp. 3721–3724.

Okuda, T., Hatano, T., Nitta, H. and Fujii, R. (1980a). Revised structure of terchebin and structure of granatin B, *Tetrahedron Lett.*, 21, pp. 4361–4364.

Okuda, T., Mori, K., and Hatano, T. (1980b). The distribution of geraniin and mallotusinic acid in the Order Geraniales, *Phytochemistry*, 19, pp. 547–551.

Okuda, T. and Seno, K. (1981). Tannins of leaf of *Mallotus japonicus*, *J. Chem. Soc. Jp.*, pp. 671–677.

Okuda, T., Yoshida, T. and Hatano, T. (1982a). Hydrated stereostructure and equilibration of geraniin, *J. Chem. Soc. Perkin Trans. 1*, pp. 9–14.

Okuda, T., Yoshida, T., Kuwahara, M., Memon, M. U. and Shingu, T. (1982b). Agrimoniin and potentillin, ellagitannin dimer and monomer having α-glucose cores, *J. Chem. Soc. Chem. Commun.*, pp. 163–164.

Okuda, T., Yoshida, T., Ashida, M. and Yazaki, K. (1982c). Casuariin, stachyurin and strictinin, new ellagitannins from *Casuarina stricta* and *Stachyrus praecox*, *Chem. Pharm. Bull.*, 30, pp. 766–769.

Okuda, T., Hatano, T. and Yazaki, K. (1982d). Dehydrogeraniin, furosinin and furosin, dehydroellagitannins from *Geranium thunbergii*, *Chem. Pharm. Bull.*, 30, pp. 1113–1116.

Okuda, T., Yoshida, T., Hatano, T., Koga, T., Toh, N. and Kuriyama, K. (1982e). Circular dichroism of hydrolysable tannins. I. Ellagitannins and gallotannins. *Tetrahedron Lett.*, 23, pp. 3937–3940.

Okuda, T., Yoshida, T., Hatano, T., Koga, T., Toh, N. and Kuriyama, K. (1982f). Circular dichroism of hydrolysable tannins. II. Dehydroellagitannins. *Tetrahedron Lett.*, 23, pp. 3941–3944.

Okuda, T., Hatano, T and Ogawa, N. (1982g). Rugosins D, E, F and G, dimeric and trimeric hydrolyzable tannins, *Chem. Pharm. Bull.*, 30, pp. 4234–4237.

Okuda, T., Yoshida, T., Hatano, T., Yazaki, K. and Ashida, M. (1982h). Ellagitannins of *Casuarinaceae*, *Stachyuraceae* and *Myrtaceae*, *Phytochemistry*, 21, pp. 2871–2874.

Okuda, T., Mori, K., Shiota, M. and Ida, K. (1982i). Reduction of heavy metal ions and solubilization of precipitates, *Yakugaku Zasshi*, 102, pp. 735–742.

Okuda, T., Mori, K. and Shiota, M. (1982j). Formation and solubilization of precipitates with alkaloids, *Yakugaku Zasshi*, 102, pp. 854–858.

Okuda, T., Yoshida, T., Ashida, M. and Yazaki, K. (1983a). Tannins of *Casuarina* and *Stachyurua* species. Part 1. Structures of pedunculagin, casuarictin, strictinin, casuarinin, casuariin and stachyurin, *J. Chem. Soc. Perkin Trans. 1*, pp. 1765–1772.

Okuda, T., Kimura, Y., Yoshida, t., Hatano, T. Okuda, H. and Arichi, S. (1983b). Inhibitory effects on lipid peroxidation in mitochondria and microsomes of liver. *Chem. Pharm. Bull.*, 31, pp. 1625–1631.

Okuda, T., Mori, K. and Hayatsu, H. (1984a). Inhibitory effect of tannins on direct-acting mutagens, *Chem. Pharm. Bull.*, 32, pp. 3755–3758.

Okuda, T., Hatano, T., Ogawa, N., Kira, R. and Matsuda, M. (1984b). Cornusiin A, a dimeric ellagitannin forming four tautomers, and accompanying new tannins in *Cornus officinalis*, *Chem. Pharm. Bull.*, 32, pp. 4662–4665.

Okuda, T., Mori, K. and Hatano, T. (1985). Relationship of the structures of tannins to the binding activities with hemoglobin and methylene blue, *Chem. Pharm. Bull.*, 33, pp. 1424–1433.

Okuda, T., Yoshida, T., Hatano, T., Yazaki, K., Kira, R. and Ikeda, Y. (1986a). Preparative fractionation of hydrolysable tannins by centrifugal partition chromatography, *J. Chromatogr.*, 362, pp. 375–381.

Okuda, T., Yoshida, T., Hatano, t., Ikeda, Y. Shingu, T. and Inoue. T. (1986b). Isolation of water-soluble tannins by centrifugal partition chromatography, and biomimetic synthesis of elaeocarpusin, *Chem. Pharm. Bull.*, 34, pp. 4075–4082.

Okuda, T., Hatano, T., Kaneda, t., Yoshizaki, M. and Shingu, T. (1987). Liquidambin, an ellagitannin from *Liquidambar formosana*, *Phytochemistry*, 26, pp. 2053–2055.

Okuda, T., Yoshida, T. and Hatano, T. (1989a). New methods of analyzing tannins, *J. Nat. Prod.*, 52, pp. 1–31.

Okuda, T., Yoshida, T. and Hatano, T. (1989b). Ellagitannins as active constituents of medicinal plants, *Planta Medica*, 55, pp. 117–122.

Okuda, T., Okuda, T., Yoshida, T. and Hatano, T. (1990). Oligomeric hydrolysable tannins, a new class of plant polyphenols, *Heterocycles*, 30, pp. 1195–1218.

Okuda, T., Yoshida, T. and Hatano, T. (1991). Chemistry and biological activity of tannins in medicinal plants. In *Economic and Medicinal Plant Research*, vol. 5, ed by Wagner, H. and Farnsworth, N. R., Academic Press: London, pp. 129–165.

Okuda, T., Yoshida, T. and Hatano, T. (1992a). Pharmacologically active tannins isolated from medicinal plants. In *Plant polyphenols*, ed. by Hemingway, R. W. and Laks, P. E., Plenum Press: New York, pp. 530–569.

Okuda, T., Yoshida, T. and Hatano, T. (1992b). Antioxidant effects of tannins and related polyphenols. In *Phenolic Compounds in Food and their Effects on Health*, vol. I, ed. by Ho, C.-T., Lee, C. Y. and Huang, M.-T., American Chemical Society: Washington, DC., pp. 87–97.

Okuda, T., Yoshida, T. and Hatano, T. (1992c). Polyphenols from Asian plants: structural diversity and antitumor and antiviral activities. In *Phenolic Compounds in Food and their Effects on Health*, vol. II, ed. by Huang, M.-T., Ho, C.-T. and Lee, C. Y., American Chemical Society: Washington, DC., pp. 160–183.

Okuda, T., Yoshida, T. and Hatano, T. (1993a). Classification of oligomeric hydrolysable tannins and specificity of their occurrence in plants, *Phytochemistry*, 32, pp. 507–521.

Okuda, T., Yoshida, T and Hatano, T. (1993b). Antioxidant phenolics in oriental medicine. In *Active Oxygen, Lipid Peroxides, and Antioxidants*, ed. by Yagi, K., CRC Press: Boca Raton, pp. 333–346.

Okuda, T. (1995a). Tannins, a new family of bioactive natural organic compounds (questions and answers), *Yakugaku Zasshi*, 115, pp. 81–100.

Okuda, T., Yoshida, T. and Hatano, T. (1995b). Hydrolyzable tannins and related polyphenols. In *Progress in the Chemistry of Organic Natural Products*, vol. 66, ed. by Herz, W., Kirby, G. W., Moore, R. E., Steglich, W. and Tamm, Ch., Springer-Verlag: Wien, pp. 1–117.

Okuda, T. (1997a). Structure-activity relationship of antioxidant and antitumor polyphenols. In *Food Factors for Cancer Prevention*, ed. by Ohigashi, H., Osawa, T., Terao, J., Watanabe, S. and Yoshikawa, T., Springer-Verlag: Wien, pp. 280–285.

Okuda, T. (1997b). Phenolic antioxidants. In *Food and Free Radicals*, ed. by Hiramatsu, M., Yoshikawa, T. and Inoue, M., Plenum Press: New York, pp 31–48.

Okuda, T. (1999a). Novel aspects of tannins – Renewed concepts and structure-activity relationships, *Curr. Org. Chem.*, 3, pp. 609–622.

Okuda, T. (1999b). Antioxidants in herbs: polyphenols. In *Antioxidant Food Supplements in Human Health*, ed. by Packer, L., Hiramatsu, M. and Yoshikawa, T., Academic Press: San Diego, pp.393–410.

Okuda, T., Yoshida, T. and Hatano, T. (2000). Correlation of oxidative transformations of hydrolysable tannins and plant evolution, *Phytochemistry*, 55, pp. 513–529.

Okuda, T. (2005). Systematics and health effects of chemically distinct tannins in medicinal plants, *Phytochemistry*, 66, pp. 2012–2031.

Quideau, S. and Feldman, K. S. (1996). Ellagitannin chemistry. *Chem. Rev.*, 96, pp. 475–503.

Sakagami, H., Kuribayashi, N., Iida, M., Sakagami, T., Takeda, M., Fukuchi, K., Gomi, K., Ohta, H., Momose, K., Kawazoe, Y., Hatano, T., Yoshida, T. and Okuda, T. (1995). Induction of DNA fragmentation by tannin- and lignin-related substances, *Anticancer Res.*, 15, pp. 2121–2128.

Sakagami, H., Satoh, K., Ida, Y., Koyama, N., Premanathan, M., Arakaki, R., Nakashima, H., Hatano, T., Okuda, T. and Yoshida, T. (1999). Induction of apoptosis and anti-HIV activity by tannin- and lignin-related substances. In *Plant Polyphenols*, vol. 2, ed. by Gross, G.G., Hemingway, R. W. and Yoshida, T., Kluwer Academic/Plenum Publishers: New York, pp. 595–611.

Schmidt, O.T. and Mayer, W. (1956). Naturliche Gerbstoffe, *Angew. Chem.*, 68, pp. 103–115.

Tanaka, T., Nonaka, G. and Nishioka, I. (1985). Revision of the structures of sanguiins H-6, H-2 and H-3, and isolation and characterization of sanguiin H-11, a novel tetrameric hydrolysable tannin, and seven related tannins from *Sanguisorba officinalis*, *J. Chem. Res. (M)*, pp. 2001–2009.

Tanaka, T., Tachibana, H., Nonaka, G., Nishioka, I. and Hsu F.-L. (1993). New dimeric trimeric and tetrameric ellagitannins, lambertianins A-D, from *Rubus lambertianus*, *Chem. Pharm. Bull.*, 41, pp. 1214–1220.

Taniguchi, S., Nakamura, N., Nose, M., Takeda, S., Yabu-uchi, R., Ito, H., Yoshida, T. and Yazaki, K. (1998). Production of macrocyclic ellagitannin oligomers by *Oenothera laciniata* callus cultures, *Phytochemistry*, 48, pp. 981–985.

Taniguchi, S., Imayoshi, Y., Hatano, T., Yazaki, K. and Yoshida, T. (2002). Hydrolyzable tannin production in *Oenothera tetraptera* shoot tissue culture, *Plant Biotechnology*, 19, pp. 357–363.

Taniguchi, S., Uechi, K., Kato, R., Ito, H., Hatano, T., Yazaki, K. and Yoshida, T. (2002). Accumulation of hyfrolyzable tannins by *Aleurites fordii* callus culture. *Planta Med.*, 68, pp. 1145–1146.

Yazaki, K.and Okuda, T. (1990). Ellagitannin formation in callus culture of *Heterocentron roseum*, *Phytochemistry*, 29, pp. 1127–1130.

Yoshida, T., Fujii, R. and Okuda, T. (1980). Revised structure of chebulinic acid and chebulagic acid, *Chem. Pharm. Bull.*, 28, pp. 3713–3715.

Yoshida, T., Koyama, S. and Okuda, T. (1981). Inhibitory effects of tannins on cupric ion-catalyzed autoxidation of ascorbic acid, *Yakugaku Zasshi*, 101, pp. 695–699.

Yoshida, T., Okuda, T., Memon, M. U and Shingu, T. (1982). Structure of gemin A, a new dimeric ellagitannin having α- and β-glucose cores, *J. Chem. Soc. Chem. Commun.*, pp. 351–353.

Yoshida, T., Memon, M. U., Shingu, T. and Okuda, T. (1985a). Gemins A, B and C, new dimeric ellagitannins from *Geum japonicum*, *J. Chem. Soc. Perkin Trans. 1*, pp. 315–321.

Yoshida, T., Maruyama, Y., Memon, M. U., Shingu, T. and Okuda, T. (1985b). Gemins D, E and F, ellagitannins from *Geum japonicum*, *Phytochemistry*, 24, pp. 1041–1046.

Yoshida, T., Tanaka, K., Chen, X.-M. and Okuda, T. (1989a). Hydrolyzable tannins with dehydrodigalloyl group from *Rosa laevigata*, *Chem. Pharm. Bull.*, 37, pp. 920–924.

Yoshida, T., Mori, K., Hatano, T., Okumura, T., Uehara, I., Komagoe, K., Fujita, Y. and Okuda, T. (1989b). Radical-scavenging effects of tannins and related polyphenols on 1,1-diphenyl-2-picrylhydrazyl radical, *Chem. Pharm. Bull.*, 37, pp. 1919–1921.

Yoshida, T., Chou, T., Nitta, A., Miyamoto, K., Koshiura, R. and Okuda, T. (1990a). Woodfordin C, a macro-ring hydrolysable tannin dimer with antitumor activity and accompanying dimers from *Woodfordia frutocosa* flowers, *Chem. Pharm. Bull.*, 38, pp. 1211–1217.

Yoshida, T., Chou, T., Maruyama, Y. and Okuda, T. (1990b). Camelliins A and B, two new dimeric hydrolysable tannins from flower buds of *Camellia japonica* L. and *C. sasanqua* Thunb., *Chem. Pharm. Bull.*, 38, pp. 2681–2686.

Yoshida, T., Namba, O., Chen, L. and Okuda, T. (1990c). Euphorbin E, a hydrolysable tannin dimer of highly oxidized structure feom *Euphorbia hirta*, *Chem. Pharm. Bull.*, 38, pp. 1113–1115.

Yoshida, T., Chou, T., Matsuda, M., Yasuhara, T., Yazaki, K., Hatano, T., Nitta, A. and Okuda, T. (1991a). Woodfordin D and oenothein A, trimeric hydrolysable tannins of macro-ring structure with antitumor activity, *Chem. Pharm. Bull.*, 39, pp. 1157–1162.

Yoshida, T., Ahmed, A.F., Memon, M.U. and Okuda, T. (1991b). New monomeric and dimeric hydrolysable tannins from *Reumuria hirtella* and *Tamarix pakistanica*, *Chem. Pharm. Bull.*, 39, pp. 2849–2854.

Yoshida, T., Hatano, T., Ahmed, A.F., Okonogi, A. and Okuda, T. (1991c). Structures of isorugosin E and hirtellin B, dimeric hydrolysable tannins having a trigalloyl group, *Tetrahedron*, 47, pp. 3575–3584.

Yoshida, T., Nakata, F., Hosotani, K., Nitta, A., Okuda T. (1992a). Three new complex tannins from *Melastoma malabathricum* L., *Chem. Pharm. Bull.*, 40, pp. 1727–1732.

Yoshida, T., Hatano, T., Kuwajima, T. and Okuda, T. (1992b). Oligomeric hydrolyzable tannins – Their ^{1}H NMR spectra and partial degradation, *Heterocycles*, 33, pp. 463–482.

Yoshida, T., Maruyama, T., Nitta, A. and Okuda, T. (1992c). Encalbanins A, B and C, monmeric and dimeric hydrolysable tannins from *Eucalypus alba*, *Chem. Pharm. Bull.*, 40, pp. 1750–1754.

Yoshida, T., Feng, W.-S. and Okuda, T. (1992d). Roshenins A-E, dimeric hydrolysable tannins from *Rosa henryi*, *Chem. Pharm. Bull.*, 40, pp. 1997–2001.

Yoshida, T., Ahmed, A. F. and Okuda, T. (1993a). New dimeric hygrolyzable tannins from *Reaumuria hirtella*, *Chem. Pharm. Bull.*, 41, pp. 672–679.

Yoshida, T., Ahmed, A.F. and Okuda, T. (1993b). Tamarixinin B and C, dimeric hydrolysable tannins from *Tamarix pakistanica*, *Phytochemistry*, 33, pp. 197–202.

Yoshida, T., Ito, H., Hatano, T., Kurata, M., Nakanishi, T., Inada, A., Murata, H., Inatomi, Y., Matsuura, N., Ono, K., Nakane, H., Noda, M., Lang, F. A. and Murata, J. (1996). New hydrolyzable tannins, shephagenins A and B, from *Shepherdia argentea* as HIV-1 reverse transcriptase inhibitors, *Chem. Pharm. Bull.*, 44, pp. 1436–1439.

Yoshida, T., Hatano, T., Ito, H. and Okuda, T. (1999). Highly oxidized ellagitannins and their biological activity. In *Plant Polyphenols*, vol. 2, ed. by Gross, G.G., Hemingway, R. W. and Yoshida, T., Kluwer Academic/Plenum Publishers: New York, pp. 127–144.

Yoshida, T., Ito, H., Hippolito, I. J. (2005). Pentameric ellagitannin oligomers in melasmataceous plants – Chemotaxonomic significance, *Phytochemistry*, 66, pp. 1972–1983.

Chapter 2

Structural Diversity and Antimicrobial Activities of Ellagitannins

Takashi Yoshida,*[a] Tsutomu Hatano[b] and Hideyuki Ito,[b] and Takuo Okuda[c]

*[a]College of Pharmaceutical Sciences, Matsuyama University, Bunkyo-cho, Matsuyama, Ehime 790-8578, Japan; [b]Department of Pharmacognosy, Okayama University Graduate School of Medicine, Dentistry and Pharmaceutical Sciences, Tsushima, Okayama 700-8530, Japan; [c]Emeritus Professor, Okayama University, Okayama 700-8530, Japan; *corresponding author:* tyoshida@cc.matsuyama-u.ac.jp

2.1 Monomeric Ellagitannins

Ellagitannins are characterized by their hexahydroxydiphenoyl (HHDP) group, which liberates ellagic acid on acid hydrolysis. They are produced by intramolecular C–C oxidative coupling of galloyl residues located nearby spatially, *e.g.*, 4,6-, 2,3-, or 3,6-galloyl units on the $^1C_4$ or $^4C_1$ D-glucopyranose core, as exemplified by tellimagrandin II (**1**) and casuarictin (**4**) (Okuda et *al.*, 1983), nupharin A (**7**, Ishimatsu *et al.*, 1989a), and corilagin (**8**, Schmidt *et al.*, 1952) (*vide infra*, see also Chapter 1). Interestingly, tellimagrandin II (**1**) has been hemisynthesized enzymatically from pentagalloylglucose (PGG, **9**) (Niemetz *et al.*, 2001).

Studies of many plant families have identified more than 500 ellagitannins with diverse structures. The structural variation among monomers arises mainly from metabolic modification of the HHDP unit(s), as well as from the number and location of polyphenolic groups in a molecule. The structural variety of metabolites of the HHDP group gives rise to many subgroups that compose the ellagitannin family, including dehydroellagitannins, modified dehydroellagitannins, *C*-glycosidic ellagitannins, and complex tannins. In addition, intermolecular C–O oxidative coupling between the HHDP group in an ellagitannin monomer and the galloyl group in another monomer produces additional structural variations leading to a large number of compounds in the oligomeric ellagitannin subclass.

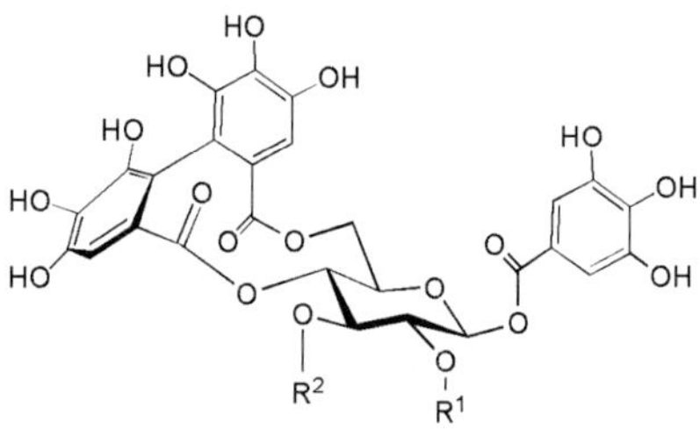

Tellimagrandin II (**1**) : R^1, R^2 = G
Strictinin (**2**) : R^1 = R^2 = H
Gemin D (**3**) : R^1 = H, R^2 = G

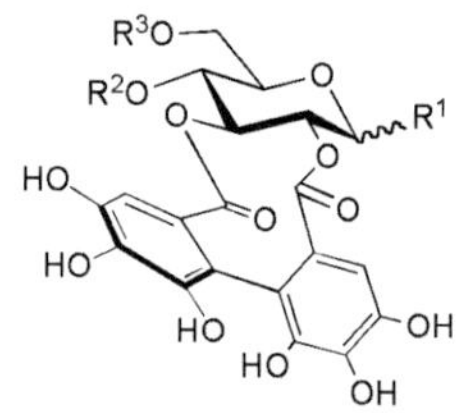

Casuarictin (**4**) : R^1 = β-OG, R^2-R^3 = (*S*)-HHDP
Pedunculagin (**5**) : R^1 = OH, R^2-R^3 = (*S*)-HHDP
Isostrictinin (**6**) : R^1 = β-OG, R^2 = R^3 = H

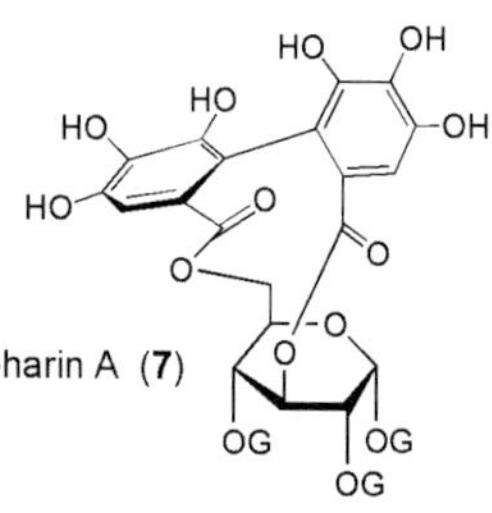

Nupharin A (**7**)

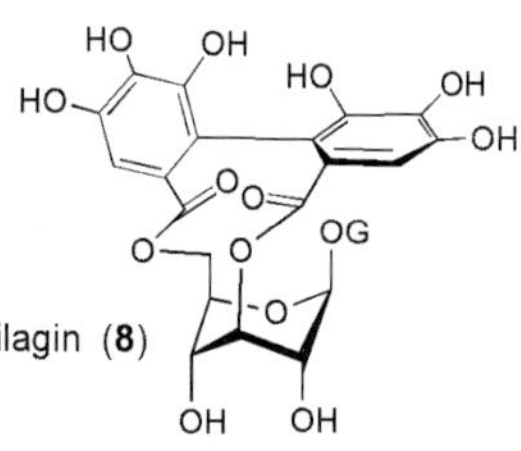

Corilagin (**8**)

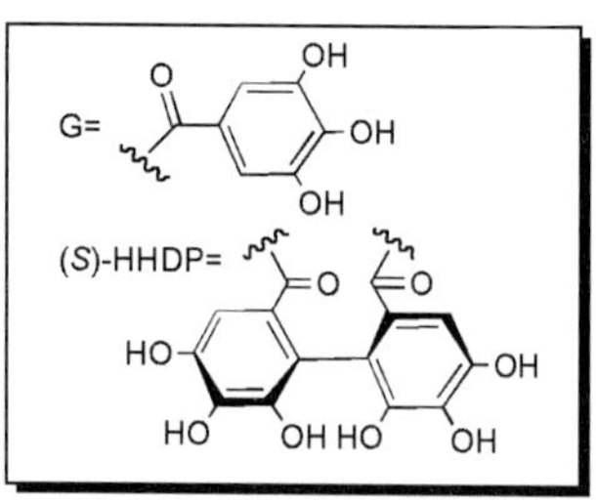

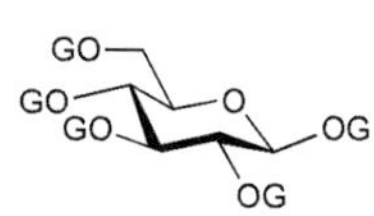

Pentagalloylglucose (**9**)

2.1.1 Primary ellagitannins (HHDP esters)

In addition to the basic ellagitannins **1** and **4**, which have a $^{4}C_{1}$ glucose core, and their deacylated congeners, strictinin (**2**), gemin D (**3**), pedunculagin (**5**), and isostrictinin (**6**) (Okuda *et al.*, 1980b, 1983, Yoshida *et al.*, 1985), many naturally occurring analogs exist, such as rugosins A (**10**) and B (**11**) (Hatano *et al.*, 1990a), tirucallin A (**12**, Yoshida *et al.*, 1991b), cornusiin B (**13**, Okuda *et al.*, 1984a), praecoxins C (**14**) and D (**15**) (Hatano *et al.*, 1991) and tergallagin (**16**, Tanaka *et al.*, 1986a), in which a HHDP or a galloyl group is replaced by a valoneoyl, a depsidone, a tergalloyl, or a gallagyl group.

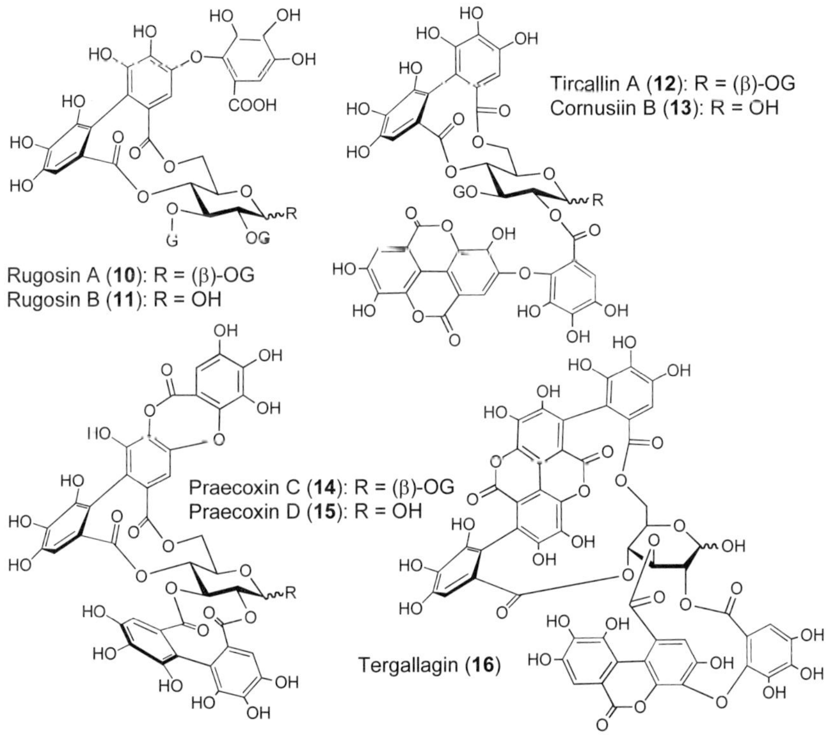

The tannins possessing a valoneoyl group or its regioisomer, the tergalloyl group, may be produced by the catabolism of oligomeric ellagitannins in a plant (*vide infra*). However, the possibility that they could be formed in an anabolic manner through oxidative coupling of the

HHDP (or galloyl) unit with an additional galloyl (or ellagic acid) group cannot be ruled out. Similarly, some ellagitannins with a digalloyl group, represented by agrimonic acid A (**17**, Okuda *et al.*, 1984b), have been obtained from species of the *Rosaceae*, *Coriariaceae* and *Tamaricaceae* plant families.

Table 2.1 Plant sources of primary ellagitannins

Compound	Plant source	Family
Tellimagrandin II (**1**)	*Tellima grandiflora*	*Saxifragaceae*
Tellimagrandin I (**114**)	*Tellima grandiflora*	*Saxifragaceae*
Strictinin (**2**)	*Casuarina stricta*	*Casuarinaceae*
Gemin D (**3**)	*Geum japonicum*	*Rosaceae*
Casuarictin (**4**)	*Casuarina stricta*	*Casuarinaceae*
Pedunculagin (**5**)	*Casuarina stricta*	*Casuarinaceae*
Isostrictinin (**6**)	*Psidium guajava*	*Myrtaceae*
Nupharin A (**7**)	*Nuphar japonicum*	*Nymphaeaceae*
Corilagin (**8**)	*Terminalia chebula*	*Combretaceae*
Rugosin A (**10**)	*Rosa rugosa*	*Rosaceae*
Rugosin B (**11**)	*Rosa rugosa*	*Rosaceae*
Tirucallin A (**12**)	*Euphorbia tirucalli*	*Euphorbiaceae*
Cornusiin B (**13**)	*Cornus officinalis*	*Cornaceae*
Praecoxin C (**14**)	*Stachyurus praecox*	*Stachyuraceae*
Praecoxin D (**15**)	*Stachyurus praecox*	*Stachyuraceae*
Tergallagin (**16**)	*Terminalia catappa*	*Combretaceae*
Agrimonic acid A (**17**)	*Agrimonia pilosa*	*Rosaceae*

As already mentioned in Chapter 1, combined chemical and circular dichroism (CD) spectroscopic data on the atropisomerism of the HHDP unit (Okuda *et al.*, 1982a, 1982b) have revealed that the 2,3-, 4,6-, and 1,6-HHDP units on the $^{4}C_{1}$ glucose core in ellagitannins have the (*S*)-configuration with only a few exceptions, exemplified by the 3,6-HHDP unit on $^{1}C_{4}$ glucose core that has the (*R*)-configuration. Table 2.1 lists the plants from which the aforementioned primary ellagitannins were first isolated; these include a wide variety of dicotyledonous species.

Ellagitannins possessing a gluconic acid as the central polyalcohol core instead of glucose, as exemplified by punigluconin (**18**), have been found in plant species of the *Punicaceae* (Tanaka *et al.*, 1986b), *Elaeagnaceae* (Yoshida *et al.*, 1991a, 1996, Ito *et al.*, 1999b) and *Lythraceae* (Tanaka *et al.*, 1992b) families.

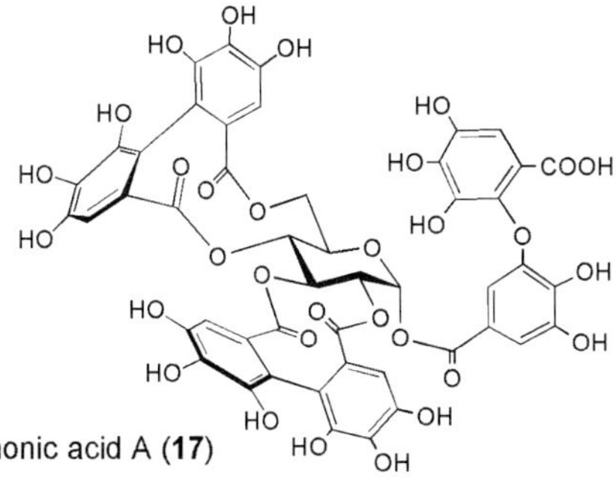

Agrimonic acid A (**17**)

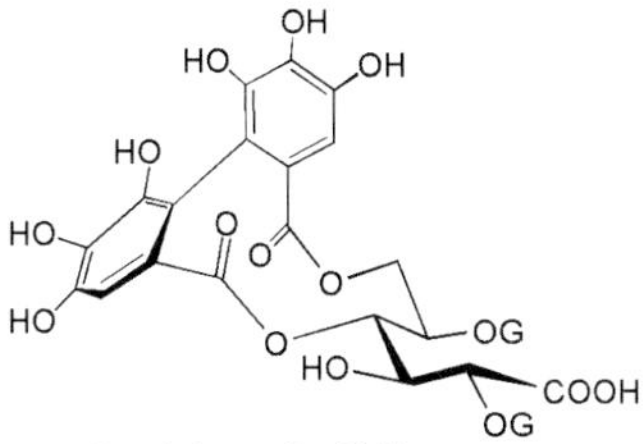

Punigluconin (**18**)

2.1.2 Dehydroellagitannins

The dehydrohexahydroxydiphenoyl (DHHDP) group of tannins belonging to the dehydroellagitannin subclass features a partially hydrated triketone moiety that forms a hemiacetal group with a nearby phenolic hydroxyl group, thus stabilizing the molecule. In aqueous or alcoholic solutions, this DHHDP hemiacetal group usually equilibrates between a six- and a five-membered ring forms. The first compound identified in this class, geraniin (**19**), a keystone compound, was shown to display the hemiacetal group under its six-membered ring form in its crystalline state (see Chapter 1). In addition to dehydrogeraniin (see Chapter 1) and amariin (**20**, from *Phyllanthus amarus*, Foo, 1993) having a 3,6-DHHDP group, and mallotusinic acid having a 3,6-valoneoyl group (from *Mallotus* species, see Chapter 1), macarinin A (**21**) and macaranin C (**22**) (from *Macaranga* species), which possess a tergalloyl and a macaranoyl group, respectively, in place of the 3,6-HHDP group, are also examples of dehydroellagitannins.

Most dehydroellagitannins are obtained as amorphous powders consisting of mixtures of the six- and five-membered hemiacetal forms of their DHHDP group, as found for dehydrogeraniin and amariin (**20**), which exist as mixtures of four isomers because of the presence of two DHHDP units in each molecule. Other dehydroellagitannins include hellioscopinin A (**23**, from *Euphorbia helioscopia*, Lee *et al.*, 1990a), carpinusin (**24**, from *Carpinus laxiflora*, Nonaka *et al.*, 1992) (1,6-HHDP; 2,4-DHHDP-type), terchebin (**25**, from *Terminalia chebula*, Okuda *et al.*, 1980a), and supinanin (**26,** from *Euphorbia spina*, Lee *et* *al.*, 1991) (3,6-digalloyl; 2,4-DHHDP-type). Isoterchebin (**27**), which was isolated from *Cytinus hypocistis* (Fürstenwerth and Schildknecht, 1976) and from *Cornus officinalis* (Okuda *et al.*, 1981), is an exceptional example of dehydroellagitannins because, unlike most compounds in this ellagitannin subclass, its glucopyranose core adopts the $^4C_1$ conformation instead of the $^1C_4$ or skewed-boat conformation.

Table 2.2 Modified dehydroellagitannins and their plant sources

Compound	Plant source	Family
Chebulagic acid (**28**)	*Terminalia chebula*	*Combretaceae*
Chebulinic acid	*Terminalia chebula*	*Combretaceae*
Macarinin B (**29**)	*Macaranga sinensis*	*Euphorbiaceae*
Macaranin A (**30**)	*Macaranga sinensis*	*Euphorbiaceae*
Eumaculin E (**31**)	*Euphorbia maculata*	*Euphorbiaceae*
Phyllanthusiin A (**32**)	*Phyllanthus flexuosus*	*Euphorbiaceae*
Phyllanthusiin B (**33**)	*Phyllanthus flexuosus*	*Euphorbiaceae*
Phyllanthusiin C (**34**)	*Phyllanthus flexuosus*	*Euphorbiaceae*
Repandusinic acid A (**35**)	*Mallotus repandus*	*Euphorbiaceae*
Mallotinin (**36**)	*Mallotus japonicus*	*Euphorbiaceae*
Bixanin (**37**)	*Euphorbia thymifolia*	*Euphorbiaceae*
Amariinic acid (**38**)	*Phyllanthus amarus*	*Euphorbiaceae*
Ascorgeraniin (**39**)	*Elaeocarpus sylvestris*	*Elaeocarpaceae*
(= Elaeocarpusin)	*Geranium thunbergii*	*Geraniaceae*
Putranjivain A (**40**)	*Macaranga sinensis*	*Euphorbiaceae*
Geraniinic acid B (**41**)	*Geranium thunbergii*	*Geraniaceae*
Euphormisin M_2 (**42**)	*Euphorbia humifusa*	*Euphorbiaceae*

2.1.3 Modified dehydroellagitannins

Chebulagic acid (**28**, Yoshida *et al.*, 1980), isolated from myrobalans (*i.e.*, fruits of *Terminalia chebula*, see Fig. 1.11 in Chapter 1) in an early study on tannin chemistry, has a chebuloyl group, which is likely a metabolite produced by a benzylic acid rearrangement-like cleavage of the cyclohexenetrione ring of the DHHDP group (see Chapter 1). The chebuloyl group is also found in the macarinin B (**29**) and macaranin A (**30**) molecules, both isolated from *Macaranga sinensis* (Lin *et al.*, 1990). Ellagitannins possessing an acyl unit related to the chebuloyl group have been found in *Geranium thunbergii* and in many species of *Euphorbiaceae* that produce geraniin (**19**) as the major tannin. It is notable that *Phyllanthus flexuosus* produces a wide variety of modified dehydroellagitannins, including **28**, phyllanthusiins A (**32**), B (**33**), and C (**34**) (Yoshida *et al.*, 1992a), repandusinic acid A (**35**, Saijo *et al.*, 1989a), ascorgeraniin (= elaeocarpusin) (**39**, Tanaka *et al.*, 1986b, Okuda *et al.*, 1986), putranjivain A (**40**, Lin *et al.*, 1990). Other analogous tannins and their plant sources are listed in Table 2.2; eumaculin E (**31**, Amakura *et al.*, 1997), mallotinin (**36**, Saijo *et al.*, 1989b), bixanin (**37**, Lee *et al.*, 1990b), amariinic acid (**38**, Foo, 1995), geraniinic acid (**41**, Ito *et al.*, 1999a) and euphormisin M_2 (**42**, Yoshida *et al.*, 1994) (see also Chapter 1).

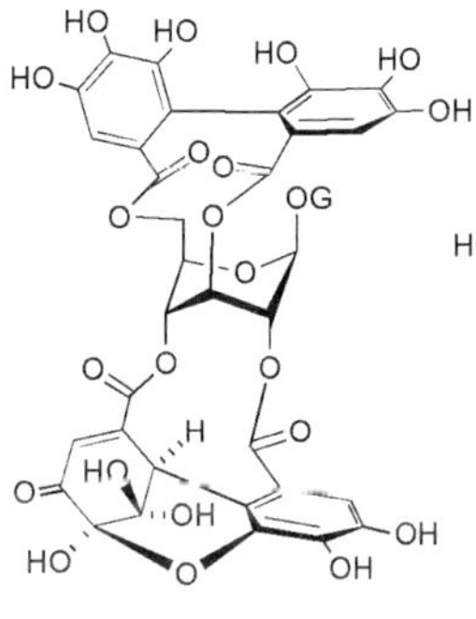

Geraniin (**19**)

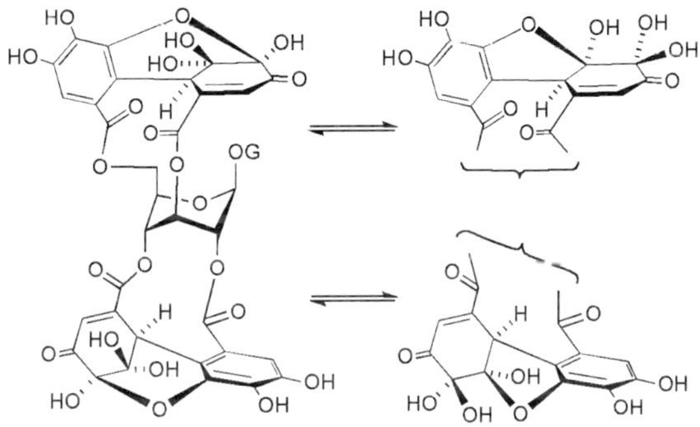

Amariin (**20**)

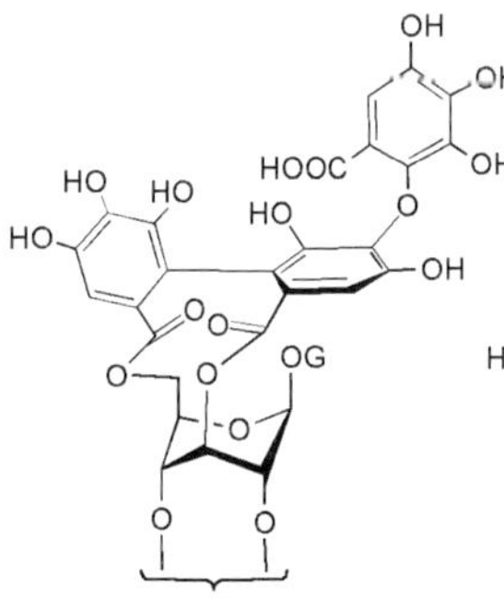

Macarinin A (**21**):
R = (1'*R*)-DHHDP

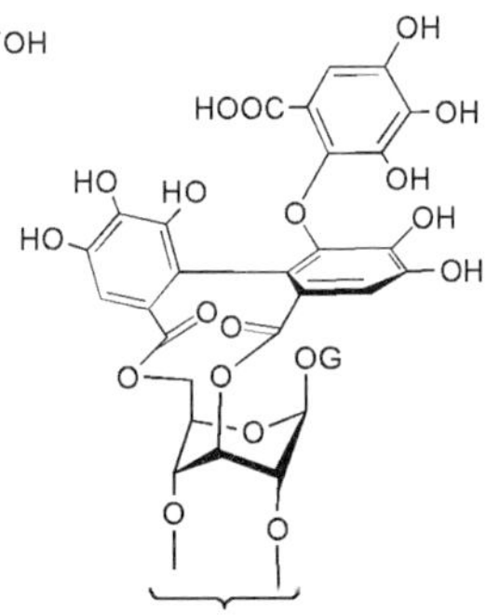

Macaranin C (**22**):
R = (1'*R*)-DHHDP

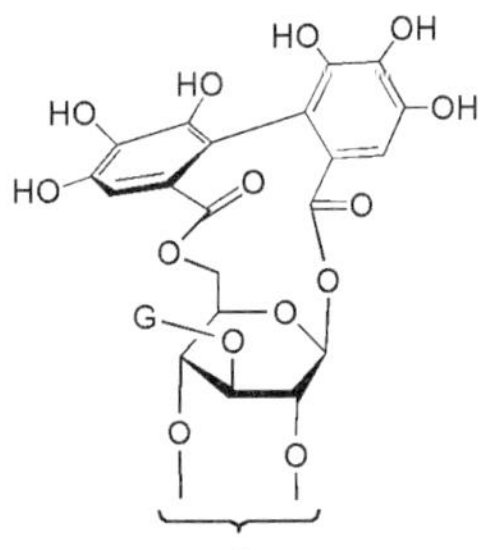

Hellioscopinin A (**23**):
R = (1"*S*)-DHHDP
Carpinusin (**24**):
R = (1'*R*)-DHHDP

(1'*R*)-DHHDP = 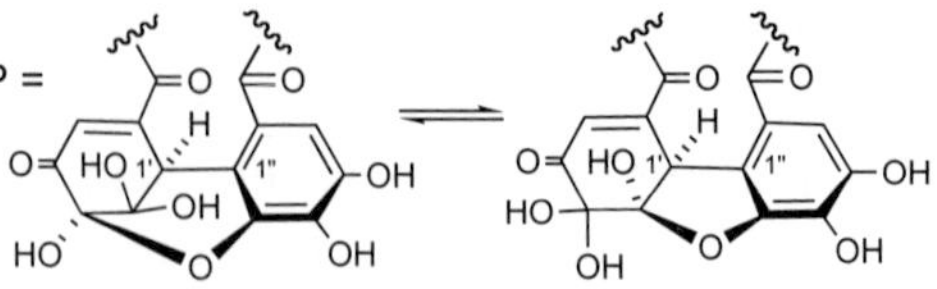

(1"*S*)-DHHDP = 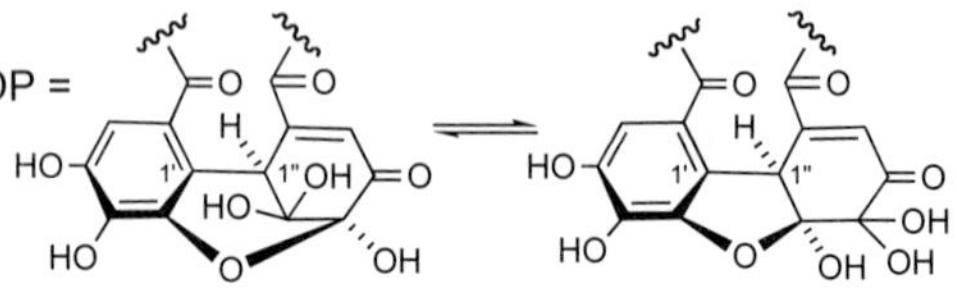

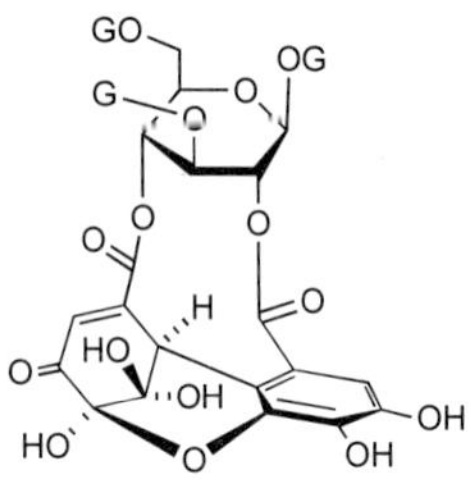

Terchebin (**25**)

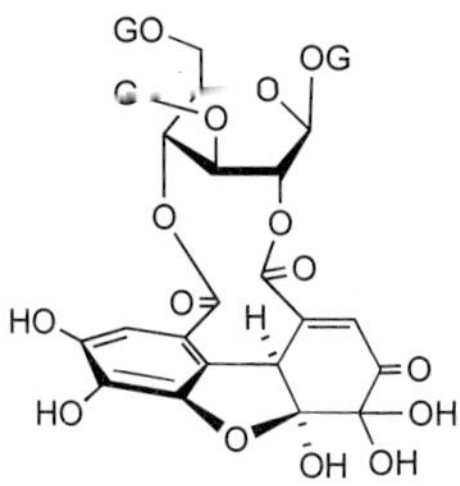

Supinanin (**26**)

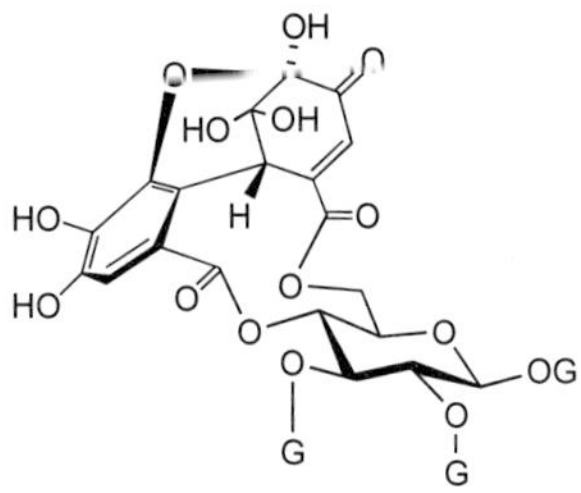

Isoterchebin (**27**)

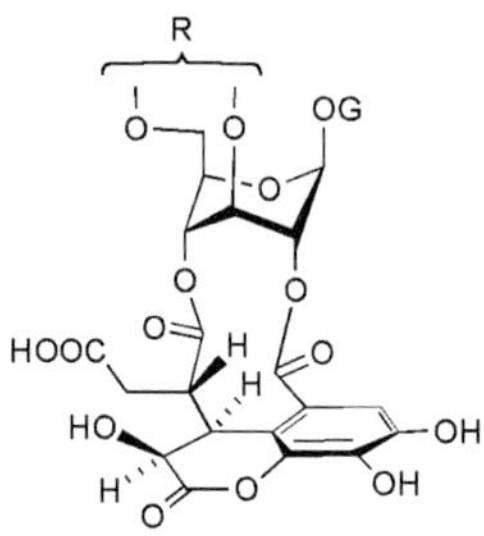

Chebulagic acid (**28**): R = (*R*)-HHDP
Macarinin B (**29**): R = tergalloyl
Macaranin A (**30**): R = macaranoyl
Eumaculin E (**31**): R = chebuloyl

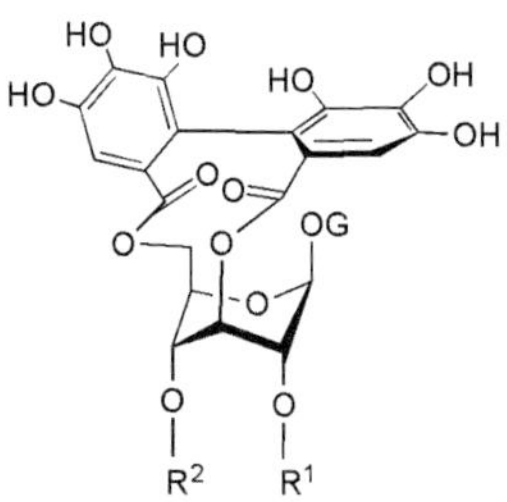

Phyllanthusiin A (**32**): R^1~R^2 = A
Phyllanthusiin B (**33**): R^1~R^2 = B
Phyllanthusiin C (**34**): R^1~R^2 = C
Repandusinic acid A (**35**): R^1 = H, R^2 = D
Mallotinin (**36**): R^1 = E, R^2 = H
Bixanin (**37**): R^1~R^2 = F
Amariinic acid (**38**): R^1~R^2 = G
Ascorgeraniin (= Elaeocarpusin) (**39**): R^1~R^2 = I
Putranjivain A (**40**): R^1~R^2 = J
Geraniinic acid B (**41**): R^1~R^2 = K
Euphormisin M_2 (**42**): R^1~R^2 = L

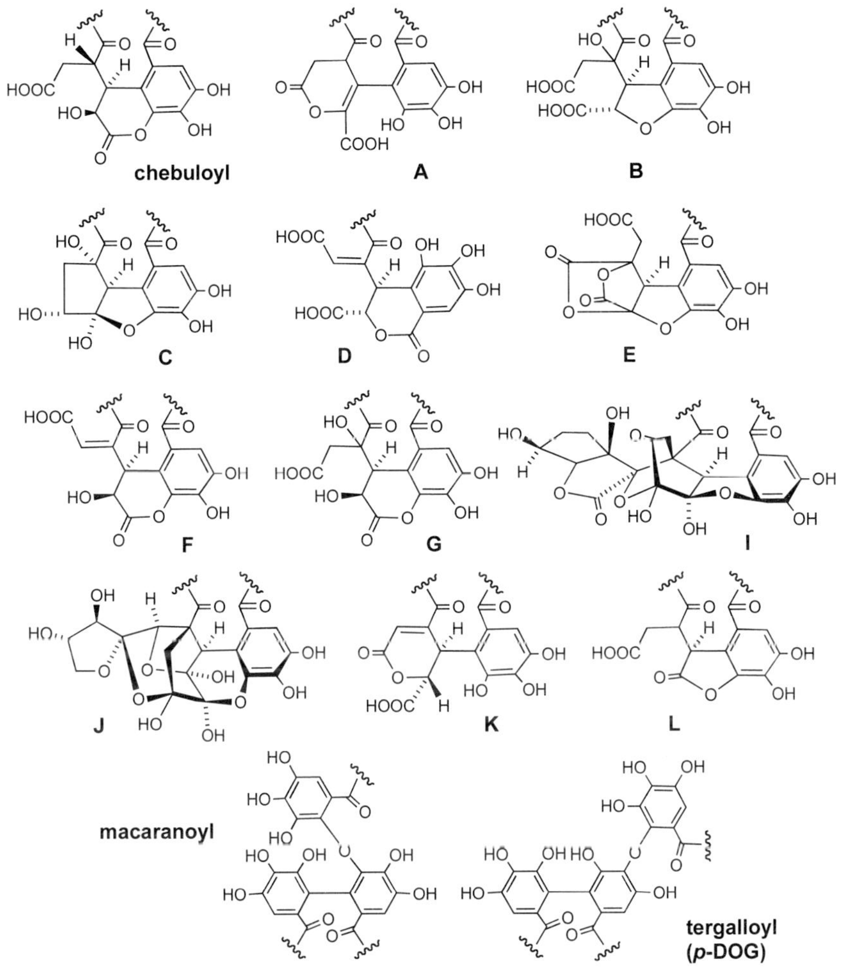

2.1.4 *C*-Glycosidic ellagitannins

Castalagin (**43**) and its C-1 epimer, vescalagin (**44**), have a flavogalloyl group participating in the *C*-glycosidic linkage, while their analogs, casuarinin (**45**) and stachyurin (**46**), have a HHDP unit at the O-2~O-4 positions of the open-chain glucose core. Although these tannins have been found in many plant families, as described in Chapter 1, their structural variation is limited to analogs such as castalin and vescalin, lacking the HHDP group of **43** and **44**, and those such as casuariin and desgalloylstachyurin, lacking the galloyl group of **45** and **46**. Pterocarinin A with a lyxose residue and hippophaenin B with a valoneoyl unit were found in *Pterocaria* species (Nonaka, *et al.*, 1989) and *Hippophae rhamnoides* (Yoshida *et al.*, 1991a), respectively. Elaeagnatin A from *Elaeagnus umbellata* (Ito *et al.*, 1999b) has both lyxose and valoneoyl units in its structure.

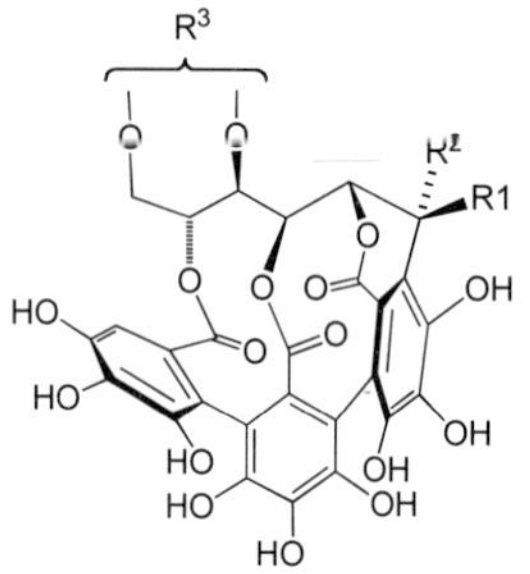

Castalagin (**43**): R^1 = H, R^2 = OH, R^3 = (S)-HHDP
Vescalagin (**44**): R^1 = OH, R^2 = H, R^3 = (*S*)-HHDP
Castalin: R^1 = H, R^2 = OH, R^3 = H, H
Vescalin: R^1= OH, R^2 = H, R^3 = H, H

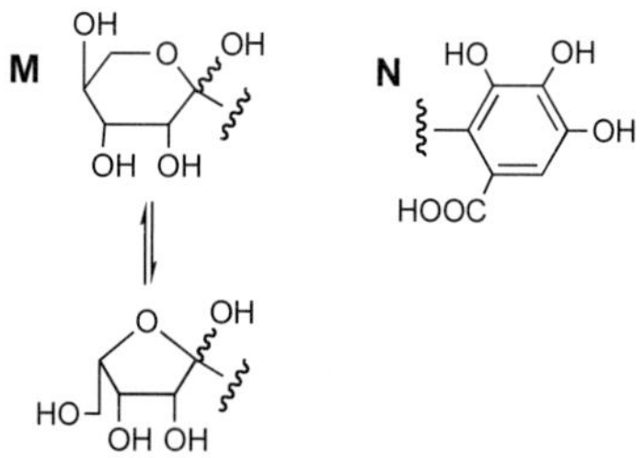

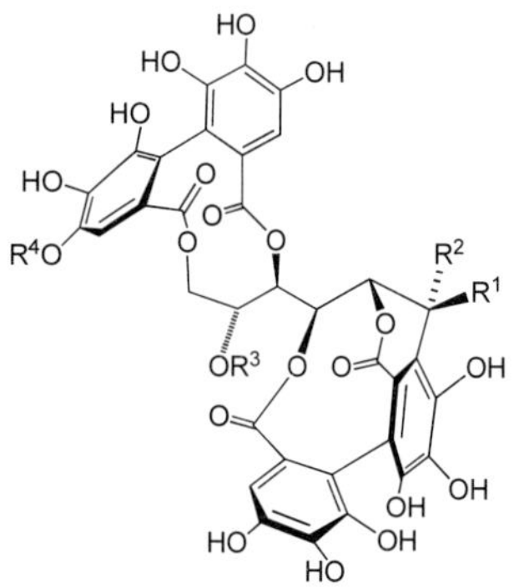

Casuarinin (**45**): R^1 = R^4 = H, R^2 = OH, R^3 = G
Stachyurin (**46**): R^1= OH, R^2 = R^4 = H, R^3 = G
Casuariin: R^1 = R^3 = R^4 = H, R^2 = OH
Desgalloylstachyurin: R^1 = OH, R^2 = R^3 = R^4 = H
Pterocarinin A: R^1 = M, R^2 = R^4 = H, R^3 = G
Hippophaenin B: R^1 = H, R^2 = OH, R^3 = G, R^4 = N
Elaeagnatin A: R^1 = M, R^2 = H, R^3 = G, R^4 = N

2.1.5 Complex tannins (hybrids bearing a flavan-3-ol moiety)

Complex tannins, also called flavano-ellagitannins, are hybrid tannins composed of a *C*-glycosidic ellagitannin [vescalagin (**44**) or stachyurin (**46**)] and a flavan-3-ol (catechin or epicatechin). These two structural units are bound through a C–C linkage between C-1 of the open-chain glucose and C-8 or C-6 of the flavan-3-ol moiety. The (+)-catechin-based complex tannins include stenophyllanins A (**47**) and B (**48**)

(Nonaka *et al.*, 1985) from *Quercus stenophylla* and acutissimins A (**49**) and B (**50**) (Ishimaru *et al.*, 1987) from *Quercus* and *Castanea* species.

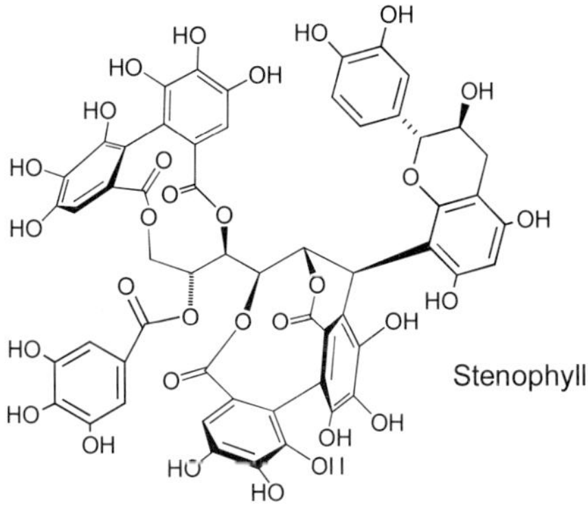

Stenophyllanin A (**47**)

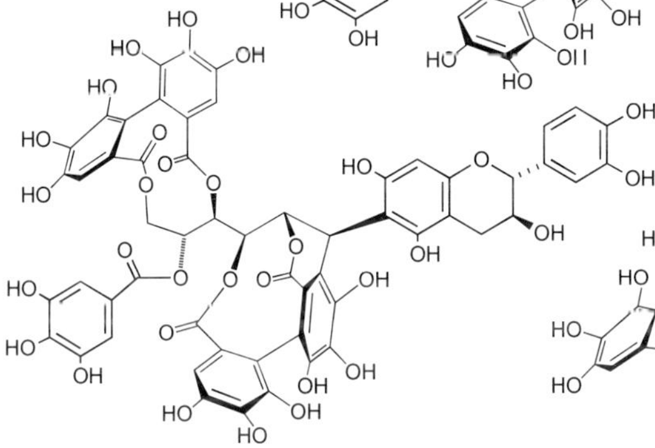

Stenophyllanin B (**48**)

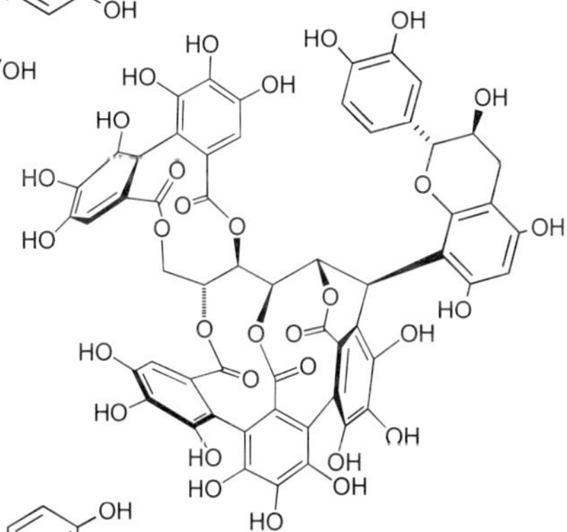

Acutissimin A (**49**)

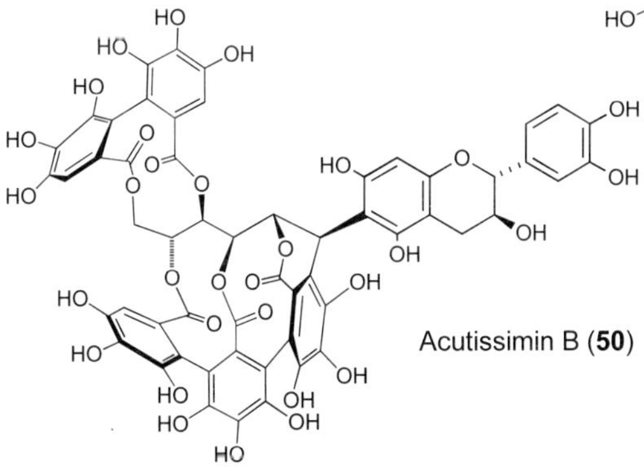

Acutissimin B (**50**)

Their oxidative analogs with a dihydrofuran structure, guavin A (**51**) (Tanaka *et al.*, 1992a) and mongolicains A (**53**) and B (**55**) (Nonaka *et al.*, 1988), were found in *Psidium guajava* and *Quercus mongolica*, respectively. The former plant also produces guavin C (**52**) (Tanaka *et al.*, 1992a) and psidinins A (**54**) and B (**56**) (Tanaka *et al.*, 1992a), which are gallocatechin-based analogs. The (–)-epicatechin-based complex tannins, such as camelliatannins A, B, and F, and malabathrins A and E, are described in Chapter 1. Mongolicins A (**57**) and B (C-6 regioisomer) from *Quercus mongolica* are rare examples of dihydroflavonol-based complex tannins (Ishimaru *et al.*, 1988a), and mongolicanin (**58**) (Ishimaru *et al.*, 1988b) is a condensate of vescalagin (**44**) and procyanidin B-3. Stachyuranins A (**59**) and B (**60**) (Han *et al.*, 1995) from *Stachyurus praecox* are regarded as precursors of stenophyllanins A (**47**) and B (**48**), respectively.

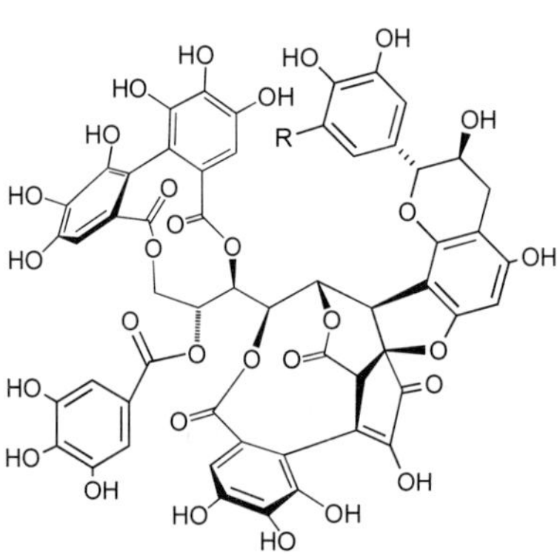

Guavin A (**51**): R = H
Guavin C (**52**): R = OH

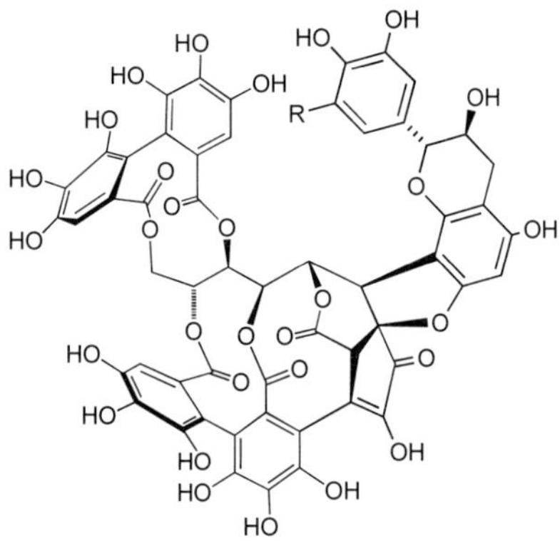

Mongolicain A (**53**): R = H
Psidinin A (**54**): R = OH

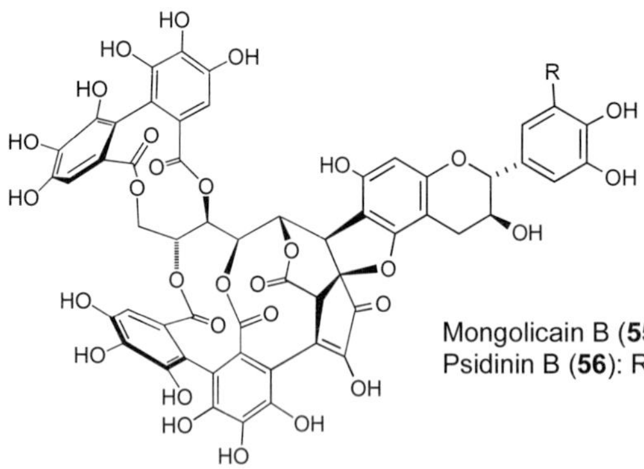

Mongolicain B (**55**): R = H
Psidinin B (**56**): R = OH

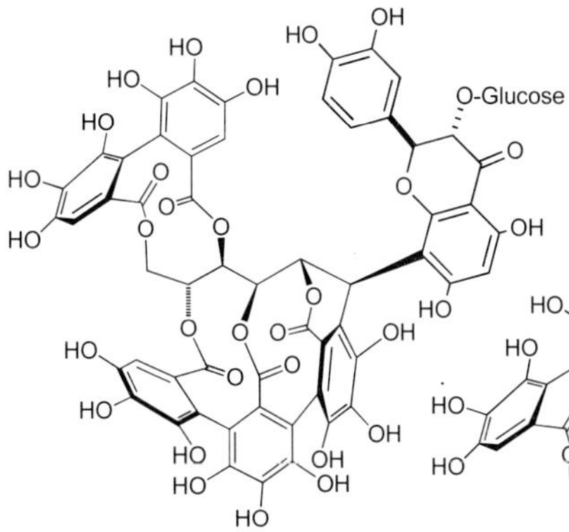

Mongolicin A (**57**)

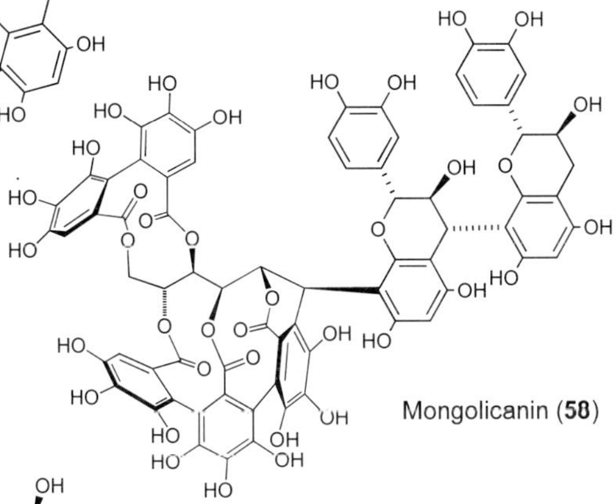

Mongolicanin (**58**)

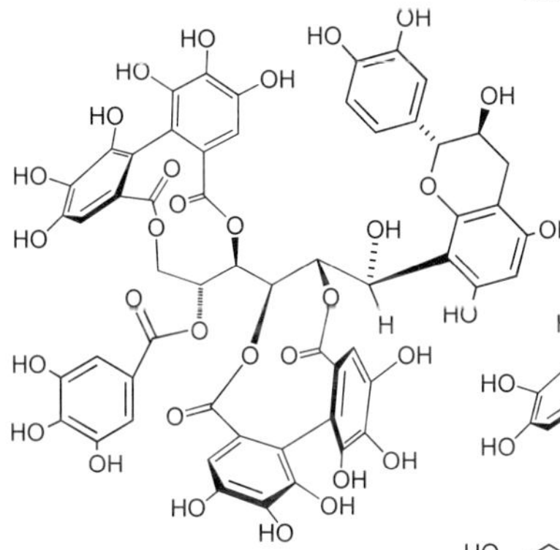

Stachyuranin A (**59**)

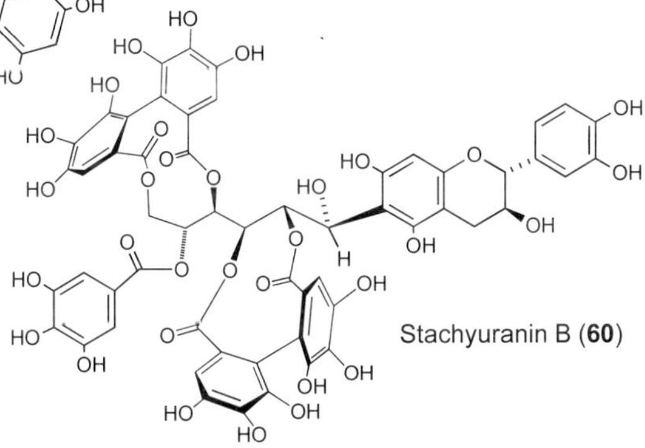

Stachyuranin B (**60**)

2.2 Oligomeric Ellagitannins

Since the 1982 discovery of agrimoniin, the first ellagitannin dimer, more than 250 ellagitannin oligomers ranging from dimers to pentamers have been found in many plant families, including the *Fagaceae*, *Betulaceae*, *Rosaceae*, *Coriariaceae*, *Melastomataceae*, *Onagraceae*, *Lythraceae*, *Theaceae*, *Myrtaceae*, *Cornaceae*, *Trapaceae*, *Combretaceae* and *Euphorbiaceae* (Table 2.3).

Table 2.3 DOG-Type oligomers and their plant sources*

1. Oligomers with 4,6-valoneoyl and 4,6-dehydrovaloneoyl linking units

Compound	Plant source	Family
Heterophylliins A, B, C (**68**)	*Corylus heterophylla*	*Betulaceae*
Heterophylliin G	*Corylus heterophylla*	*Betulaceae*
Davuriciin D_1 (**69**)	*Rosa davurica*	*Rosaceae*
Davuriciin T_1	*Rosa davurica*	*Rosaceae*
Coriariins D, E	*Coriaria japonica*	*Coriariaceae*
Nobotanins A (**70**), F (**71**)	*Tibouchina semidecandra*	*Melastomataceae*
Cornusiins A (**73**), D, E	*Cornus officinalis*	*Cornaceae*
Cornusiin C (**74**)	*Cornus officinalis*	*Cornaceae*
Tarapanin A (**75**)	*Trapa japonica*	*Trapaceae*
Camptothins A, B	*Camptotheca acuminata*	*Nyssaceae*
Calamanins B, C	*Terminalia calamansanai*	*Combretaceae*
Isorugosins D (**76**), E	*Liquidambar formosana*	*Hamamelidaceae*
Resinins A, B, C (**77**)	*Lagerstroemia flos-reginae*	*Lythraceae*
Eugeniflorin D_2 (**86**)	*Eugenia uniflora*	*Myrtaceae*
Oenotherin T_1 (**87**)	*Oenothera tetraptera*	*Onagraceae*

2. Oligomers with a 2,3-valoneoyl linking unit

Compound	Plant source	Family
Malabathrins B (**72**), C, D	*Melastoma malabathricum*	*Melastomataceae*
Nobotanins B (**78**), G, H, I	*Heterocentron roseum*	*Melastomataceae*
Resinin D (**79**)	*Lagerstoroemia flos-regina*	*Lythraceae*
Camelliatannins D (**80**), H	*Camellia japonica*	*Theaceae*

3. Oligomers with a 3,6-valoneoyl linking unit

Compound	Plant source	Family
Euphorbin A (**81**)	*Euphorbia hirta*	*Euphorbiaceae*
Euphobin F (**82**)	*Euphorbia hirta*	*Euphorbiaceae*
Euphorhellin (**83**)	*Euphorbia helioscopia*	*Euphorbiaceae*
Eumaculins A-C	*Euphorbia macrata*	*Euphorbiaceae*
Excoecarianin (**84**)	*Excoecaria kawakamii*	*Euphorbiaceae*
Bishofianin	*Bishofia javanica*	*Euphorbiaceae*
Mallotanins A (**85**), B	*Mallotus japonicus*	*Euphorbiaceae*

*Oligomers described in Chapter 1 are here omitted.

The structures of the large complex oligomeric molecules have been elucidated using spectroscopic techniques, including soft-ionizing mass spectrometry (ESIMS), 2D NMR techniques (COSY, HSQC, HMBC, NOESY), and CD spectra, and through partial chemical degradation. Evidence for the postulate (Haslam 1982, Okuda *et al.*, 1980b) that the oligomers are formed mostly through intermolecular C–O oxidative coupling between ellagitannin monomers was recently obtained with the enzymatic formation of cornusiin E from tellimagrandin II (**1**) (Niemetz and Gross, 2003). Oligomers of diverse structures are classified according to the coupling mode between the monomers involved, as described in Chapter 1.

2.2.1 Oligomers linked with a dehydrodigalloyl group

GOG-type oligomers with a dehydrodigalloyl (DHDG) unit have been found in the *Rosaceae*, *Coriariaceae*, *Nymphaeaceae*, and *Tamaricaceae*. Of these, the dimers from rosaceous and coriariaceous plants, for example, agrimoniin, gemin A, and coriariin A (see Chapter 1), have an *m*-GOG unit produced by intermolecular C–O coupling between the O-1-linked galloyl group of each monomer, while the dimers found in the *Nymphaeaceae* and *Tamaricaceae* have the *m*-GOG unit connected between the O-1 position of one monomer and the O-2 position of the other monomer, as exemplified in nupharin E (**61**) (Ishimatsu *et al.*, 1989b) and hirtellin A (**62**) (Yoshida *et al.*, 1991c).

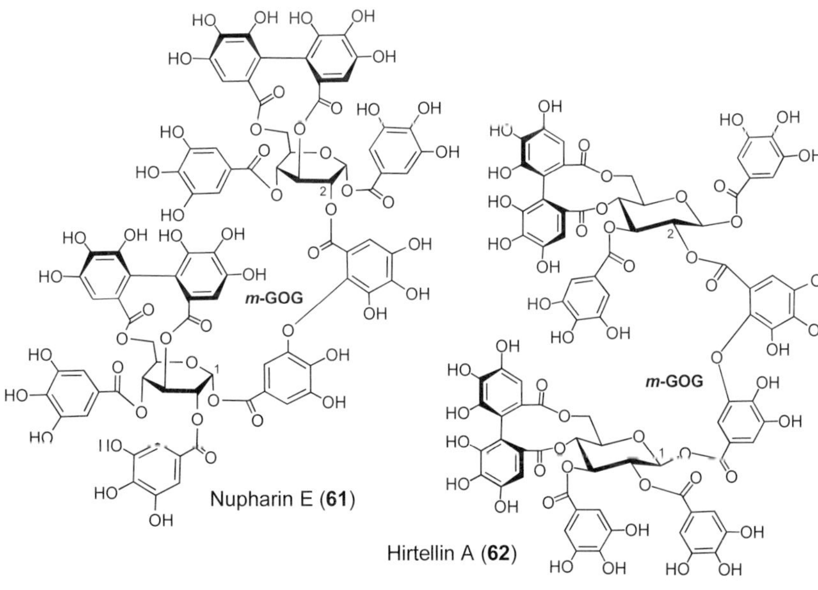

Nupharin E (**61**)

Hirtellin A (**62**)

The constituting monomeric unit of oligomers isolated from *Nuphar japonicum* is an α-glucosidic compound (*e.g.*, 3,6-HHDP-1,2,4-trigalloyl-α-D-glucose), and nupharins C (**63**) and F (**64**) (Ishimatsu *et al.*, 1989b) are unique examples of a dimer and a trimer possessing both (*S*)- and (*R*)-HHDP units in the same molecule. Conversely, tamaricaceous plants such as *Reaumuria hirtella* and *Tamarix pakistanica* produce oligomers of tellimagrandin II (**1**). Hirtellin C (**65**) (Yoshida *et al.*, 1993) is a dimer resulting from a double coupling of two molecules of **1** and featuring both the *m*-GOG and *p*-GOG linking units. Hirtellins T_1 (**66**) and Q_1 (**67**) (Ahmed *et al.*, 1994) are examples of a trimer and a tetramer of this type, formed by repetition of the same coupling mode. These oligomers might serve as chemotaxonomical markers for the plant families *Rosaceae*, *Nymphaeaceae*, and *Tamaricaceae*.

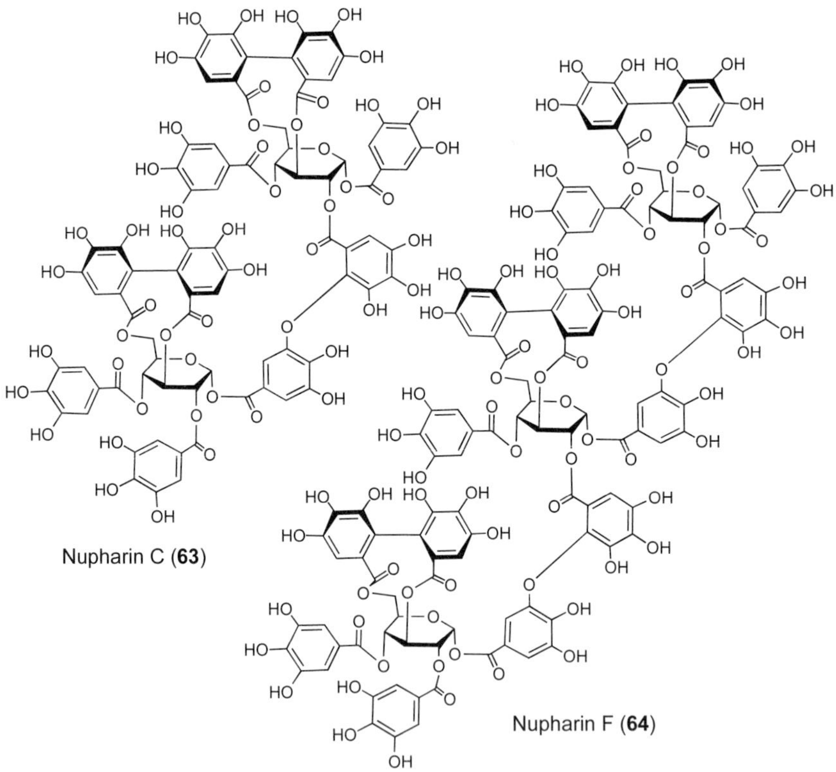

Nupharin C (**63**)

Nupharin F (**64**)

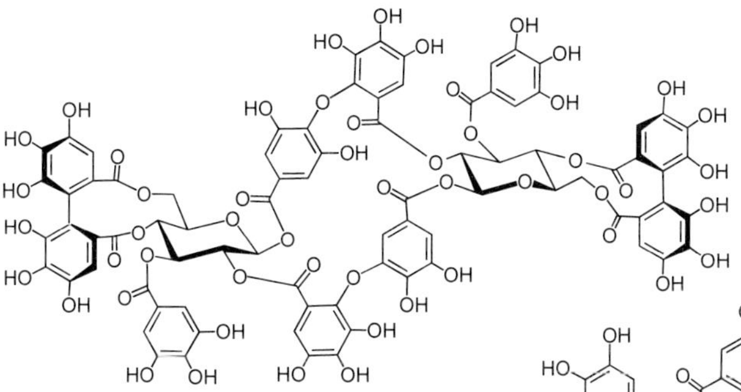

Hirtellin C (**65**)

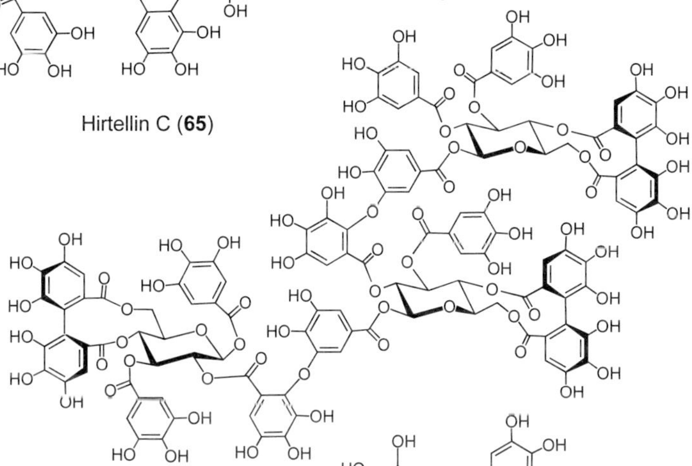

Hirtellin T_1 (**66**)

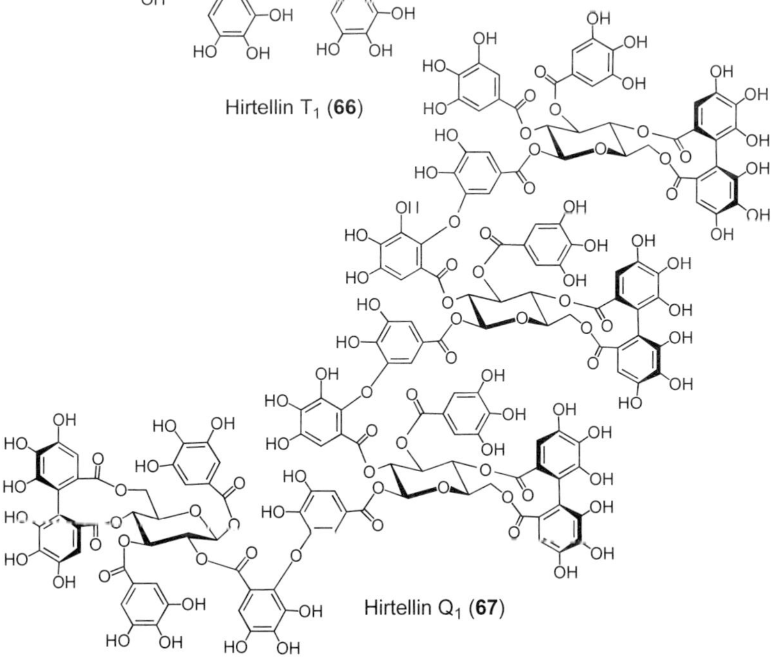

Hirtellin Q_1 (**67**)

2.2.2 Oligomers linked with valoneoyl and dehydrovaloneoyl groups

The *m*-DOG-type oligomers that bear a valoneoyl group or its oxidized variant as linking unit(s) constitute the largest class of oligomeric ellagitannins, and are distributed in a wide range of plant families. The structural diversity in these oligomers arises from the variation in the positions of the HHDP and galloyl groups that participate in the formation of a valoneoyl group through the intermolecular oxidative coupling of monomers.

Typical oligomers with 4,6-, 2,3-, or 3,6-valoneoyl groups on their glucose core include the dimers heterophylliins A, B and C (**68**, Yoshida *et al.*, 1991d) and the trimer heterophylliin G (Jin *et al.*, 1998), davuriciins D_1 (**69**) and T_1 (Yoshida *et al.*, 1991e), the dimers coriariins D and E (Hatano *et al.*, 1986), nobotanins A (**70**), B (**78**), F (**71**), G, H, I (Yoshida *et al.*, 1991f, 1991g), malabathrins B (**72**), C and D (Yoshida *et al.*, 1992), cornusiins A (**73**), D and E, and the trimers cornusiin C (**74**) (Hatano *et al.*, 1989) and trapanin A (**75**, Hatano *et al.*, 1990b), the dimers camptothins A and B (Hatano *et al.*, 1988a), calamanins B and C (Tanaka *et al.*, 1991), isorugosins D (**76**) and E (Hatano *et al.*, 1988b), camelliatannins D (**80**) and H (Han *et al.*, 1994, Hatano *et al.*, 1995), eumaculins A-C (Agata *et al.*, 1991, Amakura *et al.*, 1997), bishofianin (Tanaka *et al.*, 1995) and mallotannins A (**85**) and B (Saijo *et al.*, 1989). Plant sources of these oligomers and others discussed below are listed in Table 2.3.

Worthy of particular mention are the dimers resinins A, B, C (**77**) and D (**79**, Xu *et al.*, 1991) isolated from *Lagerstroemia flos-reginae*, as well as heterophylliins B and C (**68**) from *Corylus heterophylla*, for they contain a *C*-glycosidic ellagitannin as one of their constituting monomers. The *C*-glycosidic tannin monomer in resinin C (**77**) has a pentose residue in which the anomeric center is C–C-linked to the C-1 center of the *C*-glycosidic tannin unit. It is also worth noting that the dimers euphorbins A (**81**, Yoshida *et al.*, 1988) and F (**82**, Yoshida *et al.*, 1991h), euphorhellin (**83**, Lee *et al.*, 1991) and excoecarianin (**84**, Lin *et al.*, 1990) from plant species of the *Euphorbiaceae* family have geraniin-based structures. Since they have not been encountered in any other plant family, this type of dimer is chemotaxonomically significant.

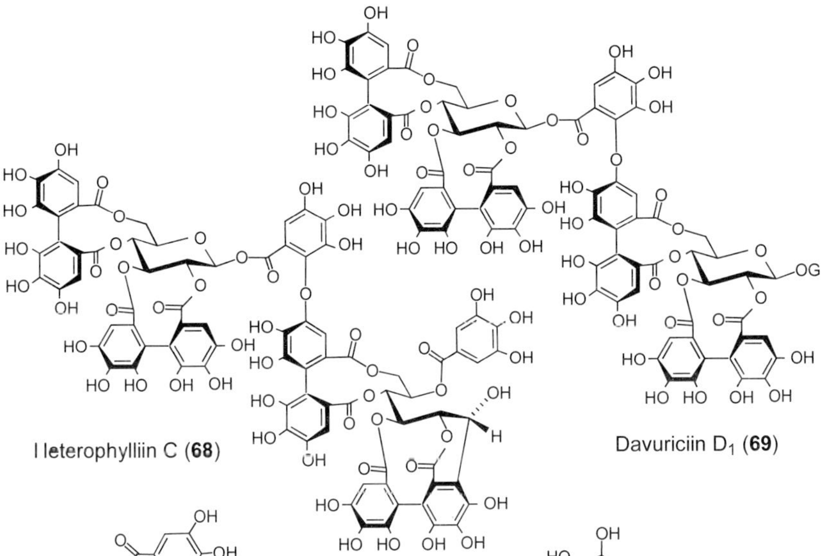

Heterophylliin C (**68**)

Davuriciin D_1 (**69**)

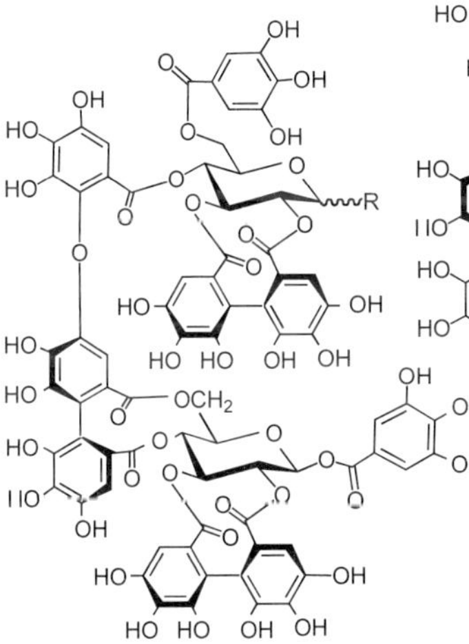

Nobotanin A (**70**) : R = OH
Nobotanin F (**71**) : R = (β)-OG

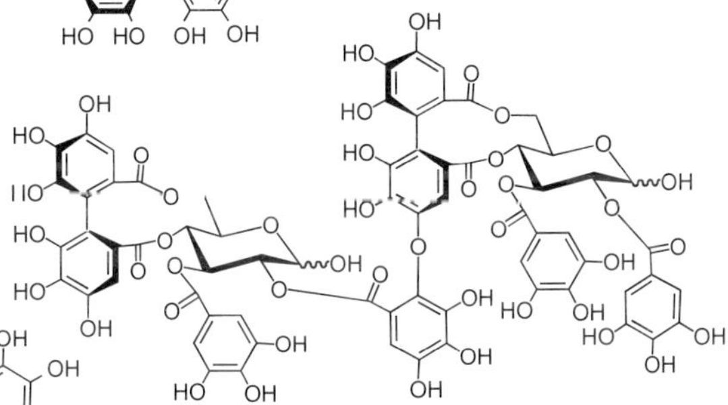

Cornusiin A (**73**)

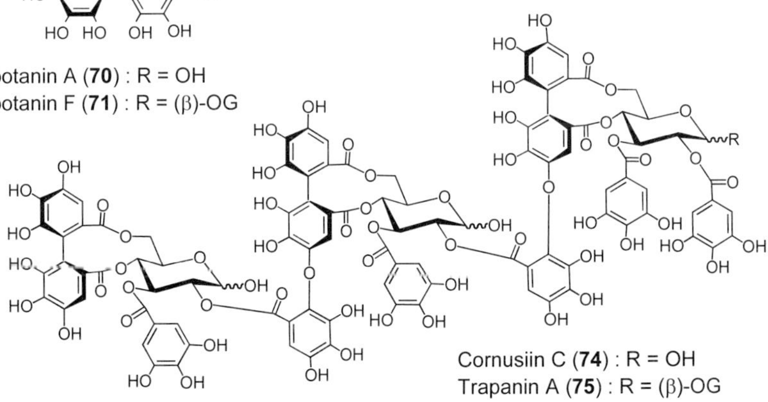

Cornusiin C (**74**) : R = OH
Trapanin A (**75**) : R = (β)-OG

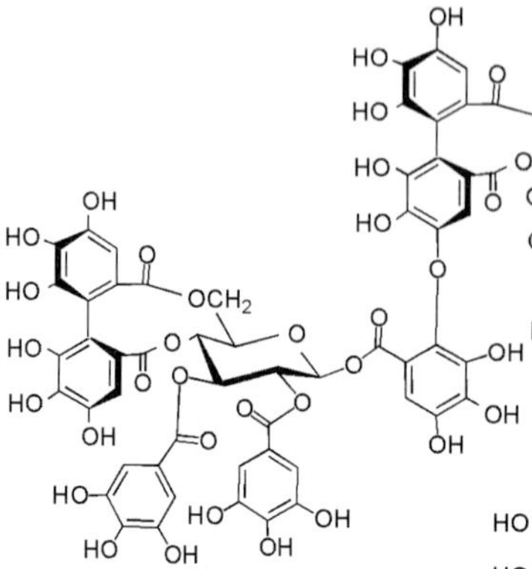

Isorugosin D (**76**)

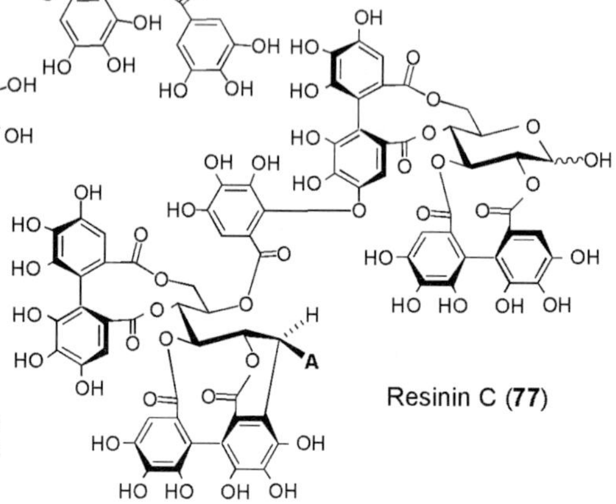

Resinin C (**77**)

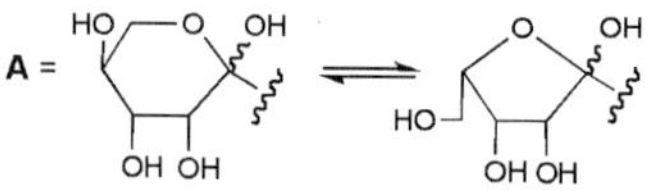

Malabathrin B (**72**) : R = OH
Nobotanin B (**78**) : R = (β)-OG

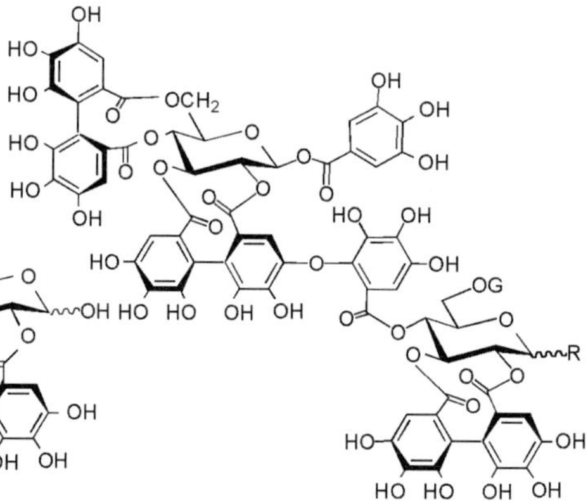

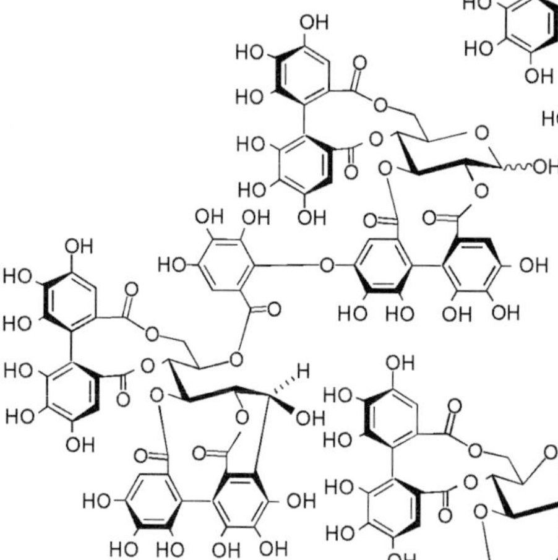

Resinin D (**79**)

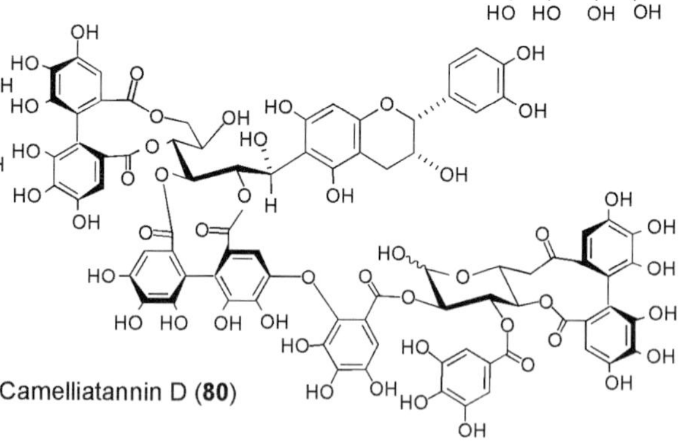

Camelliatannin D (**80**)

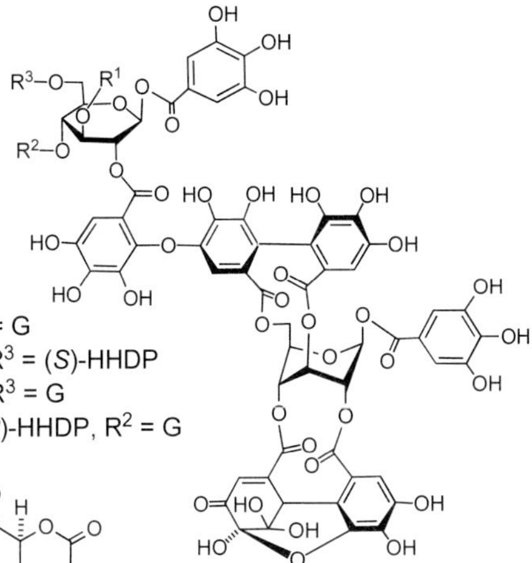

Euphorbin A (**81**): $R^1 = R^2 = R^3 = G$
Euphorbin F (**82**): $R^1 = G$, $R^2 \sim R^3 = (S)$-HHDP
Euphorhellin (**83**): $R^1 = R^2 = H$, $R^3 = G$
Excoecarianin (**84**): $R^1 \sim R^3 = (R)$-HHDP, $R^2 = G$

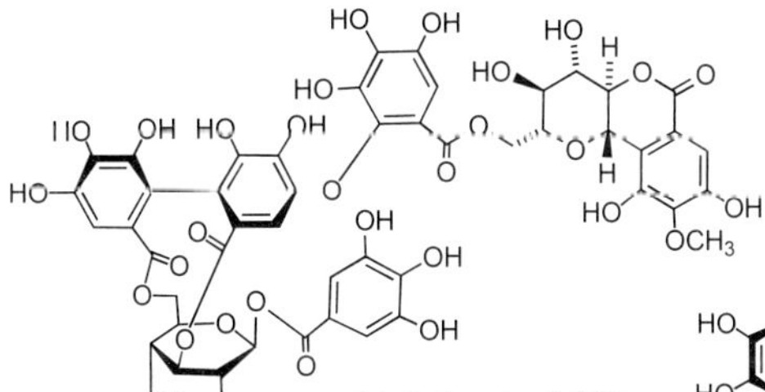

Mallotannin A (**85**)

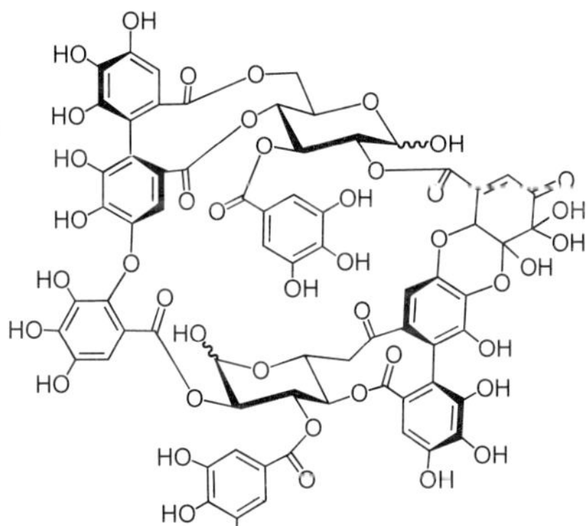

Eugeniflorin D_2 (**86**)

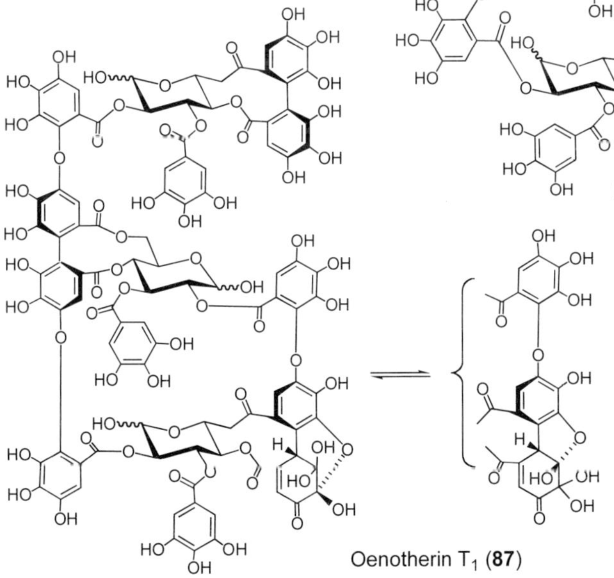

Oenotherin T_1 (**87**)

The oxidized macrocyclic dimer eugeniflorin D_2 (**86**, Lee *et al.*, 1997) also deserves special mention, for it features a dehydrovaloneoyl group. It was isolated from *Eugenia uniflora* together with eugeniflorin D_1, a regular macrocyclic dimer (see Chapter 1). Similarly, oenotherin T_1 (**87**), isolated from *Oenothera tetraptera*, is a dehydro-derivative of trimer oenothein A found in *Oenothera erythrosepala* and *Oenothera biennis* (Yoshida *et al.*, 1991c). Interestingly, the cyclohexenetrione moiety in **87** can be reduced to a valoneoyl group under mild conditions to give oenothein A (Taniguchi *et al.*, 2002).

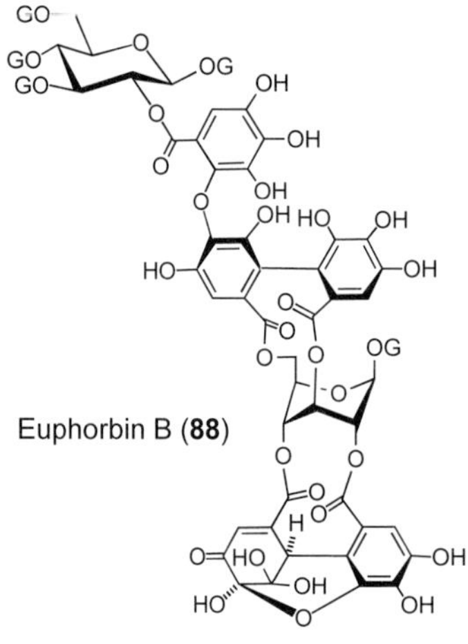
Euphorbin B (**88**)

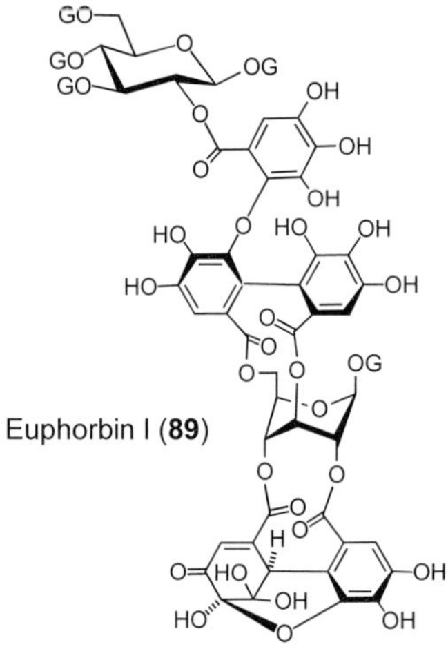
Euphorbin I (**89**)

G = 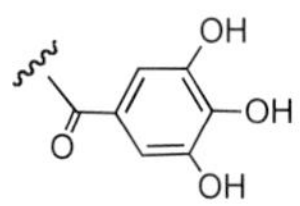

(*S*)-HHDP = 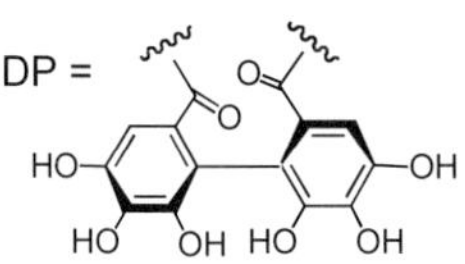

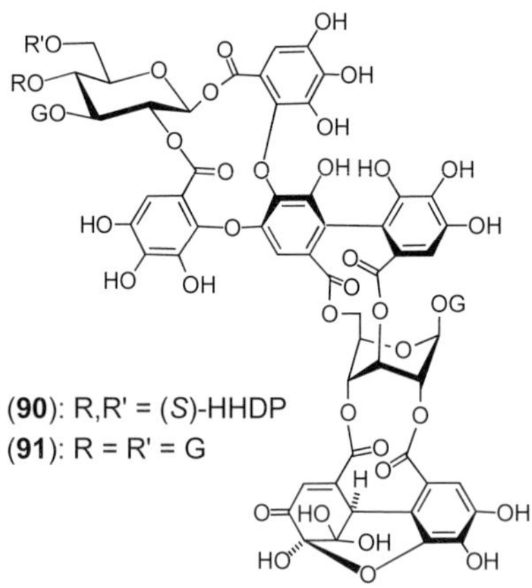
Euphorbin C (**90**): R,R' = (*S*)-HHDP
Euphorbin D (**91**): R = R' = G

2.2.3 Oligomers linked with an isomeric valoneoyl group

Sanguisorboyl, tergalloyl, and macaranoyl groups are isomeric valoneoyl groups. In addition to the oligomers possessing these groups described in Chapter 1, euphorbins B (**88**) and I (**89**) (Amakura *et al.*, 1996) from *Euphorbia* species possess (*R*)-tergalloyl and (*R*)-macaranoyl groups as their linking unit, respectively. Euphorbins C (**90**) and D (**91**) are unique dimers possessing a $D(OG)_2$-type (euphorbinoyl) linking unit in which a tergalloyl or a valoneoyl group is linked to a galloyl group.

2.2.4 *C*-Glycosidic ellagitannin oligomers

Oligomeric *C*-glycosidic ellagitannins are regarded as the products of intermolecular C–C coupling between the anomeric carbon of one monomer and the HHDP or galloyl group of the other monomer. Since 1991, approximately 30 *C*-glycosidic ellagitannin oligomers have been isolated from various plant species in the *Fagaceae*, *Casuarinaceae*, *Melastomataceae* (Yoshida *et al.*, 1992b), *Elaeagnaceae*, and *Combretaceae*. Vescalagin (**44**)-based oligomers include roburins A–D (**92-95**) (Hervé du Penhoat *et al.*, 1991), and anogeissusins A (**96**) and B (**97**) (Lin *et al.*, 1991) isolated from an European oak species (*Quercus robur*) and *Anogeissus acuminata*, respectively.

Stachyurin (**46**)-based oligomers include alienanin B (**98**) (Nonaka *et al.*, 1991; Yoshida *et al.*, 1992b) and casuglaunin A (**99**) (Shimokawa *et al.*, 1991). The leaves of *Elaeagnus umbellata* produce the analogs elaeagnatins D (**100**)-G (Ito *et al.*, 1999b). In the dimers anogeissinin (**101**) (Lin *et al.*, 1991) and casuglaunin B (**102**) (Ito *et al.*, 1999b, Shimokawa *et al.*, 1991), two molecules of a vescalagin (or stachyurin) type monomer are connected through the A-ring of catechin. Cowaniin (**103**), a putative precursor of **102**, was found in *Cowania mexicana* (Ito *et al.*, 2007) and *Melaleuca squarrosa* (Yoshida *et al.*, 2008). As for *C*-glycosidic tannin oligomers that are larger than dimers, castaneanins [dimer to pentamer (**104-107**)] (Tanaka *et al.*, 1996) from the heartwood of the Japanese chestnut tree and melasquanins B, C, and D (**108**) from *M. squarrosa* (Yoshida *et al.*, 2008) have been identified. In all of these *C*-glycosidic ellagitannin oligomers, the intermolecular linkage at the anomeric carbon participating in the oligomerization is β-oriented without any exception.

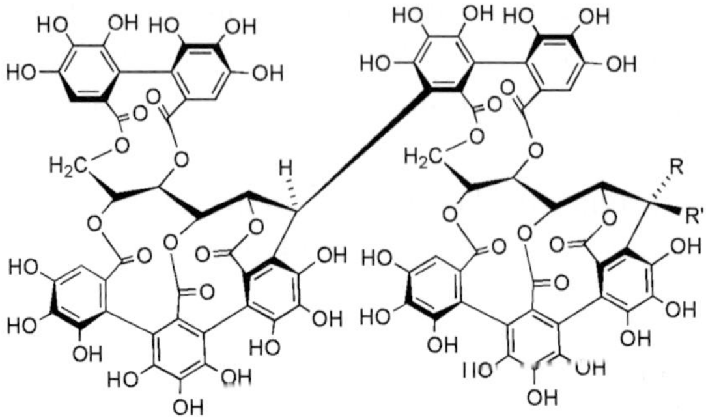

Roburin A (**92**): R = H, R' = OH
Roburin B (**93**): R = H, R' = L
Roburin C (**94**): R = H, R' = X
Roburin D (**95**): R = OH, R' = H

L =, X =

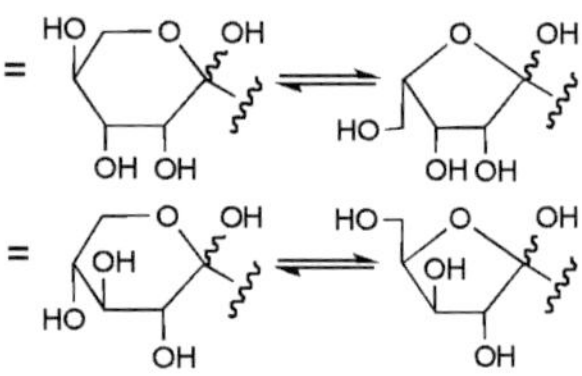

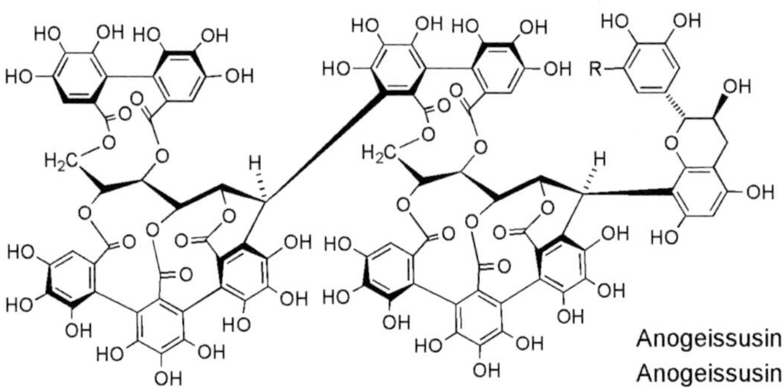

Anogeissusin A (**96**): R = H
Anogeissusin B (**97**): R = OH

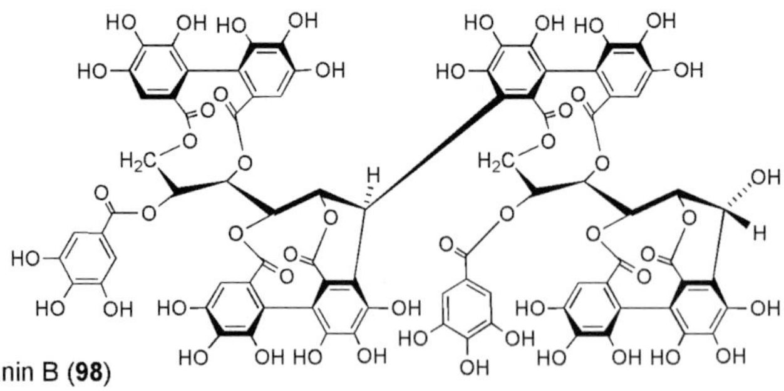

Alienanin B (**98**)

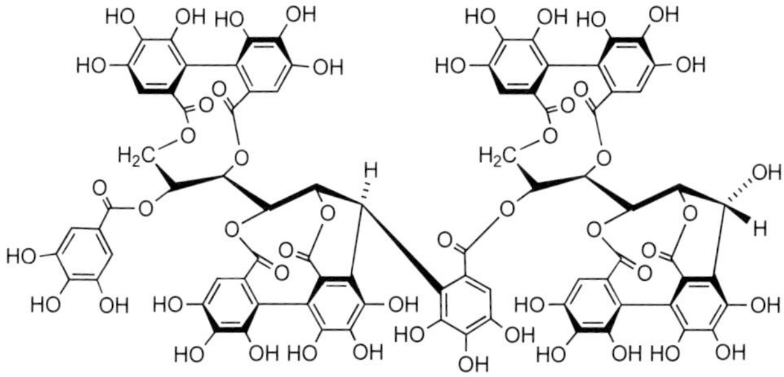

Casuglaunin A (**99**)

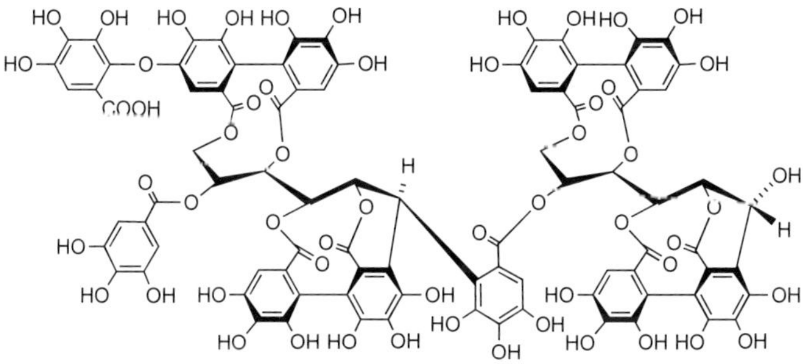

Elaeagnatin D (**100**)

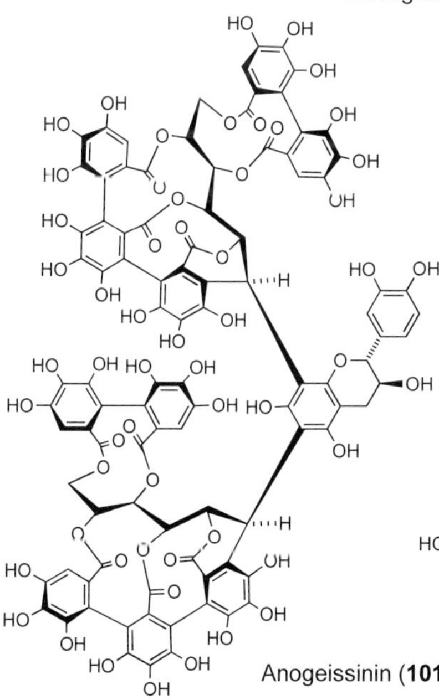

Anogeissinin (**101**)

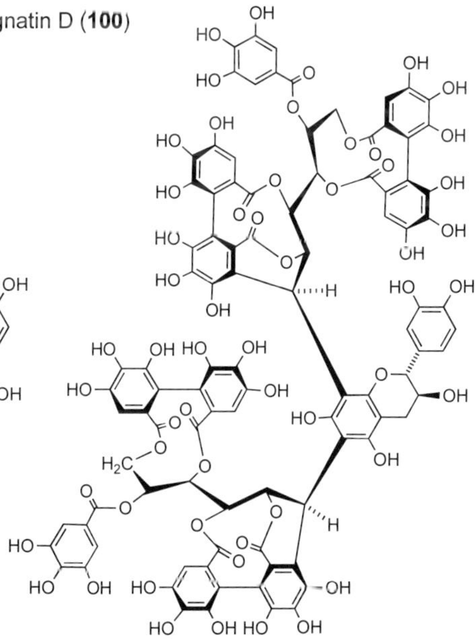

Casuglaunin B (**102**)

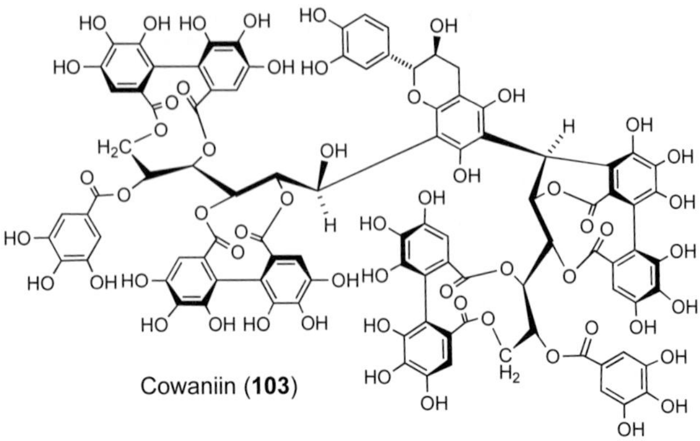

Cowaniin (**103**)

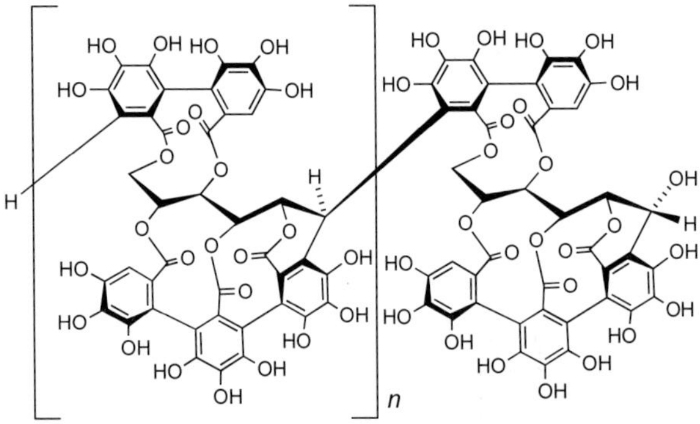

Castaneanin dimer (**104**): *n* = 1
Castaneanin trimer (**105**): *n* = 2
Castaneanin tetramer (**106**): *n* = 3
Castaneanin pentamer (**107**): *n* = 4

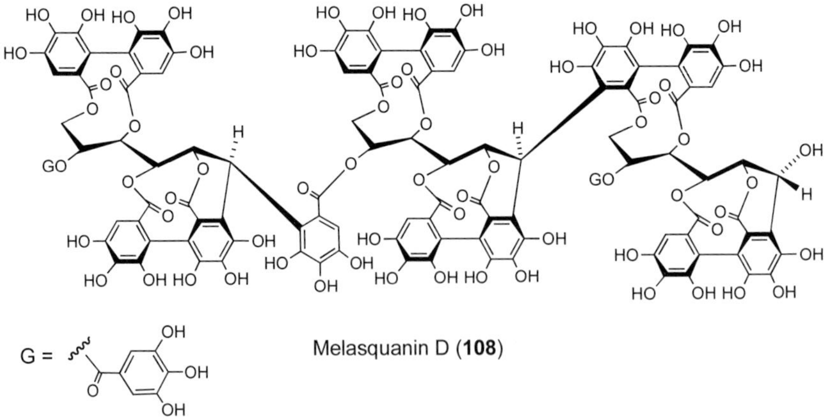

Melasquanin D (**108**)

2.3 Chemotaxonomic Significance of Oligomeric Ellagitannins

The chemotaxonomic significance of monomeric ellagitannins, which are widely distributed in dicotyledonous plants, is rather limited, except for those in certain families. The gluconic acid-based ellagitannins punigluconin (**18**) and hippophaenin A (**109**) were first isolated in *Punica granatum* (*Punicaceae*) (Tanaka, *et al.*, 1986b) and *Hippophae rhamnoides* (*Elaeagnaceae*) (Yoshida *et al.*, 1991a), respectively. However, these tannins were later found in several *Lagerstroemia* species in *Lythraceae* (Tanaka *et al.*, 1992b) and in many elaeagnaceous plants. In particular, *Elaeagnus umbellata* produces characteristic dimers with a gluconic acid-based constituting monomer, such as elaeagnatins B (**110**), C (**111**), and F (**112**) (Ito *et al.*, 1999b).

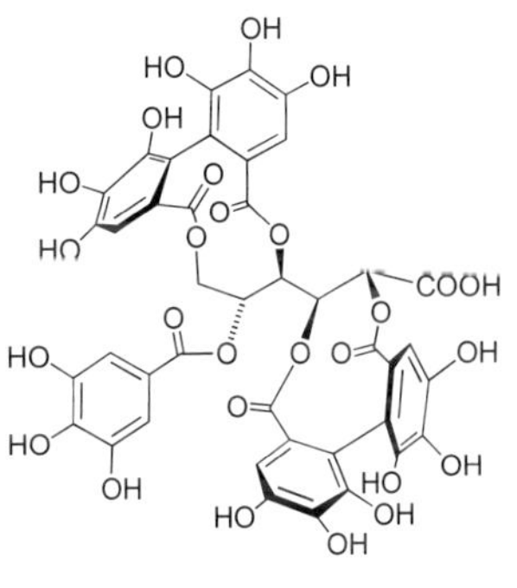

Hippophaenin A (**109**)

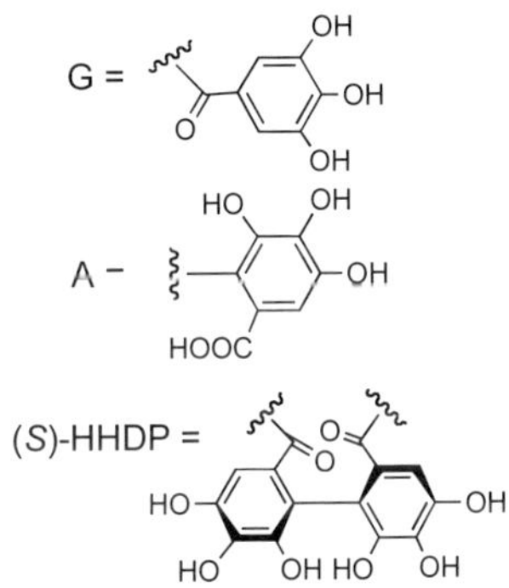

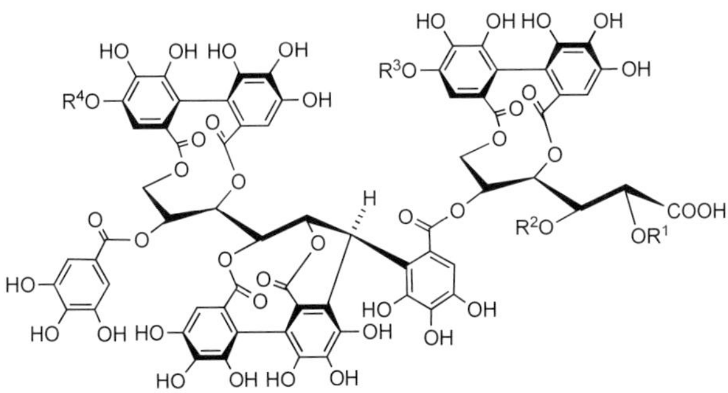

Elaeagnatin B (**110**): R^1~R^2 = (*S*)-HHDP, R^3 = R^4 = H
Elaeagnatin C (**111**): R^1~R^2 = (*S*)-HHDP, R^3 = A, R^4 = H
Elaeagnatin F (**112**): R^1 = G, R^2 = R^3 = H, R^4 = A

Unlike the monomers, many oligomeric ellagitannins can be valuable chemotaxonomic markers of plant systematic and evolutionary relationships in dicotyledonous plants because of their structural specificities. As described earlier, GOG-type oligomers with the dehydrodigalloyl (DHDG) unit are found in the *Rosaceae*, *Coriariaceae*, *Nymphaeaceae*, and *Tamaricaceae*, and the position and configuration of the DHDG unit seems to be specific to each family. The configurations of the DHDG unit on each anomeric center of the dimers in rosaceous plants (Okuda *et al.*, 1984b) are α and α, while those in coriariaceous plants (Hatano *et al.*, 1986) are α and β. Dimers such as **61** and **62** in the *Nymphaeaceae* and *Tamaricaceae*, which are produced *via* a similar coupling mode between the galloyl groups at C-1 of one monomer and C-2 of the other monomer, are made of different constituting monomers, *i.e.*, 3,6-HHDP-1,2,4-trigalloyl-α-D-glucose in the *Nymphaeaceae* and tellimagrandin II (**1**) in the *Tamaricaceae*. These structural features may be of chemotaxonomic significance.

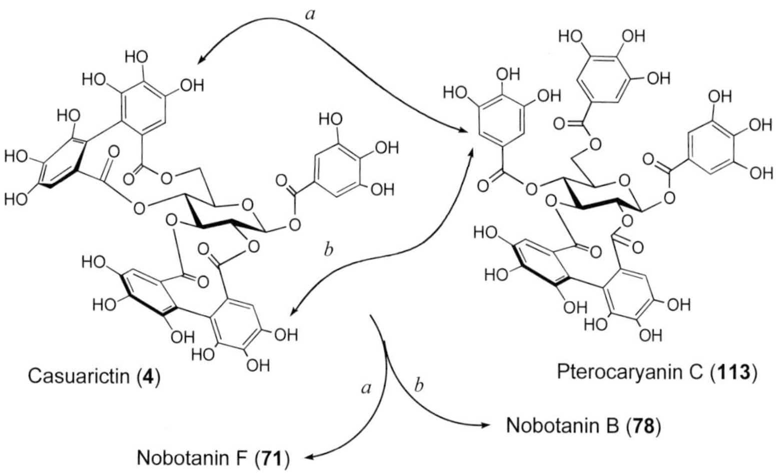

The geraniin-related dimers that bear a valoneoyl group or its isomer as the linking unit have been found only in herbaceous and woody euphorbiaceous plants, hence serving as characteristic markers of the family. The structural features of the DOG-type oligomers up to pentamers in the *Melastomataceae* are unique, for they are composed of two different monomeric units [*i.e.*, casuarictin (**4**) and pterocaryanin C

(**113**)] bound together *via* either coupling mode *a* or *b* (*vide supra*), as exemplified by the formation of either nobotanins F (**71**) or B (**78**) (Yoshida *et al.*, 2005).

2.4 Antimicrobial Effects of Ellagitannins

Despite the development of a myriad of antibiotics over the years, new and reemerging infectious diseases continue to be of major concern for human health. Antibiotic use leads to the development of drug-resistant bacteria, so that new antibacterial substances must be developed continuously. Although the antibacterial effects of various phenolic compounds, including tannin-rich extracts, have been reported, no systematic survey on the effects of tannins was conducted until a series of structurally defined tannins became available for such evaluations.

2.4.1 Variation of antibacterial effects according to bacterial species

A systematic study of the effects of chemically defined tannins using several bacterial species, along with two fungus *Candida* and *Cryptococcus* species, revealed that hydrolyzable and condensed tannins had only weak to moderate antibacterial effects irrespective of the degree of galloylation and molecular size (Kolodziej *et al.*, 1999). Among the polyphenols tested, the effects of ellagitannins such as corilagin (**8**) and phyllanthusiin C (**34**) against *Escherichia coli*, *Klebsiella pneumoniae*, *Bacillus subtilis*, *Pseudomonas aeruginosa*, and *Proteus mirabilis* were weak (MIC = 1000-2000 μg/ml). However, corilagin (**8**) showed moderate activity (*i.e.*, MIC = 250 μg/ml) against *Staphylococcus aureus*, as compared to that of penicillin G (MIC = 125 μg/ml against *Staphylococcus*). The effects of **8** and **34** against fungi such as *Candida albicans* and *Cryptcoccus neoformans* were comparable (*ca.* 125-500 μg/ml) to the effect against *Staphylococcus*. Although these data implied that the antibacterial and antifungal potencies of tannins were less pronounced than anticipated, at least for the compounds and microbial species tested, the antimicrobial effects of tannins can nevertheless be significant since they are ingested in large quantitities in traditional medicines, as noted by Kolodziej *et al.* (1999). The antimicrobial effects of ellagitannins and related polyphenols depend upon the bacterial species, as seen in their effects against *Helicobacter* (Funatogawa *et al.*, 2004), *Vibrio* species (Taguri *et al.*, 2006), and the protozoa *Leishmania donovani* (Kolodziej *et al.*, 2001).

2.4.2 Effects on *Helicobacter pylori*

Helicobacter pylori colonizing the human gastric epithelium plays a causal role in chronic gastritis and gastric ulcers in humans. Investigation of the *in vitro* anti-*pylori* effects of various polyphenols, including 11 monomeric, seven dimeric, and two trimeric ellagitannins and six proanthocyanidins, showed that all of the ellagitannins had more potent antibacterial activity with MICs of 6.25 to 50 μg/ml against four *H. pylori* strains than the B- and C-type proanthocyanidins. Among the ellagitannins, monomers exhibited stronger activity than oligomers. The strongest activity was observed for strictinin (**2**) with a MIC of 6.25 μg/ml for three of the four clinical isolates of *H. pylori*, while other monomeric ellagitannins, such as tellimagrandins I (**114**) and II (**1**), corilagin (**8**), casuarictin (**4**), and geraniin (**19**), had MICs of 12.5 to 25 μg/ml against the four strains. It was further observed that submitting tellimagrandins I (**114**) and II (**1**) to artificial gastric juice before evaluating their activity did not affect their antibacterial effects against *H. pylori*, and both of these ellagitannins had a MIC_{90} of 12.5 μg/ml, hence suggesting that they are stable and effective in the acidic gastric environment (Funatogawa *et al.*, 2004). Tellimagrandin I (**114**) also showed time- and dose-dependent bactericidal activity. A morphological study showed that *H. pylori* cells aggregated into fused clusters in the presence of tellimagrandin I (**114**), suggesting an effect on the bacterial surface structure. Conversely, the ellagitannins tested were not toxic to

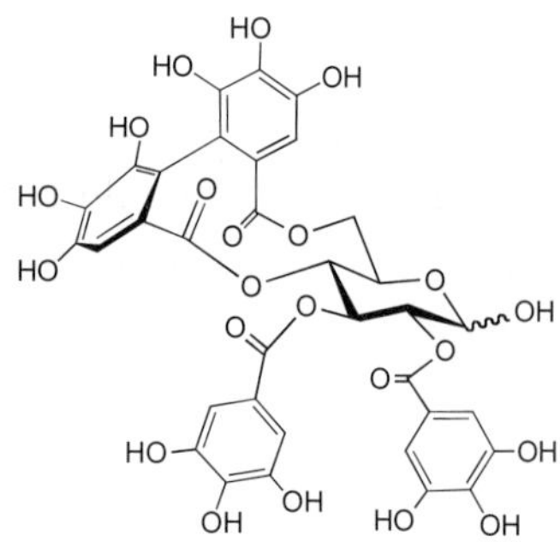

Tellimagrandin I (**114**)

E. coli, which is a normal inhabitant of the human intestinal tract, or to normal MKN-28 cells derived from gastric epithelium at concentrations up to 50 μg/ml. Therefore, ellagitannins are potentially useful agents that could be applied to suppress *H. pylori* without affecting gastric epithelial cells and nonpathogenic intestinal bacteria. *In vivo* experiments and studies on the mechanisms of their anti-*pylori* effect are desirable.

2.4.3 Effects on antibiotic-resistant bacteria

Studies have also unveiled the effect of polyphenolic compounds against methicillin-resistant *Staphylococcus aureus* (MRSA), and some ellagitannins were found to potentiate the antibacterial effect of β-lactam antibiotics. Of the diseases caused by drug-resistant bacteria, methicillin-resistant *S. aureus* (MRSA) is the one found most frequently in Japan, and over 20,000 patients with MRSA were reported in 2005 (Infectious Diseases Surveillance Center in Japan, 2007). The role of MRSA in community-associated skin diseases was also shown recently (Hasty *et al.*, 2007). The MRSA strains found in hospitals have acquired resistance not only to methicillin, but also to various antibiotics, including aminoglycosides and macrolides (Shiota *et al.*, 1999; You *et al.*, 2000).

The MICs of the β-lactam antibiotic oxacillin for four MRSA strains decreased noticeably in the presence of a small amount of (–)-epicatechin gallate (ECG) (Shiota *et al.*, 1999). An experiment examining the effect of combining ECG and oxacillin on the bacteria indicated that the effect was bactericidal. Additional studies on natural medicines have shown that the ellagitannins tellimagrandin I (**114**) and rugosin B (**11**), isolated from rose red (*i.e.*, petals of *Rosa canina*), markedly lowered the MICs of oxacillin against the MRSA strains (Shiota *et al.*, 2000). The effect of tellimagrandin I (**114**) with oxacillin was synergistic, and this tannin also had an analogous lowering effect on the MICs of tetracycline for two of the four MRSA strains. Although this tannin did not affect the MICs of aminoglycoside and macrolide antibiotics against MRSA, the MICs of tetracycline and fosfomycin against methicillin-sensitive *S. aureus* (MSSA) decreased. Corilagin (**8**), an ellagitannin isolated from the active extract of *Arctostaphylos uva-*

ursi leaves, also lowered the MICs of oxacillin, and the effect of the combination of corilagin (**8**) and oxacillin on MRSA was bactericidal (Shimizu *et al.*, 2001).

The production of penicillin-binding protein 2a (PBP2a) and drug-inactivating β-lactamase in the bacteria are regarded as important factors participating in the antibiotic-resistance mechanism of MRSA. The hydrolyzable tannins were found to decrease the production of PBP2a, to inactivate PBP2a and to suppress the β-lactamase activity (Shiota *et al.*, 2004). The macrocyclic ellagitannin dimer oenothein B (**115**), which was isolated from *Oenothera erythrosepala* (*Onagraceae*) (Hatano *et al.*, 1990c), similarly suppressed the antibiotic resistance of MRSA (Hatano *et al.*, 2006). A survey of the compounds possessing synergistic effects with antibiotics against MRSA and some other bacteria with antibiotic resistance has revealed that ellagitannins, condensed tannins, and some other polyphenolics are effective agents. The sustainability of the effect against the antibiotic resistance and the differences in the mode of action (bactericidal or bacteriostatic) were found to be dependent on the polyphenolic structure. Further research to develop polyphenol-based candidates for new medicines effective against drug-resistant bacteria or drug resistance itself is expected.

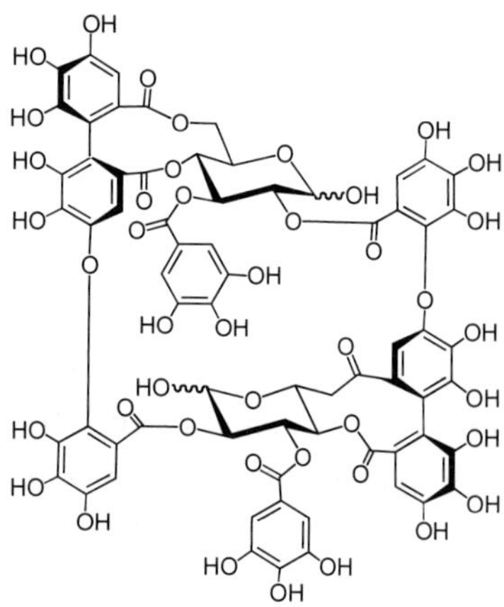

Oenothein B (**115**)

2.4.4 Effects on *Leishmania donovani*

In addition to the antibacterial and antifungal activities of ellagitannins, the antiparasitic effects of various hydrolyzable tannins against the extracellular promastigotes and intercellular amastigotes of *Leishmania donovani* were assessed (Kolodziej *et al.*, 2001). In an *in vitro* infection model of leishmaniasis using macrophage-like RAW 264.7 cells infected with the obligate intracellular parasite, all of the gallotannins and ellagitannins (monomers to trimers) showed pronounced antileishmanial effects against *L. donovani* intracellular amastigotes with EC_{50} < 0.4-7.5 μg/ml. Most of the compounds tested had low cytotoxicity (EC_{50} > 25 μg/ml) against the murine host cells.

The dehydroellagitannins geraniin (**19**), phyllanthusiin B (**33**) and elaeocarpusin (ascorgeraniin, **39**) and the gallotannins pentagalloylglucose (PGG, **9**) and tannic acid exhibited the most potent antileishmanial effect with EC_{50} < 0.4 μg/ml, which is comparable to that of the widely used therapeutic drug amphotericin B (EC_{50} = 0.3 μg/ml). Casuarinin (**45**) and castalagin (**43**) also showed potent effects with EC_{50} of 0.5 and 2.7 μg/ml, respectively. In contrast, none of the compounds were toxic to extracellular promastigotes (EC_{50} > 25 μg/ml).

The intracellular amastigote-specific effect of ellagitannins was associated with immunomodulatory effects of the tannins on macrophage functions, as shown in several functional assays. These assays evaluated the release of nitric oxide (NO), tumor necrosis factor-α (TNF-α) and interferon (IFN) in the presence of each tannin. Among the cytokines induced by tannins, the TNF-α-inducing activity evaluated in an assay using TNF-sensitive L929 fibroblasts increased in the following order: oligomeric ellagitannins (EC_{50} > 25 μg/ml; inactive) < monomeric ellagitannins and gallotannins (EC_{50} 8.5->25 μg/ml) < *C*-glycosidic ellagitannins and dehydroellagitannins (EC_{50} 0.6-2.8 μg/ml).

The aforementioned potent leishmanicidal dehydroellagitannins and *C*-glycosidic ellagitannins stimulated macrophages to release significant amounts of TNF-α, indicating that the antiparasitic effects of the tannins are due to their immunomodulatory effect, and not to a direct action. From these *in vitro* results, Kolodziej *et al.* (2001) concluded that tannins, and in particular ellagitannins, are capable of enhancing nonspecific immunity *via* macrophage activation and the release of cytokines, thereby inducing the fundamental host defense system to destroy invading pathogens. These immunological responses may account for the effectiveness of traditional polyphenol-rich medicines in infectious and other diseases.

2.5 Bibliography

Ahmed, A. F., Yoshida, T. and Okuda, T. (1994). New dimeric, trimeric and tetrameric ellagitannins from *Reaumuria hirtella*, *Chem. Pharm. Bull.*, 42, pp. 246–253.

Amakura, Y. and Yoshida, T. (1996). Euphorbin I, a new dimeric hydrolyzable tannin from *Euphorbia watanabei*, *Chem. Pharm. Bull.*, 44, pp. 1293–1297.

Amakura, Y., Kawata, K., Hatano, T., Agata, I., Sugaya, T., Nishibe, S., Okuda, T. and Yoshida, T. (1997). Four new hydrolyzable tannins and acylated flavonol glucoside from *Euphorbia maculata*, *Can. J. Chem.*, 75, pp. 727–733.

Foo, L. Y. (1993). Amariin, a di-dehydrohexahydroxydiphenoyl hydrolysable tannin from *Phyllanthus amarus*, *Phytochemistry*, 33, pp. 487–491.

Foo, L. Y. (1995). Amariinic acid and related ellagitannins from *Phyllanthus amarus*, *Phytochemistry*, 39, pp. 217–224.

Funatogawa, K., Hayashi, S., Shimomura, H., Yoshida, T., Hatano, T., Ito, H. and Hirai, Y. (2004). Antibacterial activity of hydrolyzable tannins derived from medicinal plants against *Helicobacter pylori*, *Microbiol. Immunol.*, 48, pp. 251–261.

Fürstenwerth, H. and Schildknecht, H. (1976). Pigments of *Cytinus hypocistis*, I. Isoterchebin, the yellow pigment of *Cytinus hypocistis* (*Rafflesiaceae*, parasitic plants), *Lieb. Ann. Chem.*, pp. 112–123.

Han, L., Hatano, T., Okuda, T. and Yoshida, T. (1995). Stachyuranins A, B and C, three new complex tannins from *Stachyurus praecox* leaves, *Chem. Pharm. Bull.*, 43, pp. 2109–2114.

Haslam, E. (1982). The metabolism of gallic acid and hexahydroxydiphenic acid in higher plants, *Fortschr. Chem. Org. Naturst.*, 41, pp. 1–46.

Hasty, M. B., Klasner, A., Kness, S., Denmark, T. K., Ellis, D., Herman, M. I. and Brown, L. (2007). Cutaneous community-associated methicillin-resistant *Staphylococcus aureus* among all skin and soft-tissue infections in two geographically distant pediatric emergency departments, *Acad. Emerg. Med.*, 14, pp. 35–40.

Hatano, T., Hattori, S. and Okuda, T. (1986). Coriariins A and B, new dimeric and monomeric hydrolyzable tannins, *Chem. Pharm. Bull.*, 34, pp. 4092–4097.

Hatano, T., Ogawa, N., Yasuhara, T. and Okuda, T. (1990a). Hydrolyzable tannin monomers having a valoneoyl group from flower petals of *Rosa rugosa* Thunb, *Chem. Pharm. Bull.*, 38, pp. 3308–3313.

Hatano, T., Okonogi, A., Yazaki, K., and Okuda, T. (1990b), Trapanins A and B, oligomeric hydrolyzable tannins from *Trapa japonica* Flerov., *Chem. Pharm. Bull.*, 38, pp. 2707–2711.

Hatano, T., Yasuhara, T., Matsuda, M., Yazaki, K., Yoshida, T. and Okuda, T. (1990c). Oenothein B, a dimeric hydrolysable tannin with macrocyclic structure, and accompanying tannins from *Oenothera erythrosepala*, *J. Chem. Soc., Perkin Trans. 1*, pp. 2735–2743.

Hatano, T., Yazaki, K., Okonogi, A. and Okuda, T. (1991). Praecoxins A, B, C, and D, four new hydrolyzable tannins from *Stachyurus praecox* leaves, *Chem. Pharm. Bull.*, 39, pp. 1689–1693.

Hatano, T., Tsugawa, M., Ohyabu, T., Kusuda, M., Shiota, S., Tsuchiya, T. and Yoshida, T. (2006). Effects of polyphenols in tea and foods on methicillin-resistant

Staphylococcus aureus and the sustainability of the antibacterial effects in the presence of food additives, *Assoc. J. Jpn. Soc. Med. Use Funct. Foods*, 4, pp. 43–48.

Hervé du Penhoat, C. L. M., Michon, V. M. F., Peng, S., Viriot, C., Scalbert, A. and Gage, D. (1991). The structural elucidation of new dimeric ellagitannins from *Quercus robur* L., roburin A-E, *J. Chem. Soc. Perkin Trans. I*, pp. 1653–1660.

Infectious Diseases Surveillance Center in Japan (2007). *Infectious Diseases Weekly Report*. http://idsc.nih.go.jp/idwr/ydata/report-Eb.html.

Ishimaru, K., Nonaka, G.-I. and Nishioka, I. (1987). Tannins and related compounds. LV. Isolation and characterization of acutissimins A and B, novel tannins from *Quercus* and *Castanea* species, *Chem. Pharm. Bull.*, 35, pp. 602–610.

Ishimaru, K., Ishimatsu, M., Nonaka, G., Mihashi, K., Iwase, Y. and Nishioka, I. (1988a). Isolation and characterization of mongolicins A and B, novel flavono-ellagitannins from *Quercus mongolica* var. *grosseserrata*, *Chem. Pharm. Bull.*, 36, pp. 3312–3318.

Ishimaru, K., Ishimatsu, M., Nonaka, G., Mihashi, K., Iwase, Y. and Nishioka, I. (1988b). Isolation and characterization of mongolicanin (procyanidino-ellagitannin), mongolinin A, acutissimin C and vescalagin carboxylic acid, novel tannins from *Quercus mongolica* var. grossesserrata, *Chem. Pharm. Bull.*, 36, pp. 3319–3327.

Ishimatsu, M., Tanaka, T., Nonaka, G., Nishioka, I., Nishizawa, M., and Yamagishi, T. (1989a). Isolation and characterization of novel diastereomeric ellagitannins, nupharins A and B, and their homologues from *Nuphar japonicum* DC, *Chem. Pharm. Bull.*, 37, pp. 129–134.

Ishimatsu, M., Tanaka, T., Nonaka, G., Nishioka, I., Nishizawa, M. and Yamagishi, T. (1989b). Isolation and characterization of novel dimeric and trimeric hydrolyzable tannins, nupharins C, D, E and F, from *Nuphar japonicum* DC, *Chem. Pharm. Bull.*, 37, pp. 1735–1743.

Ito, H., Hatano, T., Namba, O., Shirono, T., Okuda, T. and Yoshida, T. (1999a). Modified dehydroellagitannins, geraniinic acids B and C, and phyllanthusiin F. *Chem. Pharm. Bull.*, 47, pp. 1148–1151.

Ito, H., Miki, K. and Yoshida, T. (1999b). Elaeagnatins A–G, *C*-glucosidic ellagitannins from *Elaeagnus umbellate*, *Chem. Pharm. Bull.*, 47, pp. 536–542.

Ito, H., Miyake, M., Nishitani, E., Miyashita, K., Yoshimura, M., Yoshida, T., Takasaki, M., Konoshima, T., Kozuka, M. and Hatano, T. (2007). Cowaniin, a *C*-glucosidic ellagitannin dimer linked through catechin from *Cowania mexicana*, *Chem. Pharm. Bull.*, 55, pp. 491–494.

Kolodziej, H., Kayser, O., Latte, K. P. and Kiderlen, A. F. (1999). Enhancement of antimicrobial activity of tannins and related compounds by immune modulatory effects. In *Plant Polyphenols*, vol. 2, ed. by Gross, G.G., Hemingway, R. W. and Yoshida, T., Kluwer Academic/Plenum Publishers: New York, pp. 575–594.

Kolodziej, H., Kayser, O., Kiderlen, A. F., Ito, H., Hatano, T., Yoshida, T. and Foo, L. Y. (2001). Antileishmanial activity of hydrolyzable tannins and their modulatory effects on nitric oxide and tumor necrosis factor-α release in macrophages *in vitro*, *Planta Med.*, 67, pp. 825–832.

Lee, S. H., Tanaka, T., Nonaka, G. and Nishioka, I. (1990a). Isolation and characterization of helioscopinins and helioscopins, four new hydrolyzable tannins from *Euphorbia helioscopia* L., *Chem. Pharm. Bull.*, 38, pp. 1518–1523.

Lee, S. H., Tanaka, T., Nonaka, G. and Nishioka, I. (1990b). Hydrolysable tannins from *Euphorbia thymifolia*, *Phytochemistry*, 29, pp. 3621–3625.

Lee, S. H., Tanaka, T., Nonaka, G. and Nishioka, I. (1991). Monomeric and dimeric hydrolyzable tannins having a dehydrohexahydroxydiphenoyl group, supinanin, euphorscopin, euphorhelin and jolkianin, from *Euphorbia* species, *Chem. Pharm. Bull.*, 39, pp. 630–638.

Lin, J., Ishimatsu, M., Tanaka, T., Nonaka, G. and Nishioka, I. (1990). Structures of macaranins and macarinins, new hydrolyzable tannins possessing macaranoyl and tergalloyl ester groups, from the leaves of *Macaranga sinensis* (Baill.) Muell.-Arg., *Chem. Pharm. Bull.*, 38, pp. 1844–1851.

Lin, T.-J., Tanaka, T., Nonaka, G., Nishioka, I. and Young, T.-L. (1991). Tannins and related compounds. CVIII. Isolation and characterization of novel complex tannins (flavano-ellagitannins), anogeissinin and anogeissusins A and B, from *Anogeissus acuminata* (Roxb ex DC.) Guill. et Perr. var. *lanceolata* Wall. ex Clarke, *Chem. Pharm. Bull.*, 39, pp. 1144–1147.

Niemetz, R. and Gross, G. G. (2003). Ellagitannin biosynthesis: Laccase-catalyzed dimerization of tellimagrandin II to cornusiin E in *Tellima grandiflora*. *Phytochemistry*, 64, pp. 1197–1201.

Niemetz, R., Schilling, G. and Gross, G. G. (2001). Ellagitannin biosynthesis: Oxidation of pentagalloylglucose to tellimagrandin II by an enzyme from *Tellima grandiflora* leaves, *Chem. Commun.*, pp. 35–36.

Nonaka, G., Nishimura, H. and Nishioka, I. (1985). Isolation and structures of stenophyllanins A, B, and C, novel tannins from *Quercus stenophylla*, *J. Chem. Soc. Perkin Trans. 1*, pp. 163–172.

Nonaka, G., Ishimaru, K., Mihashi, K., Iwase, Y., Ageta, M. and Nishioka, I. (1988). Isolation and characterization of mongolicains A and B, novel tannins from *Quercus* and *Castanopsis* species, *Chem. Pharm. Bull.*, 36, pp. 847–869.

Nonaka, G., Ishimaru, K., Azuma, R., Ishimatsu, M. and Nishioka, I. (1989). Structures of novel *C*-glucosidic ellagitannins, grandinin and pterocarinins A and B, *Chem. Pharm. Bull.*, 37, pp. 2071–2077.

Nonaka, G., Sakai, K., Mihashi, K. and Nishioka, I. (1991). Isolation of alienanins A and B, novel C-C-linked ellagitannin dimers from *Quercus aliena* Blume, *Chem. Pharm. Bull.*, 39, pp. 884–888.

Nonaka, G., Akazawa, M. and Nishioka, I. (1992). Two new ellagitannin metabolites, carpinusin and carpinusnin from *Carpinus laxiflora*, *Heterocycles*, 33, pp. 597–606.

Okuda, T., Hatano, T., Nitta, H. and Fujii, R. (1980a). Hydrolysable tannins having enantiomeric dehydrohexahydroxydiphenoyl group: revised structure of terchebin and structure of granatin B, *Tetrahedron Lett.*, 21, pp. 4361–4364.

Okuda, T., Yoshida, T., Hatano, T., Yazaki, K. and Ashida, M. (1980b). Ellagitannins of the *Casuarinaceae*, *Stachyuraceae* and *Myrtaceae*, *Phytochemistry*, 21, pp. 2871–2874.

Okuda, T., Hatano, T. and Yasui, T. (1981). Revised structure of isoterchebin, isolated from *Cornus officinalis*, *Heterocycles*, 16, pp. 1681–1685.

Okuda, T., Yoshida, T., Hatano, T., Koga, T., Toh, N. and Kuriyama, K. (1982a). Circular dichroism of hydrolysable tannins. I. Ellagitannins and gallotannins, *Tetrahedron Lett.*, 23, pp. 3937–3940.

Okuda, T., Yoshida, T., Hatano, T., Koga, T., Toh, N. and Kuriyama, K. (1982b). Circular dichroism of hydrolysable tannins. II. Dehydroellagitannins, *Tetrahedron Lett.*, 23, pp. 3941–3944.

Okuda, T., Yoshida, T., Ashida, M. and Yazaki, K. (1983). Tannins of *Casuarina* and *Stachyurus* species. Part 1. Structures of pedunculagin, casuarictin, strictinin, casuarinin, casuariin, and stachyurin, *J. Chem. Soc. Perkin Trans 1*, pp. 1765–1772.

Okuda, T., Hatano, T., Ogawa, N., Kira, R. and Matsuda, M. (1984a). Cornusiin A, a dimeric ellagitannin forming four tautomers, and accompanying new tannins in *Cornus officinalis*, *Chem. Pharm. Bull.*, 32, pp. 4662–4665.

Okuda, T., Yoshida, T., Kuwahara, M., Memon, M. U. and Shingu, T. (1984b). Structures of potentillin, agrimonic acids A and B, and agrimoniin, a dimeric ellagitannin, *Chem. Pharm. Bull.*, 32, pp. 2165–2173.

Okuda, T., Yoshida, T., Hatano, T., Ikeda, Y., Shingu, T. and Inoue, T. (1986). Isolation of water-soluble tannins by centrifugal partition chromatography, and biomimetic synthesis of elaeocarpusin, *Chem. Pharm. Bull.*, 34, pp. 4075–4082.

Saijo, R., Nonaka, G. and Nishioka, I. (1989a). Isolation and characterization of four new hydrolyzable tannins from the leaves of *Mallotus repandus*, *Chem. Pharm. Bull.*, 37, pp. 2624–2630

Saijo, S., Nonaka, G. and Nishioka, I. (1989b). Isolation and characterization of five new hydrolyzable tannins from the bark of *Mallotus japonicus*, *Chem. Pharm. Bull.*, 37, pp. 2063–2070.

Shimizu, M., Shiota, S., Mizushima, T., Ito, H., Hatano, T., Yoshida, T. and Tsuchiya, T. (2001). Marked potentiation of activity of beta-lactams against methicillin-resistant *Staphylococcus aureus* by corilagin, *Antimicrob. Agents Chemother.*, 45, pp. 3198–3201.

Shimokawa, H., Nonaka, G. and Nishioka, I. (1991). Tannins and related compounds from *Casuarina glauca* Sieber, *Proc. 111th Annual Meeting of the Pharmaceutical Society of Japan*, vol. 2, Tokyo, p. 147.

Shiota, S., Shimizu, M., Mizushima, T., Ito, H., Hatano, T., Yoshida, T. and Tsuchiya, T. (1999). Marked reduction in the minimum inhibitory concentration (MIC) of beta-lactams in methicillin-resistant *Staphylococcus aureus* produced by epicatechin gallate, an ingredient of green tea (*Camellia sinensis*), *Biol. Pharm. Bull.*, 22, pp. 1388–1390.

Shiota, S., Shimizu, M., Mizusima, T., Ito, H., Hatano, T., Yoshida, T. and Tsuchiya, T. (2000). Restoration of effectiveness of beta-lactams on methicillin-resistant *Staphylococcus aureus* by tellimagrandin I from rose red, *FEMS Microbiol. Lett.*, 185, pp. 135–138.

Shiota, S., Shimizu, M., Sugiyama, J., Morita, Y., Mizushima, T. and Tsuchiya, T. (2004). Mechanisms of action of corilagin and tellimagrandin I that remarkably potentiate the activity of beta-lactams against methicillin-resistant *Staphylococcus aureus*, *Microbiol. Immunol.*, 48, pp. 67–73.

Taguri, T., Tanaka, T. and Kouno, I. (2006). Antibacterial spectrum of plant polyphenols and extracts depending upon hydroxyphenyl structure, *Biol Pharm Bull.*, 29, pp. 2226–2235.

Tanaka, T., Nonaka, G. and Nishioka, I. (1986a). Isolation and characterization of four new hydrolyzable tannins, terflavins A and B, tergallagin and tercatain from the leaves of *Terminalia catappa* L., *Chem. Pharm. Bull.*, 34, pp. 1039–1049.

Tanaka, T., Nonaka, G. and Nishioka, I. (1986b). Isolation and characterization of novel ellagitannins, punicacorteins A, B, C, D, and punigluconin from the bark of *Punica granatum* L. *Chem. Pharm. Bull.*, 34, pp. 656–663.

Tanaka, T., Nonaka, G., Nishioka, I., Miyahara, K. and Kawasaki, T. (1986c). Isolation and structure elucidation of elaeocarpusin, a novel ellagitannin from *Elaeocarpus sylvestris* var. *ellipticus*, *J. Chem. Soc., Perkin Trans. 1*, pp. 369–376.

Tanaka, T., Ishida, N., Ishimatsu, M., Nonaka, G. and Nishioka, I. (1992a). 6 new complex tannins, guajavins, psidinins and psiguavin from the bark of *Psidium guajava* L., *Chem. Pharm. Bull.*, 40, pp. 2092–2098.

Tanaka, T., Tong, H.-H., Xu, Y.-M., Ishimaru, K., Nonaka, G. and Nishioka, I. (1992b). Isolation and characterization of three new ellagitannins, lagerstannins A, B and C, having a gluconic acid core, from *Lagerstroemia speciosa* (L.) PERS, *Chem. Pharm. Bull.*, 40, pp. 2975–2980.

Tanaka, T., Ueda, N., Shinohara, H., Nonaka, G., Fujioka, T., Mihashi, K. and Kouno, I. (1996). *C*-Glucosidic ellagitannin metabolites in the heartwood of Japanese chestnut tree (*Castanea crenata* Sieb. et Zucc.), *Chem. Pharm. Bull.*, 44, pp. 2236–2242.

Yoshida, T., Fujii, R. and Okuda, T. (1980). Revised structures of chebulinic acid and chebulagic acid, *Chem. Pharm. Bull.*, 28, pp. 3713–3715.

Yoshida, T., Maruyama, Y., Memon, M. U., Shingu, T. and Okuda, T. (1985). Gemins D, E and F, ellagitannins from *Geum japonicum*, *Phytochemistry*, 24, pp. 1041–1046.

Yoshida, T., Tanaka, K., Chen, X.-M. and Okuda, T. (1991a). Tannins from *Hippophae rhamnoides*, *Phytochemistry*, 30, pp. 663–666.

Yoshida, T., Yokoyama, K., Namba, O. and Okuda, T. (1991b). Tirucallins A, B and euphorbin F, monomeric and dimeric ellagitannins from *Euphorbia tirucalli* L., *Chem. Pharm. Bull.*, 39, pp. 1137–1143.

Yoshida, T., Ahmed, A. F., Memon, M. U. and Okuda, T. (1991c). New monomeric and dimeric hydrolyzable tannins from *Reaumuria hirtella* and *Tamarix pakistanica*, *Chem. Pharm. Bull.*, 39, pp. 2849–2854.

Yoshida, T., Jin, Z.-X. and Okuda, T. (1991d). Heterophylliins A, B, C, D and E, ellagitannin monomers and dimers from *Corylus heterophylla* Fisch, *Chem. Pharm. Bull.*, 39, pp. 49–54.

Yoshida, T., Itoh, H., Matsunaga, S., Tanaka, R. and Okuda, T. (1992a). Tannins and related polyphenols of euphorbiaceous plants. IX. Hydrolyzable tannins with $^{1}C_{4}$ glucose core from *Phyllanthus flexuosus* Muell. ARG, *Chem. Pharm. Bull.*, 40, pp. 53–60.

Yoshida, T., Nakata, F., Hosotani, K., Nitta, A. and Okuda, T. (1992b). Three new complex tannins from *Melastoma malabathricum*, *Chem. Pharm. Bull.*, 40, pp. 1727–1732.

Yoshida, T., Ahmed, A. F. and Okuda, T. (1993). New dimeric hydrolyzable tannins from *Reaumuria hirtella*, *Chem. Pharm. Bull.*, 41, pp. 672–679.

Yoshida, T., Amakura, Y., Liu, Y. Z. and Okuda, T. (1994). Three new hydrolyzable tannins and a polyphenol glucoside from *Euphorbia humifusa*, *Chem. Pharm. Bull.*, 42, pp. 1803–1807.

Yoshida, T., Ito, H., Hatano, T., Kurata, M., Nakanishi, T., Inada, A., Murata, H., Inatomi, Y., Matsuura, N., Ono, K., Nakane, H., Noda, M., Lang, F. A. and Murata, J. (1996). New hydrolyzable tannins, shephagenins A and B, from *Shepherdia argentea* as HIV-1 reverse transcriptase inhibitors, *Chem. Pharm. Bull.*, 44, pp. 1436–1439.

Yoshida, T., Ito, H. and Isaza, J. H. (2005). Pentameric ellagitannin oligomers in melastomataceous plants – Chemotaxonomic significance, *Phytochemistry*, 66, pp. 1972–1983.

Yoshida, T., Ito, H., Yoshimura, M., Miyashita, K. and Hatano, T. (2008). *C*-Glucosidic ellagitannin oligomers from *Melaleuca squarrosa* Donn ex Sm., *Myrtaceae*, *Phytochemistry, in press*.

You, I., Kariyama, R., Zervos, M. J., Kumon, H. and Chow, J. W. (2000). *In-vitro* activity of arbekacin alone and in combination with vancomycin against gentamicin- and methicillin-resistant *Staphylococcus aureus*, *Diagn. Microbiol. Infect. Dis.*, 36, pp. 37–41.

Chapter 3

Biosynthesis of Ellagitannins: Old Ideas and New Solutions

Georg G. Gross*

*Molekulare Botanik, Universität Ulm, 89069 Ulm, Germany. *Retired; private address: Kirchstrasse 2, 88480 Achstetten, Germany; gg.gross@t-online.de*

3.1 Introduction

Plants have always been an indispensable part of human life, not only as important nutrients, but also as a source of valuable chemicals for manifold applications in medicine, flavoring, painting or technology. In contrast to alkaloids or essential oils, the role of plant phenolics, which certainly represent the most abundant class of natural plant products, has been widely ignored. This statement applies particularly to tannins, *i.e.*, plant polyphenols that have been used by mankind over many thousand years for the production of leather from raw animal hides. It was only at the beginning of the 20th century that intensive chemical studies on the nature of such substances were conducted, particularly by E. Fischer, K. Freudenberg and P. Karrer, just 'to mention a few prominent names. Some principal structural features of hydrolyzable tannins from sumac (Chinese gallotannin) and oak (Turkish gallotannin) were discovered in these early investigations, but the then available analytical armament was insufficient for an effective tackling of these complex plant constituents. Interest in this field waned therefore considerably, and it was not before the 1950s when, in response to significant progress in more effective separation techniques and sensitive analytical procedures, a remarkable renaissance could be observed with the seminal investigations of O. Th. Schmidt and W. Mayer (for a competent review of tannin research from the beginnings to the mid-1950s, see Schmidt and Mayer, 1956). This trend was continued by other research groups, mainly in England and Japan, including, among many others, the laboratories of E. C. Bate-Smith, T. Swain, E. Haslam, T. Okuda and I. Nishioka (see Chapters 1 and 2). Detailed structures of a myriad of hydrolyzable tannins and related compounds were compiled by these efforts (*e.g.*, Haslam, 1989) that not only served to describe their distribution and evolutionary relationships in the plant kingdom (*e.g.*, Haddock *et al.*, 1982, Okuda *et al.*, 1993), but also provoked considerations on possible routes in their biogenetic origin (Haddock *et al.*, 1982), an aspect, however, that was merely hypothetical at these times.

Concerning this latter question, it must be emphasized that biosynthetic studies were initially based on so-called 'feeding' experiments, a technique by which radioactively labeled putative precursors were administered to living plants or plant parts. Next, investigated compounds were isolated; total amount and distribution of radioactivity in these products served as a basis for the construction of feasible metabolic routes. The introduction of enzyme studies added a new dimension to such investigations, hence allowing experiments with labile or membrane-impermeable compounds and, in particular, giving access to work with "energy-rich" activated intermediates. Over the past 25 years, such enzyme studies on the challenging question of hydrolyzable tannin biosynthesis were performed in my laboratory. In retrospect, it was a rather daring endeavor to tackle this problem just by enzymological means if one considers the "tanning" property, *i.e.*, protein precipitating, and hence enzyme inactivating potential of such compounds. Fortunately, gallotannin- and ellagitannin-synthesizing enzymes proved to be remarkably resistant to their unfavorable substrates and products. In contrast, major problems were encountered in the availability of hydrolyzable tannins required as references or substrates because none of these delicate chemicals was, and still is, commercially available (Gross, 2007).

3.2 Principal Structures and Definitions

Before discussing the biochemical events described in the subsequent sections, it appears appropriate to insert a few comments on the structural principles and conventional terminology in this field. According to the classical definition formulated by Freudenberg (1920), plant tannins are typically divided into condensed tannins and hydrolyzable tannins. The first class, nowadays often referred to as proanthocyanidins due to the liberation of colored anthocyanidins upon treatment with alcoholic mineral acid (Scalbert, 1991), is of flavonoid origin and is no subject of this article.

The second class, hydrolyzable tannins, is best described as derivatives of gallic acid (3,4,5-trihydroxybenzoic acid, **1**; see Fig. 3.1), this principal phenolic unit being esterified with the hydroxyl groups of a central polyol moiety (usually β-D-glucopyranose). The simplest among these esters, β-glucogallin (1-*O*-galloyl-β-D-glucopyranose, **2**), is known as a natural product for more than a century. The fully acylated analogue, 1,2,3,4,6-penta-*O*-galloyl-β-D-glucopyranose (**3**), is regarded as the immediate precursor of the two subclasses of hydrolyzable tannins, *i.e.*, gallotannins and ellagitannins.

The latter, *i.e.*, ellagitannins, are the result of oxidative processes that act on pentagalloylglucopyranose (PGG, **3**) to form linkages between spatially adjacent galloyl residues, affording compounds like tellimagrandin II (**4**), a simple monomeric ellagitannin. After eventual hydrolytic release from the tannin core, the resulting 3,4,5,3',4',5'-hexahydroxydiphenic (HHDP) acid (**5**) rearranges spontaneously to the stable, rather insoluble dilactone, ellagic acid (**6**). It is interesting to note that this artifact became name-giving for the whole group of natural products. The second subclass, gallotannins, is characterized by additional galloyl residues that are linked to the basic pentagalloylglucose (**3**) core to afford *meta*-digalloyl residues (**7**) that can be classified as depsides. Total substitution degrees of 10-12 galloyl units attached to the central glucose moiety have been determined for Chinese gallotannin. Many variations of these fundamental structural principles have been discovered in higher plants, but their discussion is beyond the scope of this chapter.

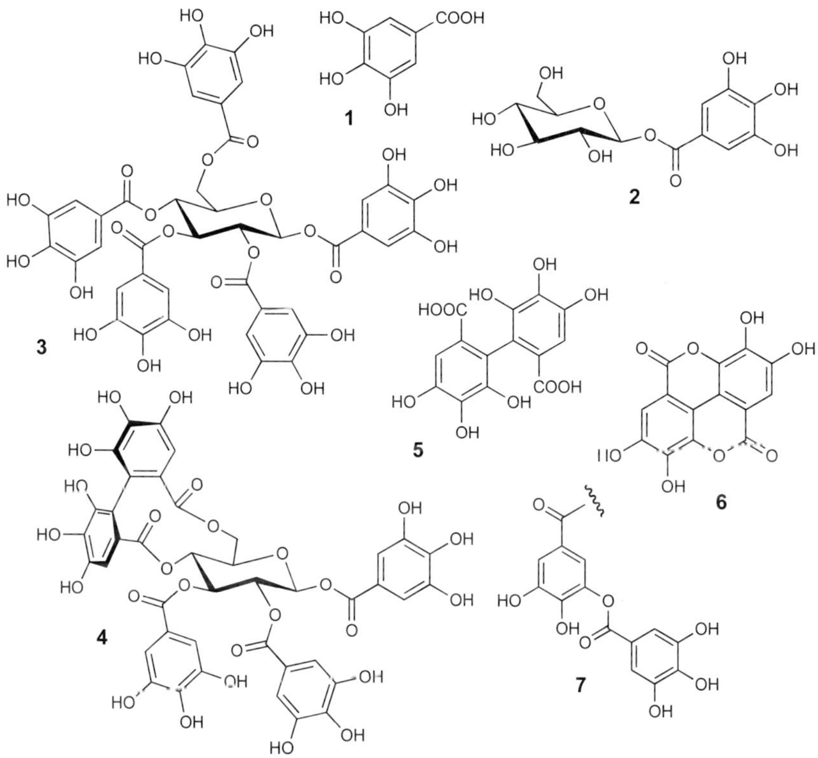

Fig. 3.1 Structures of important components related to hydrolyzable tannins: (**1**) Gallic acid (3,4,5-trihydroxybenzoic acid), the principal phenolic unit; (**2**) β-glucogallin (1-*O*-galloyl-β-D-glucopyranose; (**3**) 1,2,3,4,6-penta-*O*-galloyl-β-D-glucopyranose; (**4**) tellimagrandin II, a typical monomeric ellagitannin; (**5**) 3,4,5,3',4',5'-hexahydroxydiphenic acid (usually abbreviated as HHDP), the typical constituent of ellagitannins; (**6**) ellagic acid, the dilactone of **5**; (**7**) *meta*-digalloyl residue, the characteristic moiety of gallotannins.

It is not surprising that increasing knowledge of these chemical structures was suitable to stimulate considerations on the biogenetic routes to hydrolyzable tannins. Plausible pathways were proposed on this basis (Haddock *et al.*, 1982) which, as depicted in Fig. 3.2, can be conveniently subdivided into several principal minor challenges, aiming at (i) the biosynthetic route(s) to gallic acid (**1**) as the principal phenolic unit; (ii) the origin of β-glucogallin (**2**) as the first specific intermediate in the pathway to hydrolyzable tannins; (iii) the conversion of this monoester to pentagalloylglucopyranose (PGG, **3**) along a series of so-called "simple" galloylglucose esters; and finally the secondary transformations of this pivotal intermediate to yield (iv) gallotannins by adding galloyl *meta*-depsides groups (**7**), or to form (v) monomeric ellagitannins by oxidative intramolecular aryl C–C coupling reactions, followed by intermolecular C–O coupling leading to dimeric and oligomeric derivatives that considerably contribute to the vast structural diversity of this class of natural phenolic products.

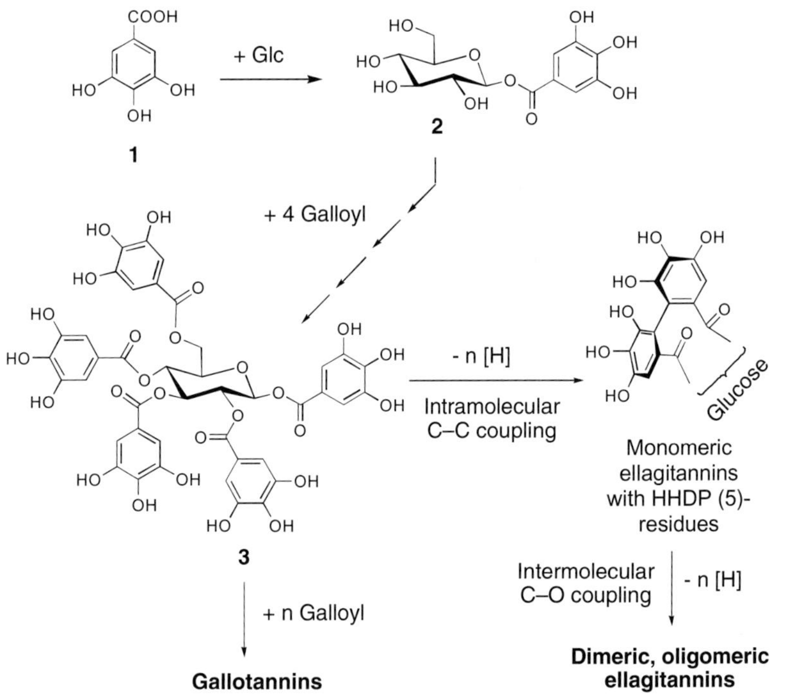

Fig. 3.2 Principal steps in the biosynthesis of gallo- and ellagitannins. Glc = glucopyranose.

Still in the 1980s, little experimental evidence was available on the natural intermediates and biochemical events of these plausible, nevertheless theoretical pathways as documented, for instance, by a review published in 1985 on the biosynthesis of tannins (Hillis, 1985). Intensive enzymatic investigations in my laboratory have significantly contributed to considerably changing this unsatisfying situation. Experiments on these questions were performed with cell-free extracts from leaves of pedunculate oak (*Quercus robur*, *syn. Q. pedunculata*) or red oak (*Q. rubra*), staghorn sumac (*Rhus typhina*) and fringe cups (*Tellima grandiflora*). The results obtained by this laborious but also highly evidential technique allowed us to describe the principles and many details of the pathways leading to these important natural plant products.

3.3 Gallic Acid, the Principal Phenolic Unit

It is now generally accepted that major anabolic routes to benzoic acids (phenylcarboxylic acids) in plants proceed via cinnamoyl-coenzyme A (CoA) thioesters, whose propenoic side-chain undergoes degradation, either oxidatively by β-oxidation to afford benzoyl-CoA's (Löscher and Heide, 1994) or in a non-oxidative sequence to free benzaldehydes (Abd El-Mawla and Berhues, 2002). Major problems, however, have always been encountered in the case of gallic acid (**1**). A rather conventional but unproven concept assuming the CoA-dependent β-oxidation of 3,4,5-trihydroxycinnamic acid (**9**; see Fig. 3.3, route A) (Zenk, 1964) suffered from the fact that this compound has never been identified as natural product, occasionally being regarded as the "missing cinnamic acid" (Haslam, 1982).

A variation of this proposal (see route B in Fig. 3.3) avoided this problem by putting the side-chain degradation reaction one step forward to caffeic acid (**10**), but remained speculative as well (El-Basyouni *et al.*, 1964). A significantly different approach was presented by others (Conn and Swain, 1961; Cornthwaite and Haslam, 1965) that favored the direct aromatization of shikimic acid or, more likely, 5-dehydroshikimic acid (**12**), the *enol*-form (**13**) which should be dehydrogenated to gallic acid (**1**; see route C in Fig. 3.3). The detour *via* C_6–C_3 compounds, as required for routes A and B, would thus be replaced by this shortcut from a very early intermediate to gallic acid.

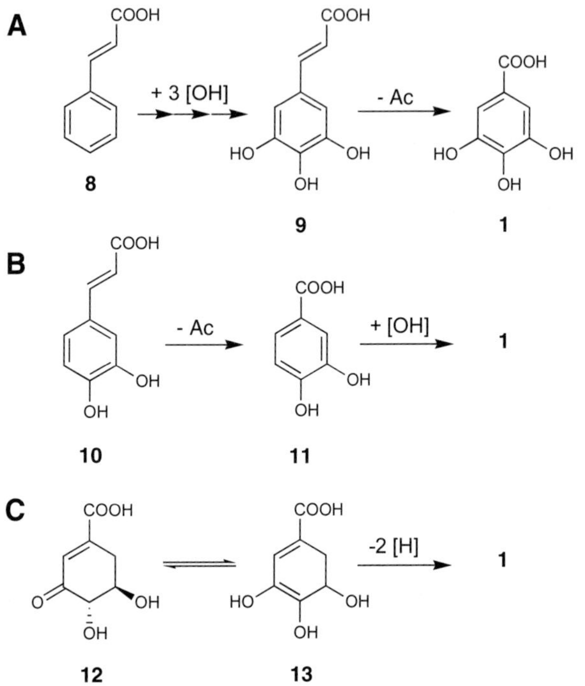

Fig. 3.3 Proposed biosynthetic pathways to gallic acid (**1**). **A**. The route from cinnamic acid (**8**) *via* 3,4,5-trihydroxycinnamicacid (**9**) that undergoes β-oxidation to gallic acid. **B**. Alternative route *via* cinnamate-derived caffeic acid (**10**) being degraded to protocatechuic acid (**11**) and hydroxylated to yield gallic acid. **C**. Direct reduction of 5-dehydroshikimic acid (**12**) via its *enol*-form (**13**) to gallic acid. Ac = acetyl residue. For details, see text.

The latter concept is currently the most plausible scheme, although it is still a matter of dispute whether this mechanism represents the exclusive route to gallic acid. First conclusive results on the old enigma of gallic acid biosynthesis were obtained by feeding [^{13}C]glucose to the fungus *Phycomyces blakesleeanus* and the dicotyledonous plant *Rhus typhina* (staghorn sumac). NMR spectroscopy of the isotope distribution of gallic acid and aromatic amino acids isolated from these species provided the basis for the interpretation of isotopomer patterns of these metabolites in a "retrobiosynthetic" approach. It was concluded that gallic acid was derived from an early intermediate of the shikimate pathway, most likely, 5-dehydroshikimic acid (**12**; Werner *et al.*, 1997). Recently, this interpretation was corroborated by determination of $\delta^{18}O$-values of gallic acid from *R. typhina* leaves that indicated its formation by dehydrogenation of 5-dehydroshikimic acid and also excluded alternative routes *via* phenylpropanoid C_6–C_3 intermediates (Werner *et al.*, 2004).

Supporting evidence for these results was published by Ossipov *et al.* (2003), who reported the *in vitro* reduction of 5-dehydroshikimic acid to gallic acid with enzyme preparations from birch (*Betula pubescens*) leaves.

3.4 Biosynthesis of β-Glucogallin

β-Glucogallin (1-*O*-galloyl-β-D-glucopyranose, **2**) was first isolated from Chinese rhubarb (*Rheum officinale*) in 1903 (Haslam, 1998) and was long ago proposed as the first specific metabolite in the biosynthesis of hydrolyzable tannins (Haddock *et al.*, 1982). For thermodynamic reasons, esterification of gallic acid (**1**) and glucose must be expected to involve the participation of an "activated" intermediate with a high group-transfer potential. By analogy to the well-known caffeoyl-CoA dependent formation of chlorogenic acid and related depsides (Stöckigt and Zenk, 1974; Ulbrich and Zenk, 1980), galloyl-CoA was considered the most likely candidate for this reaction. This unknown thioester was synthesized *via* the *N*-hydroxysuccinimidyl derivative of 4-*O*-β-D-glucosidogallic acid (Gross, 1982a), but it was soon recognized that this compound was not involved in the biosynthesis of β-glucogallin (Gross, 1983a). It was found instead that a glucosyltransferase from oak leaves catalyzed the efficient esterification of free gallic acid (**1**) and "activated" glucose, uridine-5'-diphosphate glucose, affording β-glucogallin (**2**) and related 1-*O*-acyl-β-D-glucopyranoses (Fig. 3.4; Gross, 1982b, 1983b; Weisemann *et al.*, 1988). Numerous analogous enzymes catalyzing the formation of phenolic 1-*O*-acylglucoses have been isolated in the meantime, providing evidence that this reaction type is of general importance in plants (see references in Gross, 1989).

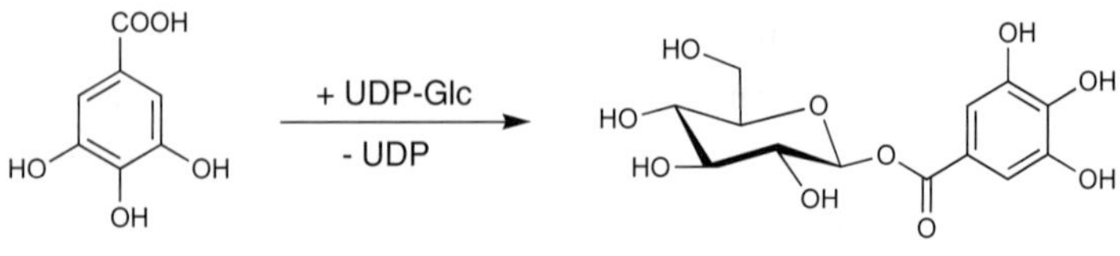

Fig. 3.4 Biosynthesis of β-glucogallin (1-*O*-galloyl-β-D-glucopyranose, **2**) from free gallic acid (**1**) and "activated" glucose, uridine-5'-diphosphate glucose (UDP-Glc). UDP = uridine-5'-diphosphate.

3.5 'Simple' Galloylglucose Esters: the Route to Pentagalloylglucose

In the above discussed investigations on the biosynthesis of β-glucogallin (**2**) with cell-free extracts from oak leaves, it was observed that these enzyme preparations also catalyzed the transformation of *in situ* generated β-glucogallin to di- and trigalloylglucopyranoses and this, to be emphasized, in the absence of any other cofactors. This unexpected result could be explained only by the assumption that β-glucogallin had a dual role in these systems, functioning not only as an acyl acceptor, but also as an efficient acyl donor to yield a higher galloylated derivative and free glucose as a by-product (Gross, 1983a). Such a behaviour would require that β-glucogallin had the unexpected characteristic of an energy-rich "activated" compound. Considering the known and comparatively low group-transfer potentials ΔG_o' of acylglucoses (*e.g.*, glucose-1-phosphate, *ca.* 21 kJ mol^{-1} or glucose-6-phosphate, *ca.* 10.5 kJ mol^{-1}; Atkinson and Morton, 1960), the existence of such a reaction was surprising. The logical classification of β-glucogallin as an activated compound was later supported by Mock and Strack (1993) that reported that a related ester, 1-*O*-sinapoyl-β-D-glucopyranose, had an unexpectedly high ΔG_o' value of 35.7 kJ mol^{-1}. This value is comparable to that of the well-known acyl-CoA thioesters, and it is reasonable to assume that the ΔG_o' of β-glucogallin is on the same order of magnitude. Consequently, glucose esters lacking the energy-rich 1-*O*-acyl group should display no donor function, a conclusion that was corroborated in later experiments. Meanwhile, similar donor functions of 1-*O*-acylglucoses have been observed for many other pathways (see references in Gross, 1999). It thus became evident that the widely neglected or underestimated phenolic 1-*O*-acylglucose esters, often previously regarded as metabolically inert compounds or waste products, occupy a central position in plant secondary metabolism, that is at least comparable to that of the generally acknowledged role of acyl-CoA esters.

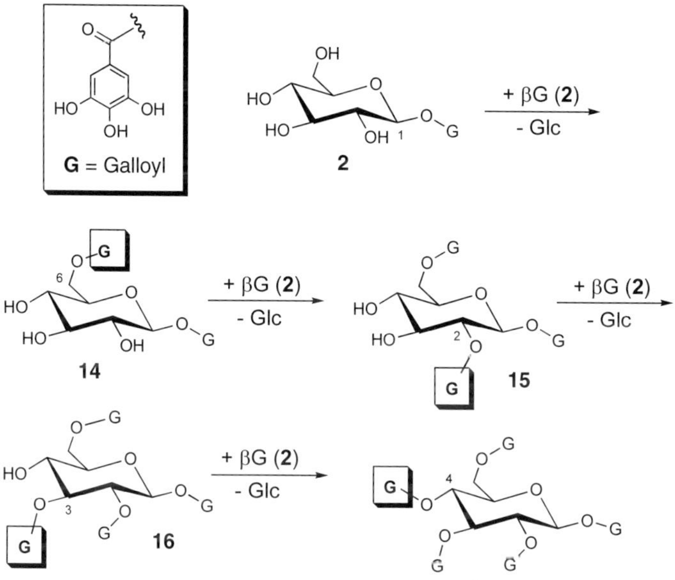

Fig. 3.5 Enzyme reactions catalyzing the pathway from β-glucogallin (βG, **2**) to 1,2,3,4,6-penta-*O*-galloyl-β-D-glucopyranose (**3**) in oak leaves. Positions of newly introduced galloyl residues are marked by square boxes and numbering of the corresponding positions on the glucopyranose core. **14**: 1,6-Digalloyl-β-D-glucopyranose; **15**: 1,2,6-trigalloyl-β-D-glucopyranose; **16**: 1,2,3,6-tetragalloyl-β-D-glucopyranose. Glc = glucose.

Further studies with enzyme preparations from oak or sumac revealed that the described mechanism, *i.e.*, galloyl transfer from β-glucogallin (**2**) to glucose hydroxyls, applied to all of the subsequent transformation reactions up to 1,2,3,4,6-pentagalloylpyranose (**3**). As depicted in Fig. 3.5, it was remarkable to find that these substitution reactions were not randomly distributed, but displayed high regiospecificity to constitute the following metabolic sequence: β-glucogallin (**2**) → 1,6-digalloylglucopyranose (**14**, Schmidt *et al.*, 1987) → 1,2,6-trigalloylglucopyranose (**15**, Gross and Denzel, 1991) → 1,2,3,6-tetragalloylglucopyranose (**16**, Hagenah and Gross, 1993) → 1,2,3,4,6-pentagalloylglucopyranose (PGG) (**3**, Cammann *et al.*, 1989). Interestingly, an identical pattern has been reported for the chemical substitution of glucose hydroxyls in experiments with 1-benzyl or 1-methyl-β-D-glucopyranose that was explained as a combination of reactivity differences in response to variations in chemical nature (*i.e.*, primary *vs.* secondary hydroxyl), neighboring group activation and steric hindrance (Williams and Richardson, 1967; Reinefeld and Ahrens, 1971).

The above discussed enzyme studies may give the impression of a straightforward directed pathway that is depending on one simple and uniform reaction mechanism. There is no doubt that this interpretation applies to the route from β-glucogallin to pentagalloylglucopyranose, however, observations on the existence of other enzyme activities, leading to side-reactions and ramifications of the main pathway, should be briefly mentioned for completeness. For instance, an acyltransferase was found in oak leaves that catalyzed an unusual galloyl exchange between β-glucogallin (**2**) and free glucose, a reaction that was detectable only with labeled substrates, *i.e.*:

β-glucogallin + *glucose ↔ *β-glucogallin + glucose

(the asterisk symbolizes the radioactive label), the physiological significance of which remaining obscure (Gross *et al.*, 1986). Another enzyme from sumac interfered with the acylation of 1,6-digalloylglucopyranose (**14**) to 1,2,6-trigalloylglucopyranose (**15**) by promoting this reaction in the absence of the established acyl donor, β-glucogallin (**2**) (see Fig. 3.6). It was recognized that this enzyme catalyzed a "disproportionation" reaction, in which two molecules of 1,6-digalloylglucopyranose (**14**) were converted into 1,2,6-trigalloylglucopyranose (**15**) and anomeric 6-*O*-galloylglucose as a partially deacylated by-product (Denzel and Gross, 1991). This finding stimulated investigations to what extent also higher substituted galloylglucoses could act as acyl donors, provided they still possessed the energetically indispensable 1-*O*-acetyl moiety. Such compounds were actually found to exert galloyl donor potentials, however, the reactivity of higher substituted analogues was drastically reduced, most likely because of increasing steric hindrance due to excessive bulkiness. Thus, 1-mono- and, at a lower degree, 1,6-diesters proved to be the predominating galloyl donors (Denzel and Gross, 1991).

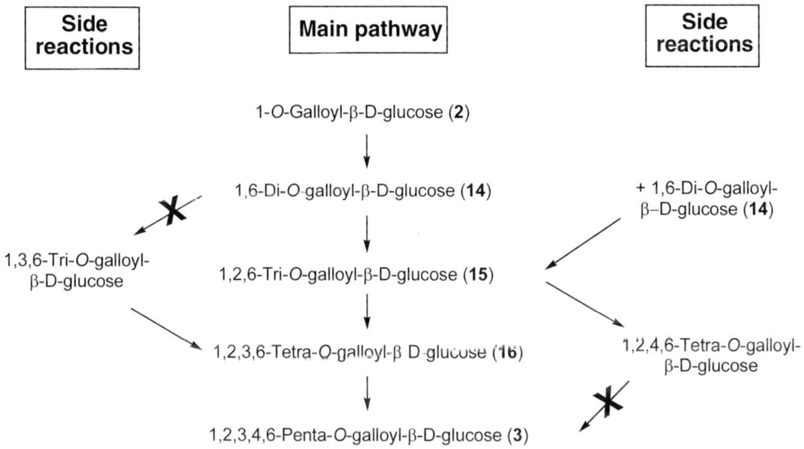

Fig. 3.6 Overview of the main pathway and side-reactions in the biosynthetic route from mono- to pentagalloylated β-D-glucopyranoses.

Some abortive side-reactions are shown in Fig. 3.6. Concerning the acylation of 1,2,6-trigalloylglucopyranose (**15**), its 1,3,6-isomer was transformed efficiently to the same product, 1,2,3,6-tetragalloylglucopyranose (**16**). However, it was impossible to assign any importance to this alternative *in vivo* due to negligible supply of the precursor from the preceding step. The reverse situation applies to 1,2,4,6-tetragalloylglucopyranose, which is formed as a by-product in the synthesis of the 1,2,3,6-isomer (**16**), but is not accepted as substrate for the subsequent conversion to pentagalloylglucopyranose (**3**). Thus, in spite of these *in vitro* detectable side-reactions, the main pathway as depicted in Figs. 3.5 and 3.6 must be considered as the exclusive route in the biosynthesis of 1,2,3,4,6-pentagalloylglucopyranose (PGG, **3**). As often observed in higher plants, it appears, also in this case, that the details of a biosynthetic pathway are regulated by the actual availability of substrates, instead of theoretical enzyme specificities that are accessible only by *in vitro* assays.

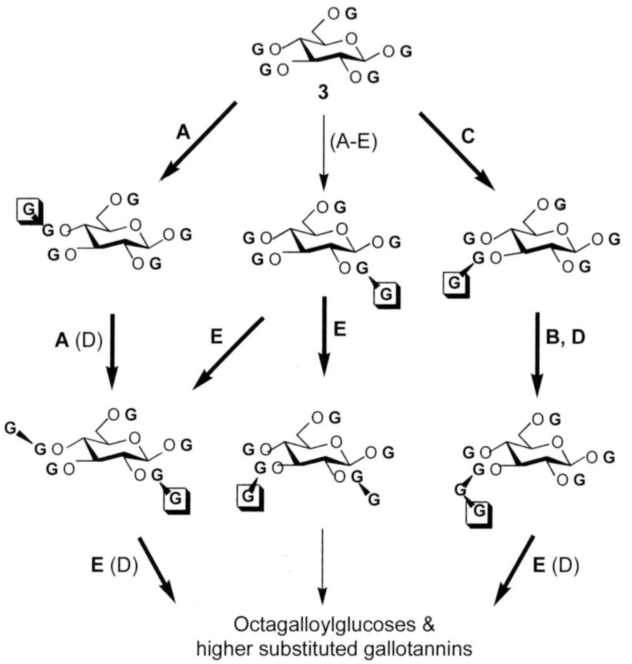

Fig. 3.7 Metabolic routes from pentagalloyl-β-D-glucopyranose (**3**) to "complex" gallotannins in sumac leaves. Main transitions are indicated by bold arrows, minor reactions by thin arrows. Newly introduced galloyl groups are marked by square boxes. Properties of galloyltransferases catalyzing these reactions have been described by Niemetz and Gross, 1999 (enzyme A), Niemetz and Gross, 2001 (enzyme B), Niemetz and Gross, 1998 (enzyme C), Fröhlich *et al.*, 2002 (enzymes D, E). Bold letters represent major enzyme activities, minor activities are symbolized by plain letters in parentheses. **G** = Galloyl; **G▶G** = *meta*-digalloyl residues.

3.6 The Side Route to Gallotannins

The polyester 1,2,3,4,6-pentagalloylglucopyranose (PGG, **3**) occupies a pivotal role in the metabolism of hydrolyzable tannins as this compound marks the point where the routes to gallotannins and ellagitannins diverge (see Fig. 3.2). Before discussing the biosynthesis of ellagitannins in detail, it appears appropriate to insert a short detour for a brief summary of the state of the art in the biochemistry of this important branch of gallic acid-derived natural products.

As mentioned earlier in this chapter, these compounds originate from the addition of *meta*-depsidically-bound galloyl groups to the pentagalloylglucopyranose (**3**) core to yield fairly large molecules, the proper gallotannins, which have eventually been described as "complex" gallotannins to discriminate them from the "simple" galloylglucoses that precede their biosynthesis. Experimental access to their formation proved rather trivial when crude enzyme extracts from sumac (*Rhus typhina*) leaves, a rich source of gallotannins, were found to catalyze the galloylation of 1,2,3,4,6-pentagalloylglucopyranose (**3**) in a manner already known for the biogenesis of "simple" galloylglucoses. Again, β-glucogallin (**2**) served as the principal acyl donor in these transformations, which, however, involved phenolic hydroxyl groups that are chemically quite different from the aliphatic hydroxyls of glucose (Hofmann and Gross, 1990). Several enzymes catalyzing the galloylation of penta- and hexagalloylglucoses have been isolated and characterized until today (for details, see the recent review article of Niemetz and Gross, 2005). A summary of the biogenetic interrelationships of gallotannins is given in Fig. 3.7.

3.7 Biosynthesis of Ellagitannins

3.7.1 Search for pentagalloylglucose oxidizing enzymes

In contrast to the limited occurrence of gallotannins in nature, ellagitannins are typical constituents of many plant families, in which they can be found in an enormous structural variety, a feature that is for the most part based on their pronounced tendency to form dimeric and oligomeric derivatives. The structural characteristics of this challenging group of natural products have attracted chemists already many decades ago and have consequently also stimulated considerations on their biogenetic origin. It was observed already in the mid-1930s that free gallic acid cannot be oxidatively coupled into ellagic acid, and it was concluded that "*ellagic acid has been formed in nature by dehydrogenation of gallic acid-esters* (*e.g.*, *depsidic tannins*)" (Erdtman, 1935). Similar conclusions were drawn 20 years later by Schmidt and Mayer (1956), stating that "*it is much more likely that hexaoxydiphenic acid is formed in situ by dehydrogenation of two gallic acids bound to sugar in suitable positions*" (both quotations are translated from the original text written in German). This view that the hexahydroxydiphenoyl (HHDP, **5**) groups of ellagitannins originated from C–C coupling of neighboring galloyl residues of a depside precursor was corroborated and refined later in several laboratories, with particular emphasis on the role of PGG (**3**) as a general precursor of these compounds (Haslam, 1982, Hatano *et al.*, 1986).

Numerous attempts to unravel the mechanism of such an oxidative C–C coupling have been carried out for decades, either using chemical oxidants (O_2, Fe^{3+}) or by relying on *in vitro* studies with fungal or plant enzymes (phenolase, laccase, peroxidase) that utilize O_2 or H_2O_2 as substrates. Free ellagic acid (**6**), the spontaneously formed dilactone of HHDP units (**5**), was occasionally obtained in experiments with gallic acid (**1**), methyl gallate, β-glucogallin (**2**), 3,6-digalloylglucopyranose or pentagalloylglucopyranose (**3**). A true ellagitannin, however, characterized by a glucose-bound HHDP group (**5**), was never found (see references in Gross, 1999). Also in my laboratory, only negative results were encountered in intensive efforts on this question, employing soluble and microsomal enzyme preparations from various plants and testing many oxidants and oxidoreductase cofactors. It was finally concluded that inadequate analytical techniques presented a major problem, being caused by a combination of extremely low enzyme reaction rates and the possible formation of numerous similar products and by-products that could hardly be discriminated even with the best available HPLC techniques. Another concern was the possible contamination of samples with *in vivo* formed ellagitannins.

The necessity of a radically altered approach to these questions was evident. In this connection, it was our idea to reduce the number of possible reaction products to be identified to a minimum, preferentially to only one variable parameter. As summarized in Fig. 3.8, such a requirement could be met by hydrolysis of *all* possibly formed ellagitannins in the reaction mixtures to release free HHDP units (**5**) from *any* reaction product. All these derivatives, in turn, would spontaneously rearrange to only *one* product, the corresponding dilactone, ellagic acid (**6**). This latter compound could thus serve as the sole, general probe for numerous oxidative reactions. Moreover, working with a pentagalloylglucopyranose (**3**) substrate labelled either uniformly

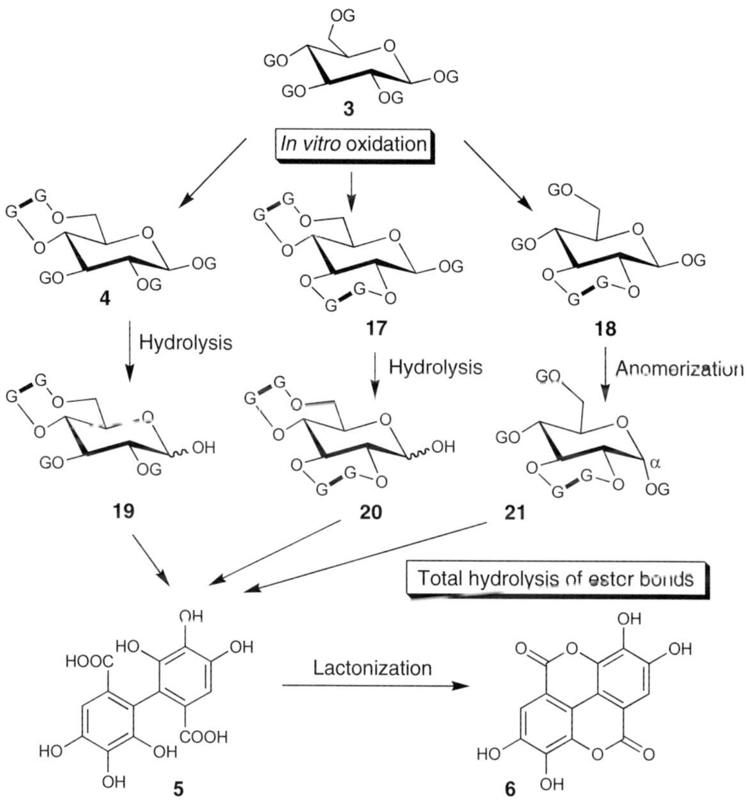

Fig. 3.8 Strategies for elucidating the *in vitro* synthesis of ellagitannins. Oxidation of the principal precursor, pentagalloyl-β-D-glucopyranose (**3**, shown in the $^4C_1$ conformation) and subsequent secondary reactions can afford numerous structurally related products. Energetically less favored glucose conformations (*e.g.*, $^1C_4$) can provide additional products. Also oligomerization reactions might occur. Total hydrolysis finally affords ellagic acid (**6**) as a sole and general indicator of any preceding ellagitannin formation. (**4**) Tellimagrandin II; (**17**) casuarictin; (**18**) pterocaryanin C; (**19**) tellimagrandin I; (**20**) pedunculagin; (**21**) sanguiin H-4.

or in its galloyl moieties would enormously enhance the sensitivity and specificity of such a test system.

3.7.2 Biosynthesis of a monomeric ellagitannin, tellimagrandin II

Extended screening programs based on the above strategy, employing [*U*-^{14}C]pentagalloylglucopyranose (**3**, Rausch and Gross, 1996) as enzyme substrate, finally led to the discovery of a novel enzyme in leaves of *Tellima grandiflora*, a weed that is known as a rich source of ellagitannins. As shown in Fig. 3.9 (panel A), several radioactive products were formed from this substrate. Degradation of prominent products among these afforded [^{14}C]labeled glucose, [^{14}C]gallic acid (**1**) and [^{14}C]ellagic acid (**6**, see Fig. 3.9, panel B). The major product was identified as the monomeric ellagitannin, tellimagrandin II (**4**, Niemetz *et al.*, 2001).

Detailed investigations following this principal finding (Niemetz and Gross, 2003a) led to the characterization of a soluble enzyme that catalyzed the stereoselective oxidative coupling of the galloyl residues at the 4,6-position of pentagalloylglucopyranose (**3**) to the (*S*)-HHDP unit (**5**) of tellimagrandin II (**3**, see Fig. 3.10). The reaction was oxygen dependent and did not require any other cofactor. Hydrogen peroxide was a strong inhibitor, suggesting that the enzyme was not a peroxidase. In contrast, carbon monoxide had no effect, for neither requirement for NADPH nor association with the microsome fraction was observed. It was thus concluded that the enzyme did not belong to cytochrome P450 monooxygenases, but was a member of the vast class of O_2-dependent phenol oxidases. Inhibitor studies led to the assumption that it had the characteristics of a laccase, a class of enzymes that is common to fungi, but is less distributed in plants (Mayer and Staples, 2002).

3.7.3 Biosynthesis of a dimeric ellagitannin, cornusiin E

Analysis of some unexpected side-products in these experiments, obtained with crude enzyme preparations from *T. grandiflora*, indicated the rather specific formation of another oxidation product derived from

[^{14}C]pentagalloylglucopyranose (**3**, see Fig. 3.11, panel A). Time-course experiments revealed that a first product, *i.e.*, tellimagrandin II (**4**), was rapidly converted to a second metabolite that was later identified as the dimeric ellagitannin cornusiin E (**22**, see Fig. 3.11, panel B). Subsequent enzyme studies employing **4** as substrate showed that this compound was indeed specifically oxidized into **22** (Fig. 3.12, Niemetz *et al.*, 2003).

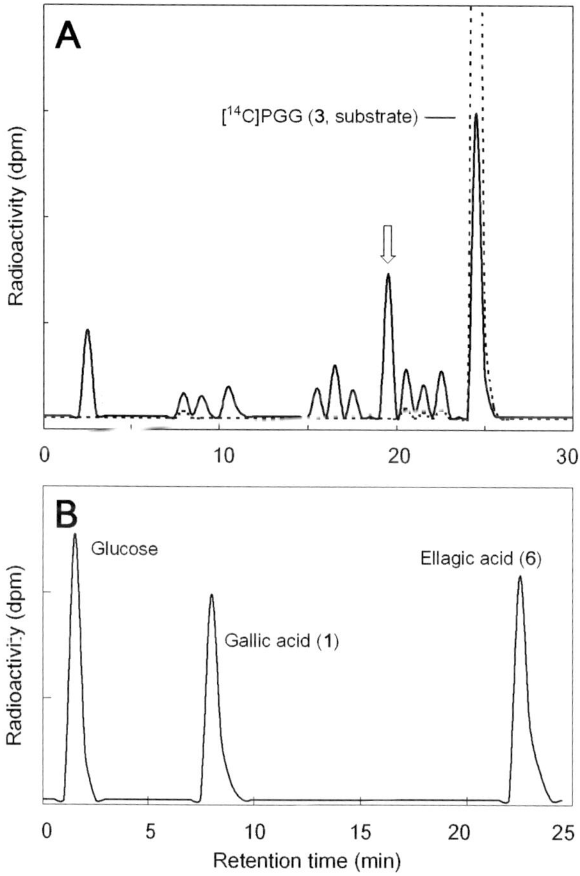

Fig. 3.9 *In vitro* oxidation of [*U*-^{14}C]pentagalloylglucopyranose (**3**) to ellagitannins with cell-free extracts from *Tellima grandiflora* leaves. **A**: Reversed phase (RP-18) HPLC of reaction products. (—) complete enzyme assay; (····) control with acid-denatured enzyme. **B**: RP-18 HPLC of hydrolysis products of marked (⇓) main product in **A**, yielding [^{14}C]glucose, [^{14}C]gallic acid (**1**) and [^{14}C]ellagic acid (**6**) as degradation products. Radioactivity was monitored by fractionation of eluates and liquid-scintillation counting.

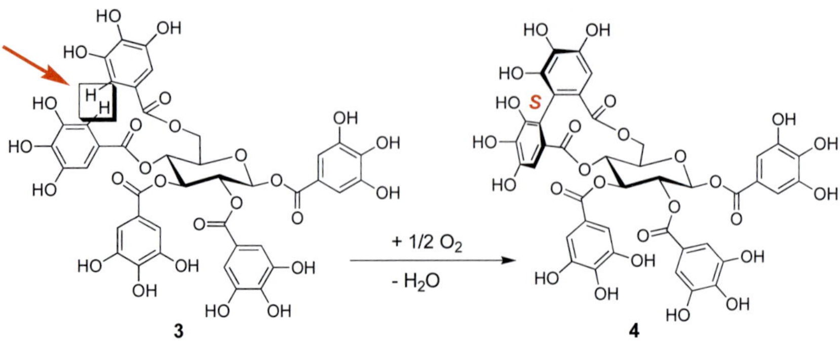

Fig. 3.10 *In vitro* oxidation of 1,2,3,4,6-pentagalloylpyranose (**3**) to tellimagrandin II (**4**) by a laccase-type phenolase from leaves of *Tellima grandiflora* (fringe cups). Oxidation sites in the substrate are marked by a square box and a red arrow.

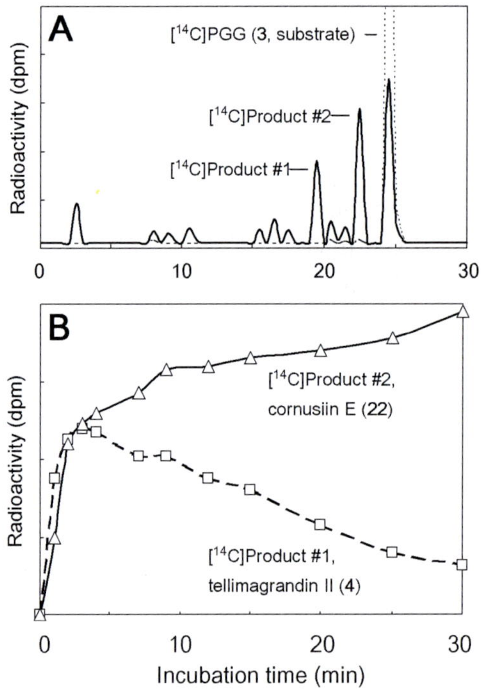

Fig. 3.11 Analysis and time course of reaction products of [*U*-^{14}C]pentagalloylglucopyranose (**3**) obtained using crude enzyme preparations from *Tellima grandiflora* leaves. **A**: RP-18 HPLC of reaction products. (—) complete enzyme assay; (····), control with acid-denatured enzyme. **B**: Time course of formation of reaction product #1 (tellimagrandin II, **4**; - -□- -) and its transformation to product #2 (cornusiin E, **22**; –Δ–). Radioactivity was monitored by fractionation of eluates and liquid-scintillation counting.

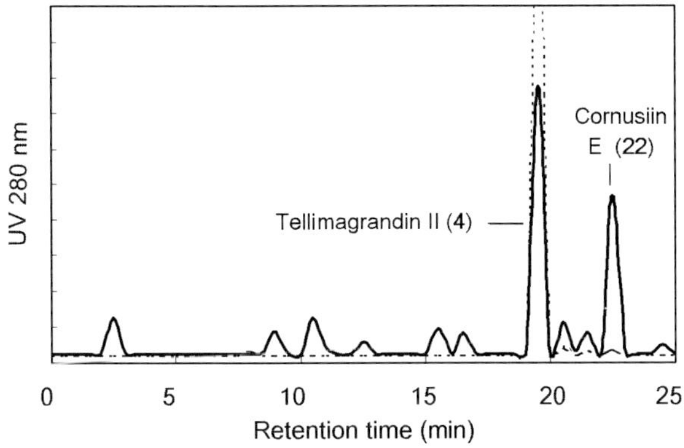

Fig. 3.12 RP-18 HPLC analysis of the enzymatic oxidation of tellimagrandin II (**4**) into the dimeric ellagitannin cornusiin E (**22**) using an enzyme preparation from *Tellima grandiflora* leaves. (—) complete enzyme assay; (····) control with acid-denatured enzyme.

The cornusiin E (**22**)-forming enzyme was also purified and characterized. Its catalytic properties were found to be similar to the tellimagrandin II (**4**)-forming enzyme, but the molecular characteristics of the two proteins were clearly different as shown by polyacrylamide gel electrophoresis experiments (Niemetz and Gross, 2003b). The reaction catalyzed by the enzyme is depicted in Fig. 3.13 to show how the newly introduced 2,3'4'-valoneoyl bridge is formed by oxidative intermolecular C–O coupling.

3.8 Conclusions and Perspectives

This chapter was written to demonstrate how efficiently extended biochemical pathways can be elucidated by application of enzyme studies, and also to show that such techniques can be easily practiced even when using substrates as complex as hydrolyzable tannins. Besides the presentation of scientific facts that had been gained in the past decades, it would be satisfying if this article could have a positive impact on present and future research by eliminating unfounded anxieties or reservations to use such methods and experimental approaches. Moreover, I hope — or better, I am confident — that the reported views and strategies will be augmented, not replaced, by embedding them into the arsenal of recently developed and emerging advanced technologies, from genomics and proteomics to comprehensive metabolomics.

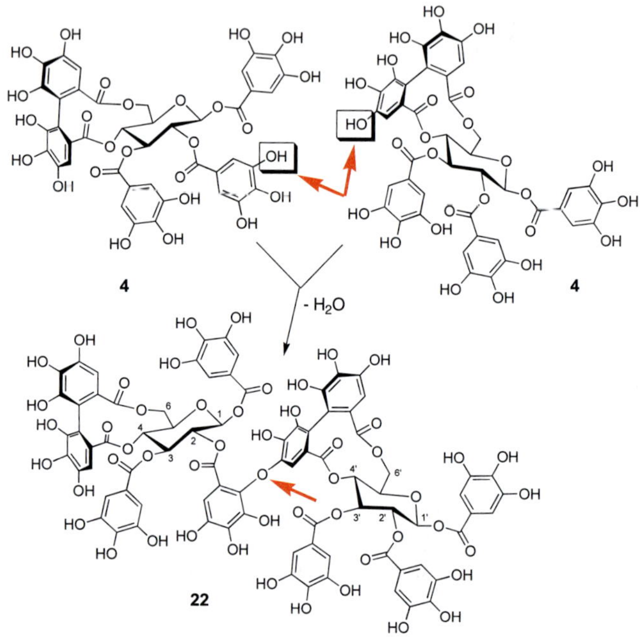

Fig. 3.13 Enzymatic dimerization of tellimagrandin II (**4**) into cornusiin E (**22**) by a laccase-type phenolase from *Tellima grandiflora* (fringe cups) leaves, affording an (*S*)-2,4',6'-valoneoyl bridge between the two monomeric units. Reaction sites are marked by square boxes and red arrows.

Acknowledgements

I am grateful to the many colleagues and co-workers that have contributed to the research reported in this chapter and deeply thank them for their input of advice, enthusiasm and skill, or for having provided rare chemicals. Generous financial support provided for many years by the Deutsche Forschungsgemeinschaft (Bonn) and the Fonds der Chemischen Industrie (Frankfurt) is gratefully acknowledged.

3.9 Bibliography

Abd El-Mawla, A. M. and Beerhues, L. (2002). Benzoic acid biosynthesis in cell cultures of *Hypericum androsaemum*, *Planta*, 214, pp. 727–733.

Atkinson, M. L. (1960). Free energy and the biosynthesis of phosphates. In *Comparative Biochemistry – Free Energy and Biological Function*, vol. II, ed. by Florkin, M. and Mason, H. S., Academic Press: New York, London, pp. 1–95.

Cammann, J., Denzel, K., Schilling, G. and Gross, G. G. (1989). Biosynthesis of gallotannins: β-glucogallin-dependent formation of 1,2,3,4,6-pentagalloylglucose by enzymatic galloylation of 1,2,3,6-tetragalloylglucose. *Arch. Biochem. Biophys.*, 73, pp. 58–63.

Conn, E. E. and Swain, T. (1961). Biosynthesis of gallic acid in higher plants. *Chem. Ind.*, pp. 592–593.

Cornthwaite, D. C. and Haslam, E. (1965). Gallotannins. IX. The biosynthesis of gallic acid in *Rhus typhina*, *J. Chem. Soc.*, pp. 3008–3011.

Denzel, K. and Gross, G. G. (1991). Biosynthesis of gallotannins. Enzymatic 'disproportionation' of 1,6-digalloylglucose to 1,2,6-trigalloylglucose and 6-galloylglucose by an acyltransferase from leaves of *Rhus typhina* L., *Planta*, 184, pp. 285–289.

El-Basyouni, S. Z., Chen, D., Ibrahim, R. K., Neish, A. C. and Towers, G. H. N (1964). The biosynthesis of hydroxybenzoic acids in higher plants, *Phytochemistry*, 3, pp. 485–492.

Erdtman, H. (1935). Phenoldehydrierungen VI. Dehydrierende Kupplung einiger Guajakolderivate, *Svensk. Kem. Tidskr.*, 47, pp. 225–230.

Freudenberg, K. (1920). *Die Chemie der natürlichen Gerbstoffe*, Springer Verlag: Berlin.

Fröhlich, B., Niemetz, R. and Gross, G. G. (2002). Gallotannin biosynthesis: two new galloyltransferases from *Rhus typhina* leaves preferentially acylating hexa- and heptagalloylglucoses, *Planta*, 216, pp. 168–172.

Gross, G. G. (1982a). Synthesis of galloyl-coenzyme A thioester, *Z. Naturforsch.*, 37c, pp. 778–783.

Gross, G. G. (1982b). Synthesis of β-glucogallin from UDP-glucose and gallic acid by an enzyme preparation from oak leaves., *FEBS Lett.*, 148, pp. 67–70.

Gross, G. G. (1983a). Synthesis of mono-, di- and trigalloyl-β-D-glucose by β-glucogallin-dependent galloyltransferases from oak leaves, *Z. Naturforsch.*, 38c, pp. 519–523.

Gross, G. G. (1983b). Partial purification and properties of UDP-glucose: vanillate 1-*O*-glucosyl transferase from oak leaves, *Phytochemistry*, 22, pp. 2179–2182.

Gross, G. G. (1989). Enzymology of gallotannin biosynthesis. In *Plant Cell Wall Polymers: Biogenesis and Biodegradation*, ed. by Lewis, N. G. and Paice, M. G., ACS Symp. Ser., vol. 399, Washington, DC, pp. 108–121.

Gross, G. G. (1999). Biosynthesis of hydrolyzable tannins. In *Comprehensive Natural Products Chemistry – Carbohydrates and Their Derivatives Including Tannins, Cellulose, and Related Lignins*, vol. 3, ed. by Pinto, B. M., Elsevier: Amsterdam, pp. 799–826.

Gross, G. G. (2008). From lignins to tannins. Forty years of enzyme studies on the biosynthesis of phenolic compounds, *Phytochemistry*, *in press* (doi:10.1016/j.phytochem.2007.04.031).

Gross, G. G. and Denzel, K. (1991). Biosynthesis of gallotannins. β-Glucogallin-dependent galloylation of 1,6-digalloylglucose to 1,2,6-trigalloylglucose, *Z. Naturforsch.*, 46c, pp. 389–394.

Gross, G. G., Schmidt, S. W. and Denzel, K. (1986). β-Glucogallin-dependent acyltransferase from oak leaves. I. Partial purification and characterization, *J. Plant Physiol.*, 126, pp. 173–179.

Haddock, E. A., Gupta, R. K., Al-Shafi, S. M. K., Layden, K., Haslam, E. and Magnolato, D. (1982). The metabolism of gallic acid and hexahydroxydiphenic acid in plants: biogenetic and molecular taxonomic considerations, *Phytochemistry*, 21, pp. 1049–1062.

Hagenah, S. and Gross, G. G. (1993). Biosynthesis of 1,2,3,6-tetra-*O*-galloyl-β-D-glucose, *Phytochemistry*, 32, pp. 637–641.

Haslam, E. (1982). The metabolism of gallic and hexahydroxydiphenic acid in higher plants, *Fortschritte Chem. Org. Naturstoffe*, 41, pp.1–46.

Haslam, E. (1989). *Plant Polyphenols: Vegetable Tannins Revisited*, Cambridge University Press: Cambridge.

Haslam, E. (1998). Practical Polyphenols: From Structure to Molecular Recognition and Physiological Action, Cambridge University Press: Cambridge.

Hatano, T., Kira, R., Yoshizaki, M. and Okuda, T. (1986). Seasonal changes in the tannins of *Liquidambar formosana* reflecting their biogenesis, *Phytochemistry*, 25, pp. 2787–2789.

Hillis, W. E. (1985). Biosynthsis of tannins. In *Biosynthesis and Biodegradation of Wood Components*, ed. by Higuchi, T., Academic Press: Orlando, pp. 325–347.

Hofmann, A. S. and Gross, G. G. (1990). Biosynthesis of gallotannins: Formation of polygalloylglucoses by enzymatic acylation of 1,2,3,4,6-penta-*O*-galloylglucose, *Arch. Biochem. Biophys.*, 283, pp. 530–532.

Löscher, R. and Heide, L. (1994). Biosynthesis of *p*-hydroxybenzoate from *p*-coumarate and *p*-coumaroyl-coenzyme A in cell-free extracts of *Lithospermum erithrorhizon* cell cultures, *Plant Physiol.*, 106, pp. 271–279.

Mayer, A. M. and Staples, R. C. (2002). Laccase: new functions for an old enzyme, *Phytochemistry*, 60, pp. 551–565.

Mock, H. P. and Strack, D. (1993). Energetics of the uridine-5'-diphosphoglucose: hydroxycinnamic acid acyl-glucosyltransferase reaction, *Phytochemistry*, 32, pp. 575–579.

Niemetz, R. and Gross, G. G. (1998). Gallotannin biosynthesis. Purification of β-glucogallin: 1,2,3,4,6-pentagalloyl-β-D-glucose galloyltransferase from sumac leaves, *Phytochemistry*, 49, pp. 327–332.

Niemetz, R. and Gross, G. G. (1999). Gallotannin biosynthesis: a new β-glucogallin-dependent galloyltransferase from sumac leaves acylating gallotannins at positions 2 and 4, *J. Plant Physiol.*, 155, pp. 441–444.

Niemetz, R. and Gross, G. G. (2001). Gallotannin biosynthesis: β-glucogallin: hexagalloyl 3-*O*-galloyltransferase from *Rhus typhina* leaves, *Phytochemistry*, 58, pp. 657–661.

Niemetz, R. and Gross, G. G. (2003a). Oxidation of pentagalloylglucose to the ellagitannin, tellimagrandin II, by a phenol oxidase from *Tellima grandiflora* leaves, *Phytochemistry*, 62, pp. 301–306.

Niemetz, R. and Gross, G. G. (2003b). Ellagitannin biosynthesis: laccase-catalyzed dimerization of tellimagrandin II to cornusiin E in *Tellima grandiflora*, *Phytochemistry*, 64, pp. 1197–1201.

Niemetz, R. and Gross, G. G. (2005). Enzymology of gallotannin and ellagitannin biosynthesis, *Phytochemistry*, 66, pp. 2001–2011.

Niemetz, R., Schilling, G. and Gross, G. G. (2001). Ellagitannin biosynthesis: oxidation of pentagalloylglucose to tellimagrandin II by an enzyme from *Tellima grandiflora* leaves, *Chem. Commun.*, pp. 35–36.

Niemetz, R., Schilling, G. and Gross, G. G. (2003). Biosynthesis of the dimeric ellagitannin, cornusiin E, in *Tellima grandiflora*, *Phytochemistry*, 64, pp. 109–114.

Okuda, T., Yoshida, T. and Hatano, T. (1993). Classification of oligomeric hydrolysable tannins and specificity of their occurrence in plants, *Phytochemistry*, 32, pp. 507–521.

Ossipov, V., Salminen, J., Ossipova, S., Haukioja, E. and Pihlaja, K. (2003). Gallic acid and hydrolyzable tannins are formed in birch leaves from an intermediate compound of the shikimate pathway, *Biochem. Syst. Ecol.*, 31, pp. 3–16.

Rausch, H. and Gross, G. G. (1996). Preparation of [^{14}C]-labelled 1,2,3,4,6-penta-*O*-galloyl-β-D-glucose and related gallotannins, *Z. Naturforsch.*, 51c, pp. 473–476.

Reinefeld, E. and Ahrens, D. (1971). Der Einfluß der Konfiguration auf die partielle Veresterung von D-Glucopyranosiden, *Liebigs Ann. Chem.*, 747, pp. 39–44.

Scalbert, A. (1991). Antimicrobial properties of tannins, *Phytochemistry*, 30, pp. 3875–3883.

Schmidt, O. Th. and Mayer, W. (1956). Natürliche Gerbstoffe. *Angew. Chem.*, 68, pp 103–115.

Schmidt, S. W., Denzel, K., Schilling, G. and Gross, G. G. (1987). Enzymatic synthesis of 1,6-digalloylglucose from β-glucogallin by β-glucogallin: β-glucogallin 6-*O*-galloyltransferase from oak leaves, *Z. Naturforsch.*, 42c, pp. 97–92.

Stöckigt, J. and Zenk, M. H. (1974). Enzymatic synthesis of chlorogenic acid from higher plants, *Phytochemistry*, 19, pp. 1625–1629.

Ulbrich, B. and Zenk, M. H. (1980). Partial purification and properties of *p*-hydroxycinnamoyl-CoA: shikimate *p*-hydroxycinnamoyl transferase from higher plants, *Phytochemistry*, 19, pp. 1625–1629.

Weisemann, S., Denzel, K., Schilling, G. and Gross, G. G. (1988). Enzymatic synthesis of 1-O-phenylcarboxyl-β-D-glucose esters, *Bioorg. Chem.*, 16, pp. 29–37.

Werner, I., Bacher, A. and Eisenreich, W. (1997). Retrobiosynthetic NMR studies with ^{13}C-labelled glucose. Formation of gallic acid in plants and fungi, *J. Biol. Chem.*, 272, pp. 25474–25482.

Werner, R. A., Rossmann, A., Schwarz, C., Bacher, A., Schmidt, H.-L. and Eisenreich, W. (2004). Biosynthesis of gallic acid in *Rhus typhina*: discrimination between alternative pathways from natural oxygen isotope abundance, *Phytochemistry*, 65, pp. 2809–2813.

Williams, J. M. and Richardson, A. C. (1967). Selective acylation of pyranosides. I. Benzoylation of methyl α-D-glycopyranosides of mannose, glucose and galactose, *Tetrahedron*, 23, pp. 1369–1378.

Zenk, M. H. (1964). Zur Frage der Biosynthese der Gallussäure, *Z. Naturforsch.*, 19b, pp. 485–492.

Chapter 4

Physicochemical Properties and Biomimetic Reactions of Ellagitannins

Takashi Tanaka

Graduate School of Biomedical Sciences, College of Pharmaceutical Sciences, Nagasaki University, 1-14 Bunkyo-machi, Nagasaki 852-8521, Japan; t-tanaka@nagasaki-u.ac.jp

This chapter approaches ellagitannins from two different perspectives. The first one has to do with the intriguing relationship between their chemical structures and their water solubility. The oxidative metabolism of hydrophobic gallotannins produces various ellagitannins of increasing hydrophilicity. The hydrophobicity of ellagitannin monomers is particularly relevant to their interaction with co-existing molecules and to the biogenesis of their dimeric congeners. The second perspective concerns the reactivity of a group of ellagitannins bearing highly oxidized galloyl-derived units. Of particular interest is the special chemical reactivity expressed by dehydrogenated hexahydroxydiphenoyl ester units, which has been exploited in biomimetic synthesis, structural determination and chromatographic analysis of some of the most complex members of this fascinating family of natural products.

4.1 Introduction

Plant polyphenols, including mainly hydrolyzable and condensed tannins, are widely distributed in the plant kingdom and thought to be biosynthesized and accumulated as defensive compounds against microorganisms, herbivores, and oxidative stress (Barbehenn *et al.*, 2005, Haslam, 1996, Scalbert, 1991, Robbins *et al.*, 1987). Among plant polyphenols, tannins have the characteristic property of precipitating proteins, which leads for example to the unpleasant astringent taste and inhibition of digestive enzymes in the digestive tract of plant predators. Although the toxicity of tannins is lower than that of qualitative defense compounds, such as toxic alkaloids and terpenoids, tannins are accumulated in much higher concentration in many cases and are designated as quantitative defense compounds (Zucker, 1983). The inhibition of food digestion by tannins is sometimes severe for animals (Mehansho *et al.*, 1983, 1987). However, paradoxically, it is regarded as being beneficial for human health in developed countries, because the inhibition of digestive enzymes decreases the excess absorption of sugars and triglycerides from the intestine, hence lowering plasma sugar and lipid levels (Hara and Honda, 1990, Nakai *et al.*, 2005). The physicochemical nature of the interaction of tannins with proteins, especially with salivary proline-rich proteins, has been extensively studied and the results unveiled the importance of the hydrophobic association in aqueous media (Charlton, *et al.*, 1996, 2002a, 2002b, Baxter *et al.*, 1997, Luck *et al.*, 1994, Cai *et al.*, 1990). The hydrophobicity of tannin molecules strongly depends on their structure, and it was shown that the molecular flexibility rather than the number of phenolic hydroxyl groups is the most determinant factor.

Compared to gallo- and condensed tannins, ellagitannins exhibit a great structural diversity. Although the overwhelming majority of ellagitannins feature β-D-glucose as their core unit, ellagitannins based on other cyclic polyols (Nishimura *et al.*, 1984a, 1986), *O*- or *C*-glucosides (Nishimura *et al.*, 1984b, Tanaka *et al.*, 1993a, 2005], 1-*O*-caffeoyl glucose (Jiang *et al.*, 2001) and even hydroxytriterpenes (Ageta *et al.*, 1988; Chen *et al.*, 1993) have also been identified. As for the acyl groups of ellagitannins, their characteristic hexahydroxydiphenoyl

(HHDP) ester groups, which are produced by oxidative C–C coupling between two galloyl groups (Haslam and Cai, 1994), are further metabolized to form complex molecules via dehydrogenation or intermolecular oxidative coupling with another galloyl group (Okuda, 2005). The products thus generated by these oxidative processes are accumulated in high concentrations in some plants, such as *Euphorbia*, *Punica* and *Geranium* species (Okuda *et al.*, 1980, 2000]. The ellagitannins found in these plant species feature dehydrohexahydroxydiphenoyl (DHHDP) ester groups that exhibit very interesting reactivity, hence contributing further to the structural diversity of the ellagitannin natural products family.

4.2 Water Solubility and Hydrophobic Association of Ellagitannins

4.2.1 Structures of ellagitannins and their water solubility

The relation between the structures of hydrolyzable tannins and their water solubility has been thoroughly examined (Tanaka *et al.*, 1997a). The partition coefficients of a series of galloylated glucoses or gallotannins between *n*-octanol and water are shown in Fig. 4.1 (*i.e.*, log P *vs*. number of galloyl groups). The water solubility of galloyl glucoses is inversely correlated with the number of galloyl groups. Pentagalloylglucose (PGG, **8**) was the most hydrophobic gallotannin examined in this experiment. The increase of the number of galloyl esters increases the global hydrophobic area of the molecule. However, when two galloyl groups are connected via a biaryl bond, such as in eugeniin (**9**, 1,2,3-tri-*O*-galloyl-4,6-(*S*)-HHDP-β-D-glucose, Nonaka *et al.*, 1980), the water solubility is much higher than that of PGG (**8**, see Fig. 4.2). It was then observed that the number of biaryl bonds in each ellagitannin molecule is positively correlated with the water solubility. Furthermore, sanguiin H-6 (**21**), an example of a dimeric ellagitannin formed by intermolecular C–O coupling (Tanaka *et al.*, 1985), was also found much more hydrophilic than the monomeric tannin. Overall, these results thus seem to indicate that the oxidative metabolism of ellagitannins increases their water solubility.

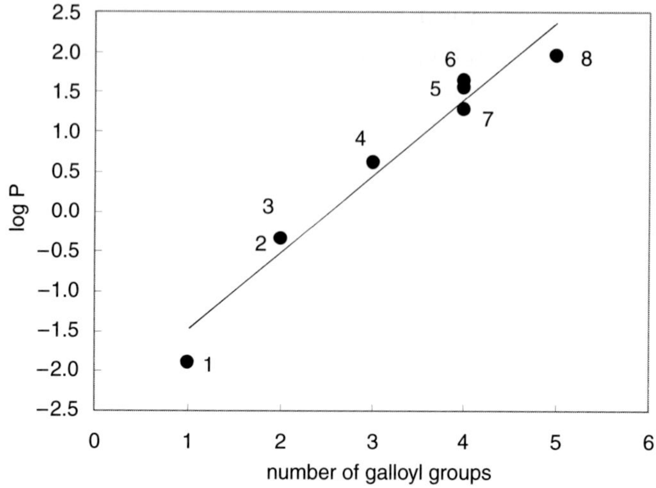

Fig. 4.1 Partition coefficients of gallotannins. **1**: 1-*O*-galloyl-β-D-glucose (1); **2**: 1,6-di-*O*-galloyl- β-D-glucose (2); **3**: hamamelitannin (2); **4**: 1,2,6-tri-*O*-galloyl-β-D-glucose (3); **5**: 1,2,3,6-tetra-*O*-galloyl-β-D-glucose (4); **6**: methyl 2,3,4,6-tetra-*O*-galloyl-β-D-glucoside (4); **7**: 2,3,4,6-tetra-*O*-galloyl-β-D-glucose (4); **8**: 1,2,3,4,6-penta-*O*-galloyl-β-D-glucose (5). The number between parenthesis represents the number of galloyl groups in the molecule.

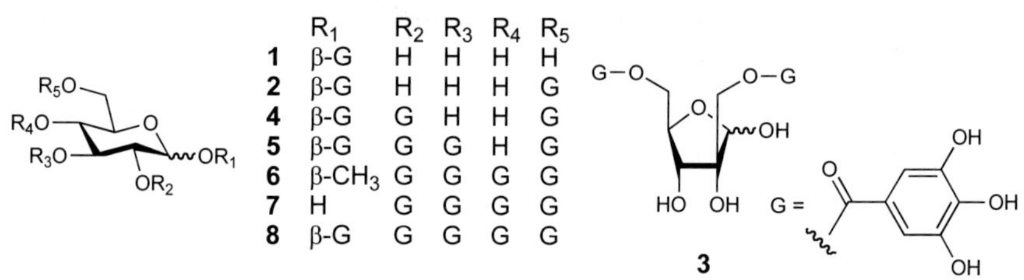

4.2.2 Hydrophobic association of pentagalloylglucose and ellagitannins with co-existing compounds

The most hydrophobic galloylated glucose, 1,2,3,4,6-pentagalloyl-β-D-glucose (PGG, **8**), forms a gel in an aqueous solution at room temperature (Fig. 4.3), indicating that the molecules are physically cross-linked by weak hydrophobic and hydrogen bonding (Cai *et al.*, 1990). The gel formation of **8** (PGG) was however disturbed by the presence of paeoniflorin, which is a water-soluble monoterpene glycoside having a hydrophobic benzoyl group (see structure in Fig. 4.4). These compounds co-exist in the root of *Paeonia lactiflora*. Since the presence of sucrose did not affect the gel formation, it was deduced that the benzoyl group of paeoniflorin interrupts the self-association of the PGG molecules.

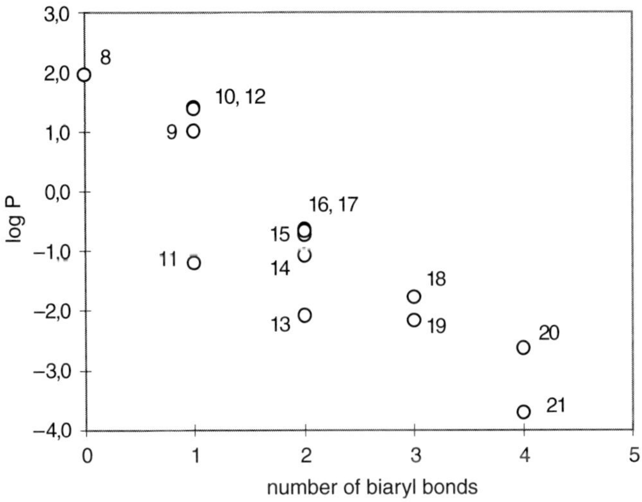

Fig. 4.2 Partition coefficients of ellagitannins. **9**: eugeniin (1); **10**: phillyraeoidin A (1); **11**: corilagin (1); **12**: des-DHHDP-bischofianin (1); **13**: pedunculagin (2); **14**: 1(β)-O-galloyl pedunculagin (2); **15**: casuarinin (2); **16**: geraniin (2); **17**: bischofianin (2); **18**: vescalagin (3); **19**: terflavin A (3); **20**: punicalagin (4); **21**: sanguiin H-6 (4). The number between parentheses represents the number of biphenyl bonds in the molecule.

In the ^{1}H-NMR spectrum, when paeoniflorin was added, the anomeric proton signal and the aromatic singlet attributable to the galloyl group attached to the anomeric position of pentagalloylglucose (PGG, **8**) were shifted to the upper field (Tanaka *et al.*, 1997a). As for paeoniflorin, the signals of the benzoyl moiety were largely shifted to the upper field, while the shifts of other proton signals were much smaller. This observation suggested that the benzoyl group of paeoniflorin is associated with the anomeric position of **8** via some hydrophobic interactions, which was supported by intermolecular NOE's between the benzoyl protons of paeoniflorin and the anomeric proton of **8**.

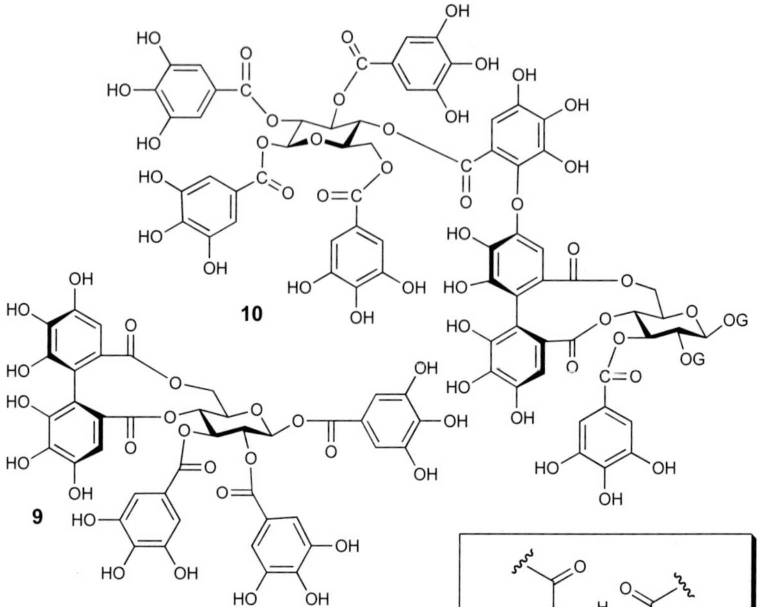

10

9

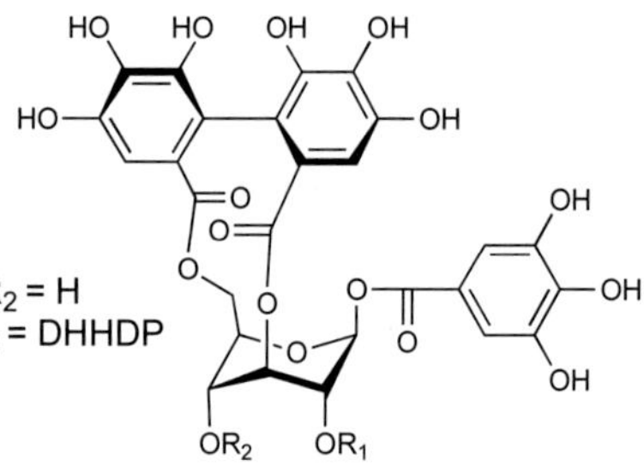

11 $R_1 = R_2 = H$
16 R_1, R_2 = DHHDP

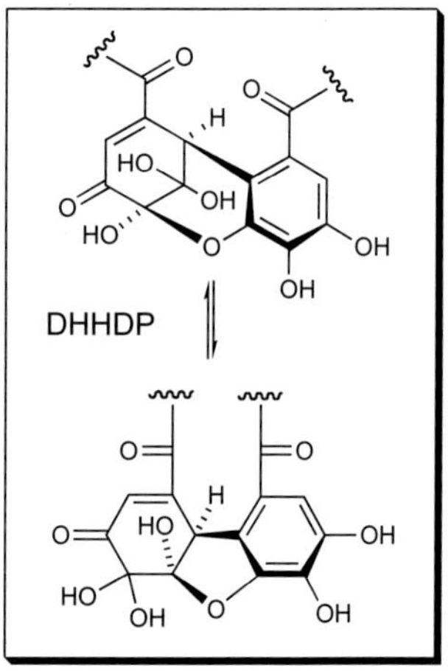

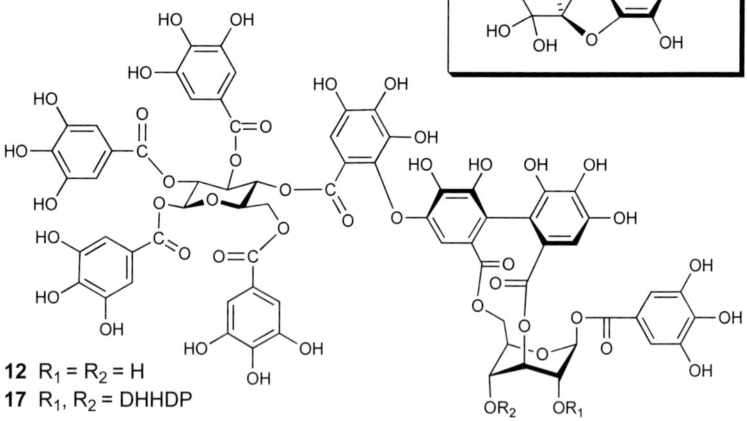

12 $R_1 = R_2 = H$
17 R_1, R_2 = DHHDP

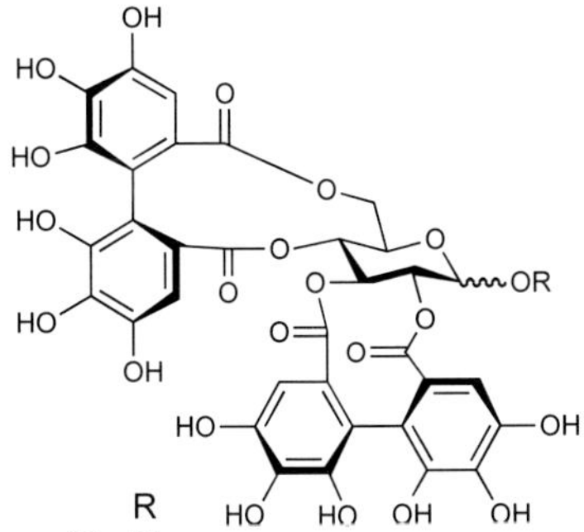

	R
13	H
14	β-G

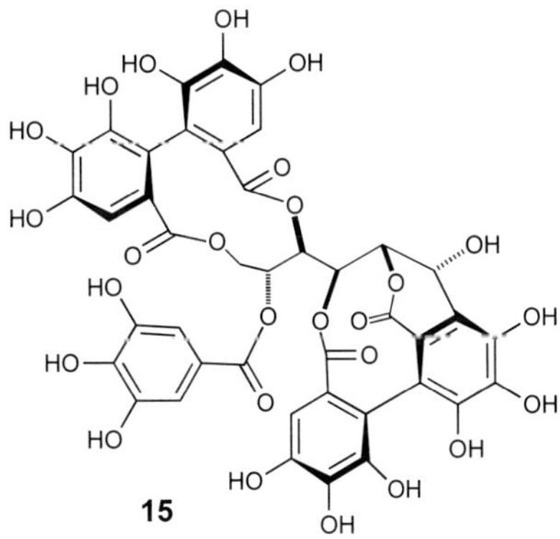

15

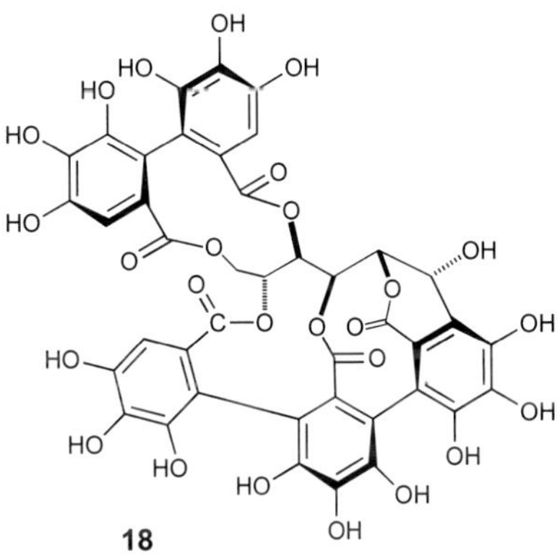

18

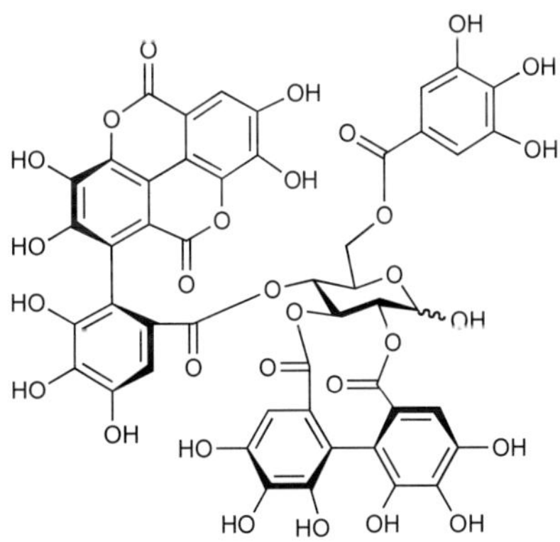

19

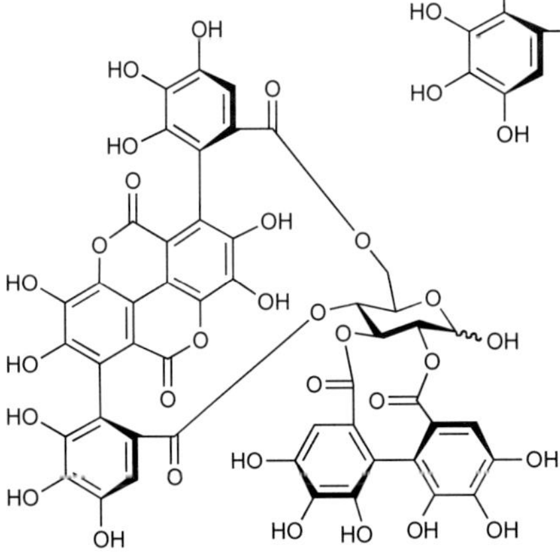

20

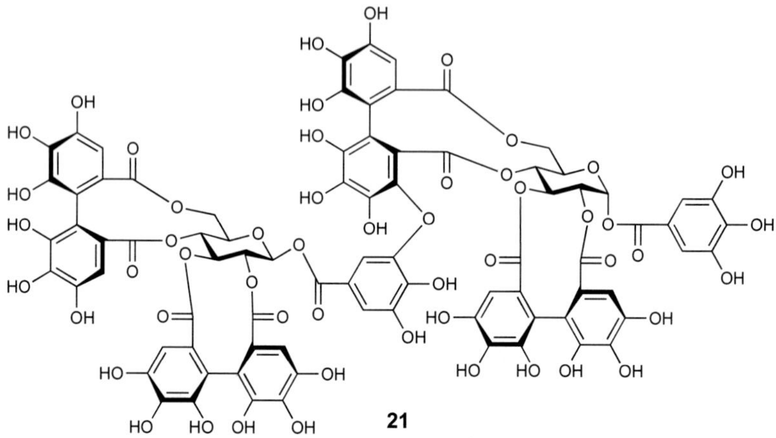

21

The acylated glucopyranose ring of pentagalloylglucose (PGG, **8**) is strongly hydrophobic and the anomeric position is the most sterically unhindered site; therefore, the hydrophobic benzoyl group of paeoniflorin selectively associates with the anomeric position of **8** (Fig. 4.4). The association of paeoniflorin with **8** was also visualized by dissolution of precipitates of the caffeine-pentagalloylglucose complex by the addition of paeoniflorin (Cai *et al.*, 1990).

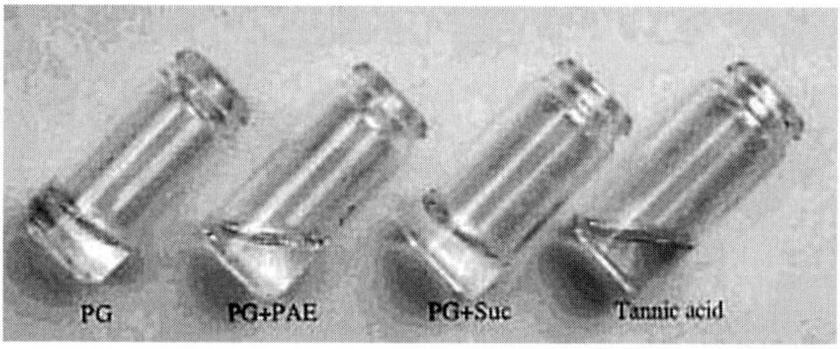

Fig. 4.3 Gel formation of pentagalloylglucose (PGG, **8**) in an aqueous solution and dissolution by addition of paeoniflorin; PG: pentagalloylglucose (10mg/ml), PAE: paeoniflorin (10mg/ml), Suc: sucrose (17mg/ml), tannic acid (20mg/ml)

Tannic acid, a mixture of polygalloylated glucoses bearing additional depsidically-linked galloyl groups, does not form any gel. This is probably because the additional galloyl groups interrupt the hydrophobic self-association. Since ellagitannins are less hydrophobic than PGG (**8**, see Fig. 4.2), their ability to engage in hydrophobic associations with co-existing substances is weaker than that of **8**. This weakness was evidenced by a lower regioselectivity of proton up-field shifts in interactions with gramicidin S, a cyclic peptide having β-turn structures (Fig. 4.5) (Zhang *et al.*, 2002). In the presence of **8** (PG = PGG), large chemical shift changes of the proline and phenylalanine moieties located at the β-turn structure were observed, indicating that selective association takes place. In contrast, addition of the galloylated pedunculagin, 1-*O*-galloyl-2,3,4,6-bis-(*S*)-HHDP-β-D-glucose (PED), caused non-selective up-field shifts of the amino acid protons.

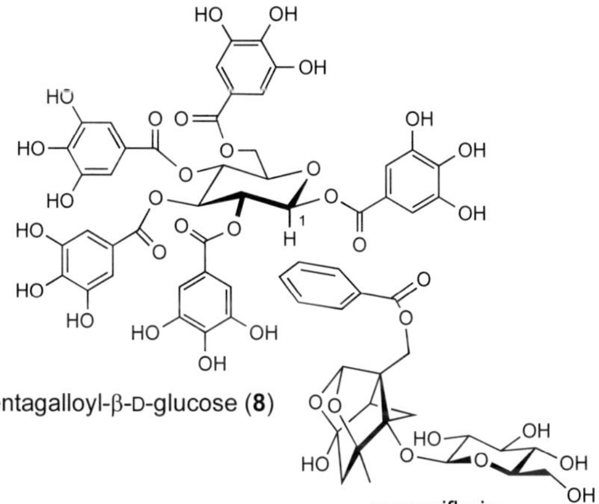

Fig. 4.4 Hydrophobic association of paeoniflorin and pentagalloylglucose (PGG, **8**).

The chemical shift changes caused by the addition of castalagin (CAST), a further oxidized and highly water-soluble ellagitannin, were much smaller than those observed with galloylated pedunculagin (PED). These results indicated that the rigid and spherical structure of castalagin decreased the number of interactions with the peptide molecule. However, on solvent partition between water and *n*-octanol, the extraction of gramicidin S into the *n*-octanol layer was strongly inhibited at pH 3-6 by the presence of sanguiin H-6 (**21**), and no inhibition was observed at pH 1-2, indicating that ionic interactions are also important in peptide-ellagitannin interactions (Fig. 4.6) (Zhang *et al.*, 2000).

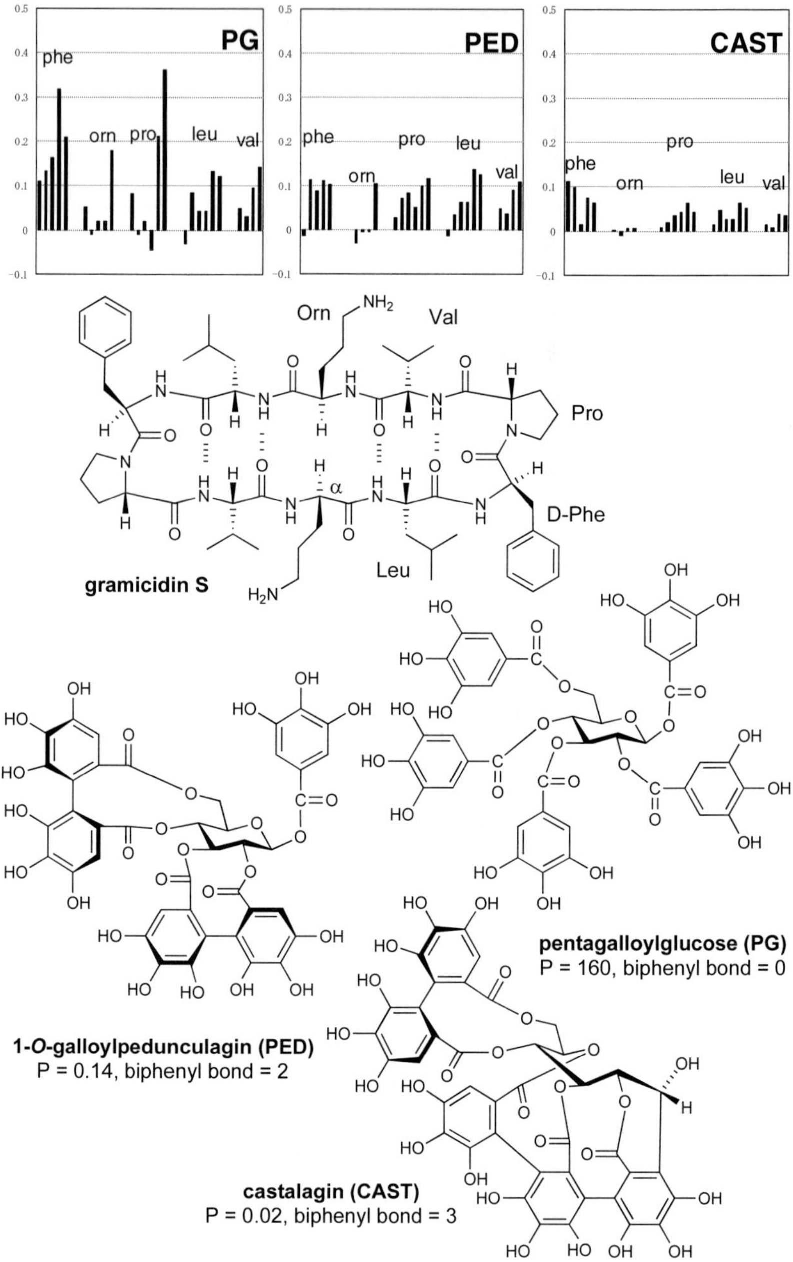

Fig. 4.5 Chemical shift change of gramicidin S in the presence of tannins; P = partition coefficient.

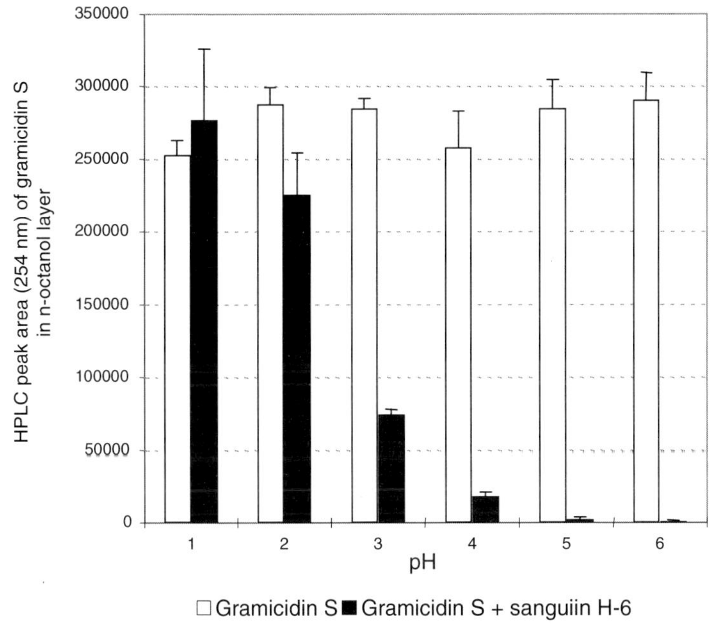

Fig. 4.6 Partition of gramicidin S in the presence of sanguiin H-6. □ : gramicidin S in 5% DMSO-buffer solution (2 mg/ml) was partitioned with *n*-octanol at 26 °C and the *n*-octanol layer was analyzed by HPLC. ■ : same experiment performed in the presence of sanguiin H-6 (21, 2 mg/ml).

4.2.3 Self-association and biogenesis of dimeric ellagitannins

The self-association of 1-*O*-galloylpedunculagin in an aqueous solution was found to be regioselective (Tanaka *et al.*, 1997b). The corresponding ^{1}H-NMR spectra measured in D_2O showed a large up-field shift of the glucose H-5 signal when the concentration was increased (Fig. 4.7). At the same time, the aromatic proton of the 2,3-HHDP pyrogallol ring connected to the 2-position of the glucose core was shifted to a lower field. The $\Delta\delta$ values were concentration-dependent at concentrations lower than 0.01 M. This result is indicative of the occurrence of a regioselective self-association, and the large $\Delta\delta$ values of H-5 suggested that this proton is shielded by the aromatic ring of another molecule. From an analogy with the case of pentagalloylglucose (PGG, **8**, see Fig. 4.4), it was deduced that the O-1 galloyl group of 1-*O*-

galloylpedunculagin overhangs the H-5 of another molecule. The low-field shift observed for the 2,3-HHDP pyrogallol ring proton can be rationalized in terms of a deshielding effect induced by the O-1 galloyl group of another molecule. Interestingly, if an oxidative coupling between these galloyl and hexahydroxydipenoyl groups occurred according to the C–C coupling mode A, it would give rise to rhoipteleanin A (see Fig. 4.8), a major dimeric ellagitannin isolated from *Rhoiptelea chiliantha* (Jiang *et al.*, 1995).

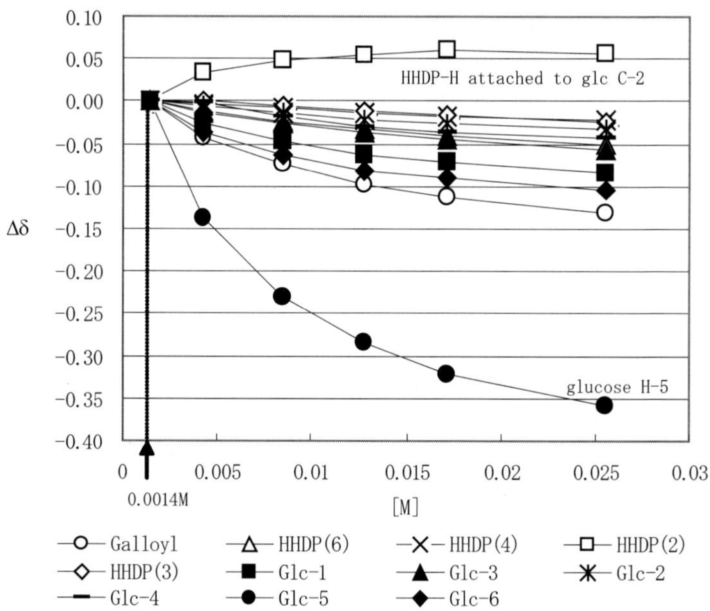

Fig. 4.7 Chemical shift change of 1-*O*-galloylpedunculagin in D_2O at different concentrations.

Rhoipteleanin A has a rather unusual structure, in which the two ellagitannin monomeric units are linked together through a (*S*,*S*)-flavogallonyl group. The stereoselectivity of the intermolecular *S*-biphenyl bond formation can be explained on the basis of the aforementioned hydrophobic association. This ellagitannin was the first and still unique example of an ellagitannin dimer generated via an intermolecular oxidative C–C coupling. It is hence quite tempting to speculate that *R. chiliantha* have specific enzymes that selectively oxidize the self-associated galloylpedunculagin dimeric complex. Additionally, if an oxidative C–O coupling occurred between one of the phenolic *meta*-hydroxyl group of the O-1 galloyl group of one molecule and the 4,6-HHDP pyrogallol ring connected to the 4-position of the glucose core of another molecule in the same self-associated complex (see coupling mode B in Fig. 4.8), a sanguisorboyl group would be formed. This group is one of the characteristic interlinking units in ellagitannin oligomers, such as lambertianin A (Tanaka *et al.*, 1993b), especially in those ellagitannins isolated from *Sanguisorba* and *Rubus* species (Tanaka *et al.*, 1985, Li *et al.*, 2007a). Since plants biosynthesize dimeric ellagitannins that are characteristic of their species and/or family to which they belong (see Chapters 1 and 2), regioselectivity and chemoselectivitity (*i.e.*, C–C *versus* C–O coupling) in oligomerizing oxidative coupling, even if facilitated by preliminary formations of self-associated and pre-arranged complexes, is probably under enzymatic control.

4.3 Biomimetic Transformation of Ellagitannins

4.3.1 Dehydroellagitannins and related compounds

Dehydroellagitannins constitute a group of ellagitannins that feature the dehydrohexahydroxydiphenoyl (DHHDP) bisester unit (Schmidt *et al.*, 1967a and b, Haddock *et al.*, 1982). This bisacyl group usually exists as an equilibrium mixture of two hydrated forms of a cyclohexenetrione structure (Fig. 4.9), displays a chiral methine carbon, and both of its enantiomeric variants are known (Lee *et al.*, 1991). It is biosynthesized by oxidation of the HHDP group, although the details of its production mechanism have not yet been clarified. In many dehydroellagitannins, the DHHDP group is bridged between the 2- and 4-positions of a glucopyranose core in its $^1C_4$-conformation. This unusual conformation of the sugar moiety and the bisester connection of the DHHDP unit with axially-oriented hydroxyl groups constitute additional characteristic features of this sub-class of ellagitannins.

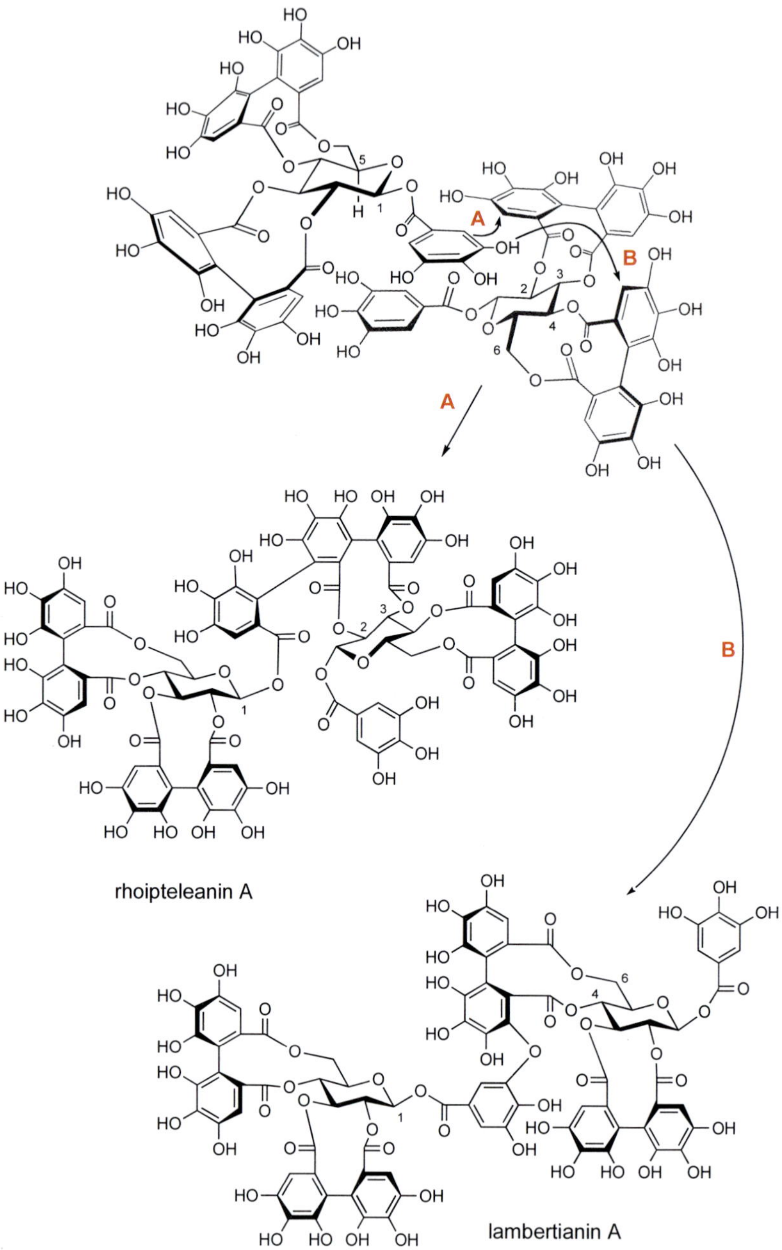

Fig. 4.8 Self-association and plausible regio- and chemoselective dimerization of 1-*O*-galloyl pedunculagin.

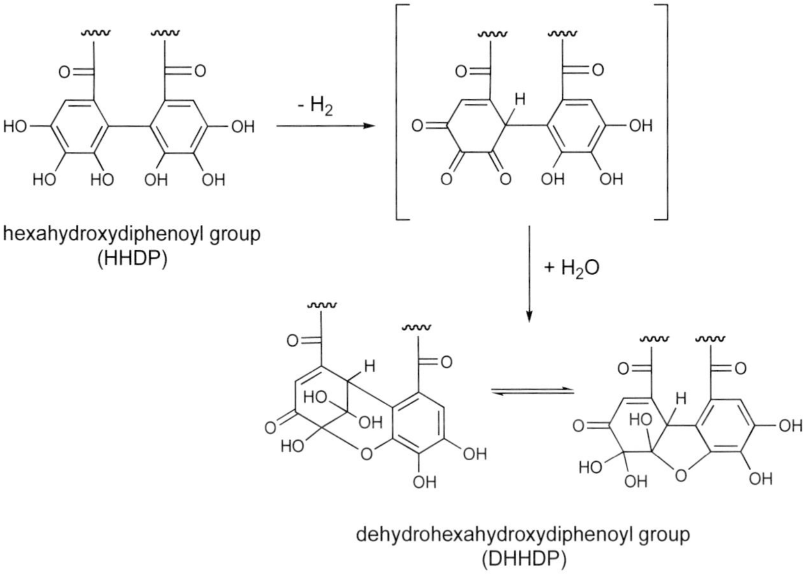

Fig. 4.9 Structures of HHDP and DHHDP groups.

Some specific dehydroellagitannins, such as geraniin (**16**, 1-*O*-galloyl-2,4-(*R*)-DHHDP-3,6-(*R*)-HHDP-β-D-glucose) (Okuda *et al.*, 1982), are accumulated in high concentration in some plants, such as those of the families *Euphorbiaceae*, *Geraniaceae* (Okuda *et al.*, 1980), and *Elaeocarpaceae* (Tanaka *et al.*, 1986). In many cases, the tannins are accompanied by structurally-related ellagitannins having their acyl groups biogenetically derived from the DHHDP group. Typical examples are the chebuloyl (Schmidt and Mayer, 1951, Yoshida *et al.*, 1980, Yoshida *et al.*, 1982, Lin *et al.*, 1990a, Nonaka *et al.*, 1992), brevifolin carboxyl, dehydrochebuloyl (Saijo *et al.*, 1989a, Lin *et al.*, 1990b), tetrahydroxydibenzofuran dicarboxyl (Saijo *et al.*, 1989b), biscyclohexenetrionyl (Nonaka *et al.*, 1990), elaeocarpusinoyl (Tanaka *et al.*, 1986), putranjivainoyl (Lin *et al.*, 1990a) and jolkinoyl (Lee *et al.*, 2004) groups (Fig. 4.10). The unique and attractive chemical structures of these acyl groups reveal the complexity of the oxidative metabolism of the HHDP groups. Of particular note is the fact that the oxidation states of the tetrahydroxydibenzofuran dicarboxyl and chebuloyl groups are the same as those of the HHDP and DHHDP groups, respectively.

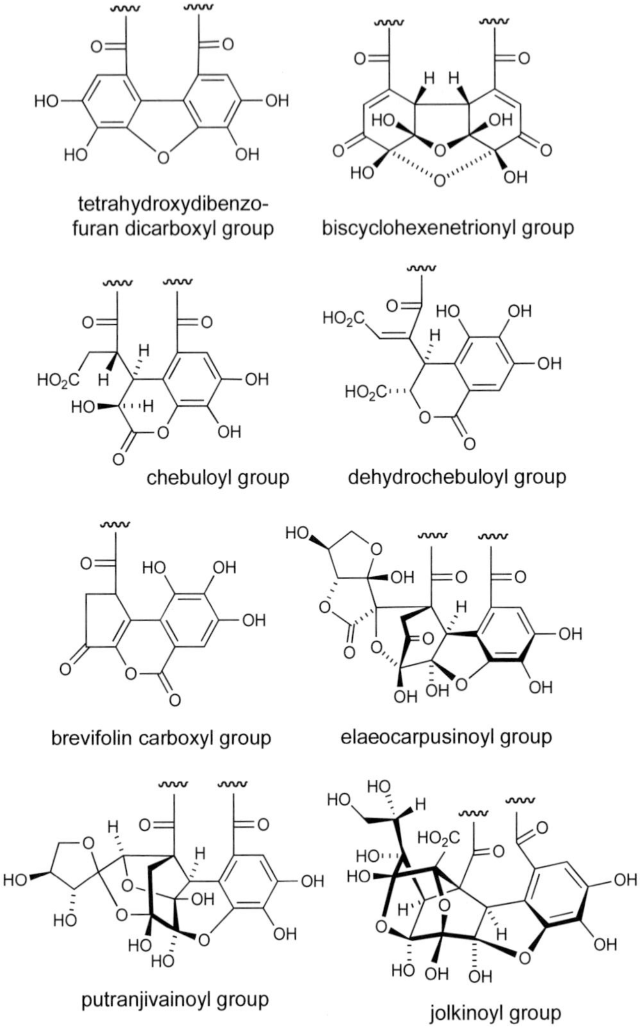

Fig. 4.10 Some acyl groups derived from the DHHDP group.

The biscyclohexenetrionyl group was found in ellagitannins isolated from *Carpinus* sp. (*Betulaceae*). The corresponding bisester is presumably derived from further oxidation of the aromatic ring of the DHHDP group. The elaeocarpusinoyl, putranjivainoyl and jolkinoyl groups comprise an ascorbic acid moiety, and the putative biogenesis of these acyl groups is depicted in Fig. 4.11.

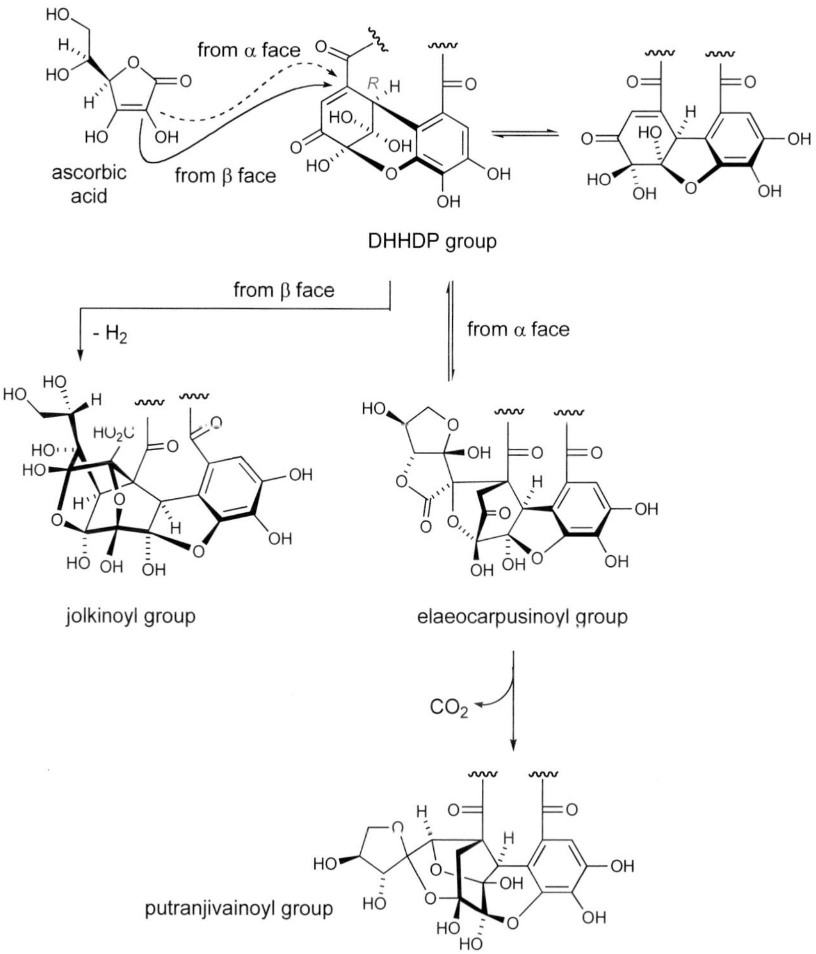

Fig. 4.11 Biogenesis of elaeocarpusinoyl, putranjivainoyl and jolkinoyl groups.

4.3.2 Similarities and differences between the oxidative dimerization of the galloyl group and the pyrogallol-type catechin B-ring

In the biosynthesis of ellagitannins, it is generally accepted that the HHDP group is derived from an oxidative coupling of two galloyl groups (Gupta *et al.*, 1982). The presence of enzymes catalyzing the reaction was recently evidenced (Niemetom *et al.*, 2005). Similar C–C coupling of two pyrogallol rings also occurs during black tea production

(Nonaka *et al.*, 1983, Hashimoto *et al.*, 1988). Enzymatic oxidation of (–)-epigallocatechin-3-*O*-gallate affords theasinensin A, which possesses a structure similar to that of the ellagitannin HHDP group (Hashimoto *et al.*, 1992). However, the production of the catechin-derived dimer does not occur through direct coupling of the pyrogallol rings. Instead, epigallocatechin is first oxidized to give a B-ring *ortho*-quinone, then dimerization spontaneously and stereoselectively occurs to give dehydrotheasinensin A (Fig. 4.12) (Tanaka *et al.*, 2002, 2003).

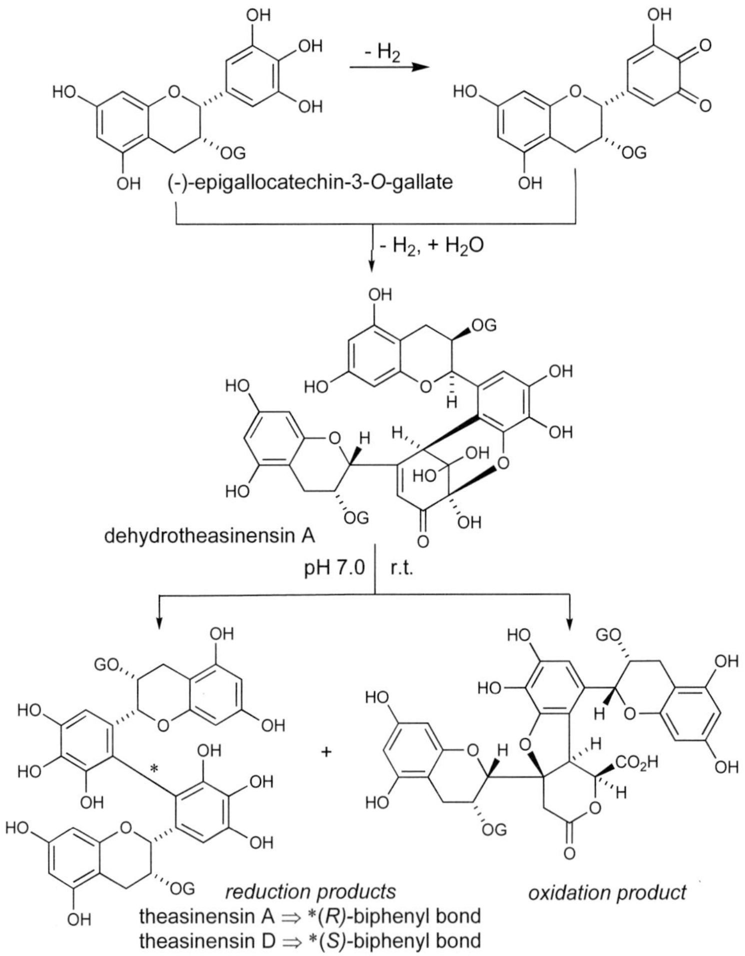

Fig. 4.12 Oxidation of epigallocatechin and degradation of dehydrotheasinensin A.

The dehydrotheasinensin A has the same partial structure as that of the DHHDP group. This dimeric product is unstable and gradually decomposed under neutral conditions to yield theasinensins A and D as major products, which are diastereomers differing only by the atropisomerism of the biphenyl bond. The reaction also produced oxidation products, including a product having two carboxyl groups. Therefore, the decomposition reaction is an oxido-reductive dismutation. The oxidation product has a carbon skeleton similar to that of the chebuloyl and dehydrochebuloyl groups (Fig. 4.10). Despite the structural similarities between dehydrotheasinensin A and the DHHDP group, the production of the ellagitannin DHHDP unit probably proceeds via a different mechanism. Under similar conditions, PGG (**8**) did not yield the products derived from quinone intermediates, although the production of a very small amount of ellagic acid was observed. In the oxidation of epigallocatechin-3-*O*-gallate, only a limited amount of the galloyl group was oxidized, and only coupling reactions between the galloyl group and the B-ring were observed, because the redox potential of the galloyl group is higher than that of the pyrogallol B-ring (Li *et al.*, 2007b).

4.3.3 Reactions of dehydroellagitannins and biomimetic conversions

The DHHDP group exists as hydrated forms of its cyclohexenetrione moiety. Therefore, the dehydroellagitannins are susceptible to suffer from alkaline degradation (Tanaka *et al.*, 1990). In fact, after a brief treatment with a basic aqueous solution, the DHHDP group of geraniin was converted to both dehydrochebuloyl and brevifolin carboxyl groups (see Figs 4.10 and 4.13). Also, heating in pyridine afforded the tetrahydroxydibenzofuran dicarboxyl group via an oxido-reductive dismutation. The reaction with triethylamine gave the brevifolin carboxyl group, and the aromatic ester of the DHHDP group was selectively hydrolyzed during the reaction. This reaction was used to determine the location of the DHHDP group, and it was shown that the cyclohexenetrione ring of the (*S*)-DHHDP group in granatins A and B (Okuda *et al.*, 1980) and in hellioscopinin A (Lee *et al.*, 1990) was attached to the glucose 2-position, which is different from the orientation of the (*R*)-DHHDP group in geraniin (Fig. 4.13).

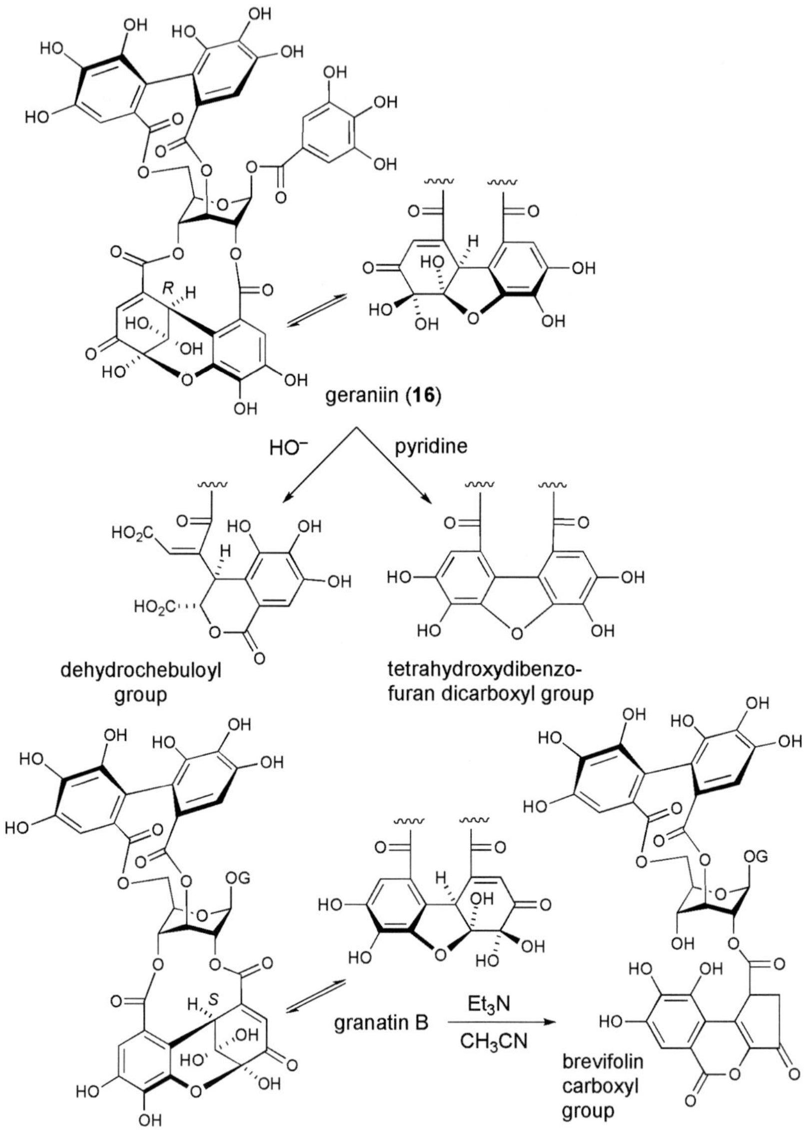

Fig. 4.13 Structures of two dehydroellagitannins and reactions under basic conditions.

In addition, the reaction of geraniin with thiol compounds, such as glutathione or *N*-acetyl cysteine, gave thioethers of the chebuloyl group, which were further accompanied by a second thioether of the cyclopentane carboxyl moiety (Fig. 4.14) (Tanaka *et al.*, 1996). The former product was reductively converted into chebulagic acid (1-*O*-galloyl-3,6-(*R*)-HHDP-2,4-chebuloyl-β-D-glucose), a major ellagitannin of *Terminalia chebula* (Schmidt and Nieswandt, 1950) and *Phyllanthus embrica* (Zhang *et al.*, 2001). Another thioether derivative with a cyclopentane carboxyl structure is spontaneously decomposed to give an ellagitannin with α-ketocarboxyl groups. This product was structurally related to a naturally occurring ellagitannin, euphormisin M_2, which is isolated from *Euphorbia* sp. (Yoshida *et al.*, 1994).

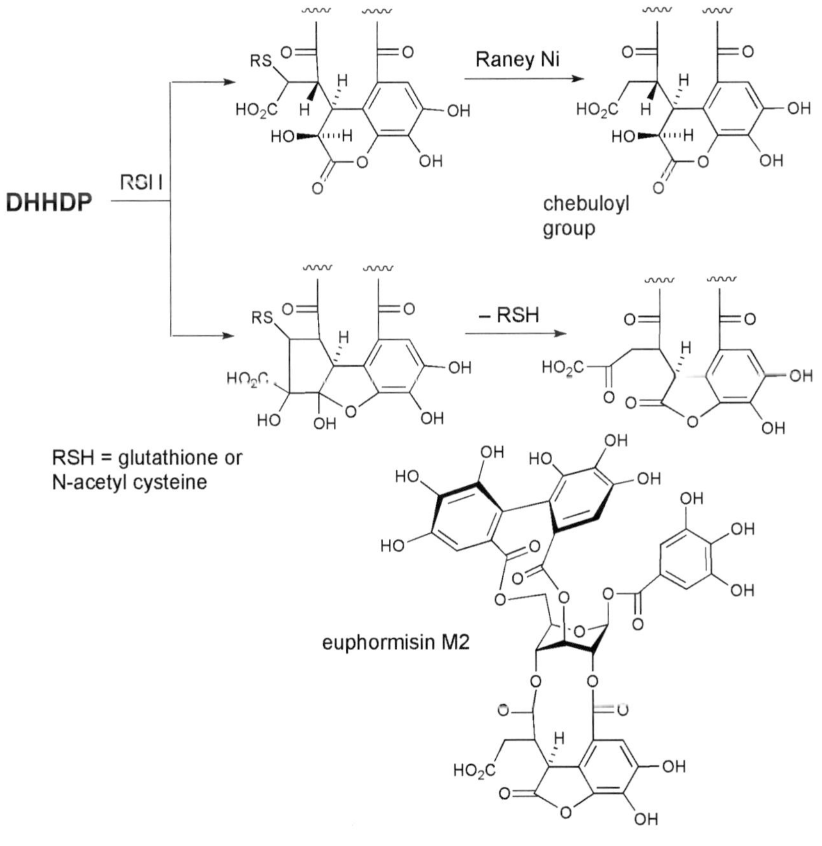

Fig. 4.14 Conversion of the DHHDP group into a chebuloyl group.

4.3.4 Application of the reactions of the DHHDP bisester

As mentioned before, the DHHDP bisester group exists as an equilibrium mixture of two hemiacetal structures, and this phenomenon sometimes increases the difficulty of spectral interpretation and HPLC analysis. Moreover, in alcoholic solutions, the acetal hydroxyl groups are substituted by the alcohol molecules, such that the mixture becomes even more complex. To avoid this complexity, it is necessary to convert the DHHDP units into phenazine derivatives by condensation with *o*-phenylenediamine (Schmidt *et al.*, 1967a, Okuda *et al.*, 1982, see also Chapter 1). The NMR signals of the resulting sugar moieties are largely shifted after derivatization, because the anisotropic effect of the phenazine ring caused significant up-field shifts of the sugar proton signals. These properties are often useful, because the observed chemical shift changes correlate to the stereochemistry of the molecule. However, the derivatives are usually unstable and gradually hydrolyzed in solution.

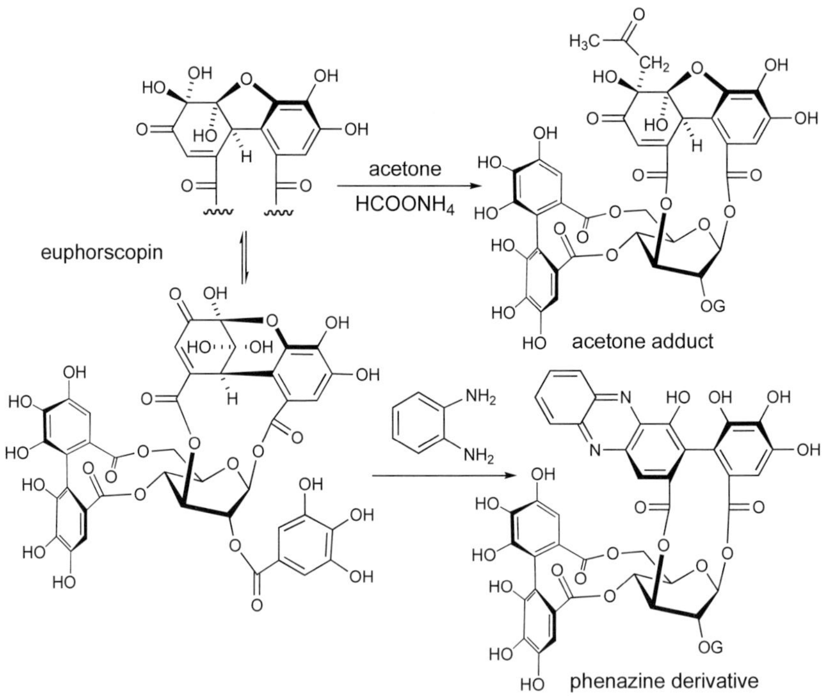

Fig. 4.15 Formation of phenazine and acetone adducts of euphorscopin.

In the course of chemical studies on dehydroellagitannins, acetone adducts are sometimes isolated as artifacts of the extraction procedure using aqueous acetone. The formation of these acetone adducts of the DHHDP units has nevertheless some benefits, because these adducts are sufficiently stable as compared to the phenazine derivatives. Moreover, with the exception of the signals due to CH_2COCH_3, the NMR spectra closely resembled those of the five-membered ring hemiacetal structures. Furthermore, the absence of the additional aromatic ring renders the ^{13}C-NMR spectra much simpler than those of the phenazine derivatives.

The addition of acetone to DHHDP bisester groups is efficiently catalyzed by ammonium ions; an alternative and simple derivatization method was thus developed. An example applied to euphorscopin is shown in Fig. 4.15 (Lee *et al.*, 1991, Tanaka *et al.*, 1992a). This derivatization is very simple: ammonium formate is added as a catalyst to an aqueous acetone solution of tannins or plant extracts that is left standing at room temperature overnight or at 50 °C for 1 to 2 h.

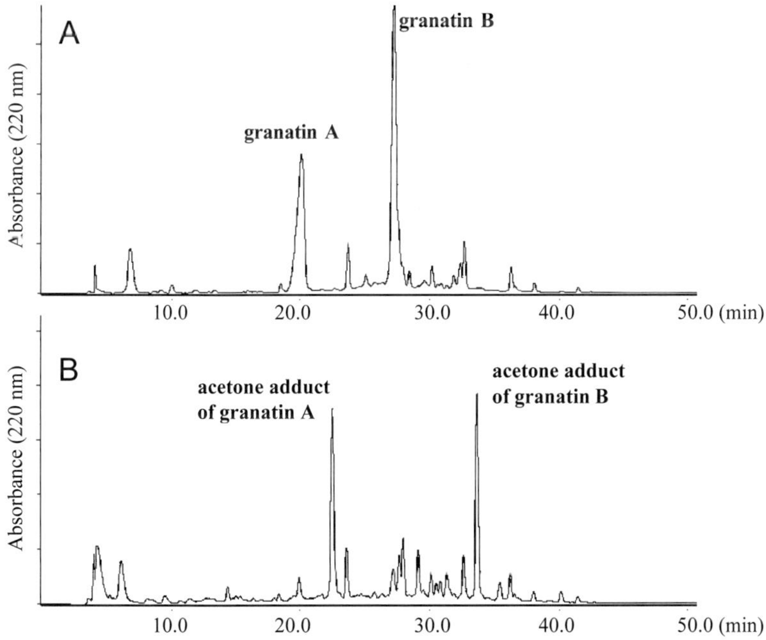

Fig. 4.16 HPLC analyses (220 nm) of the leaves of *Punica granatum*. **A**: aqueous acetone extract; **B**: aqueous acetone extract treated with $HCOONH_4$.

Application of this method to the HPLC analysis of plant extracts discriminates the dehydroellagitannins from other tannins (Fig. 4.16). The HPLC peaks attributed to dehydroellagitannins disappear after treatment of the extracts with aqueous acetone containing ammonium formate, and instead, sharp peaks arising from the acetone adducts appear at longer retention times.

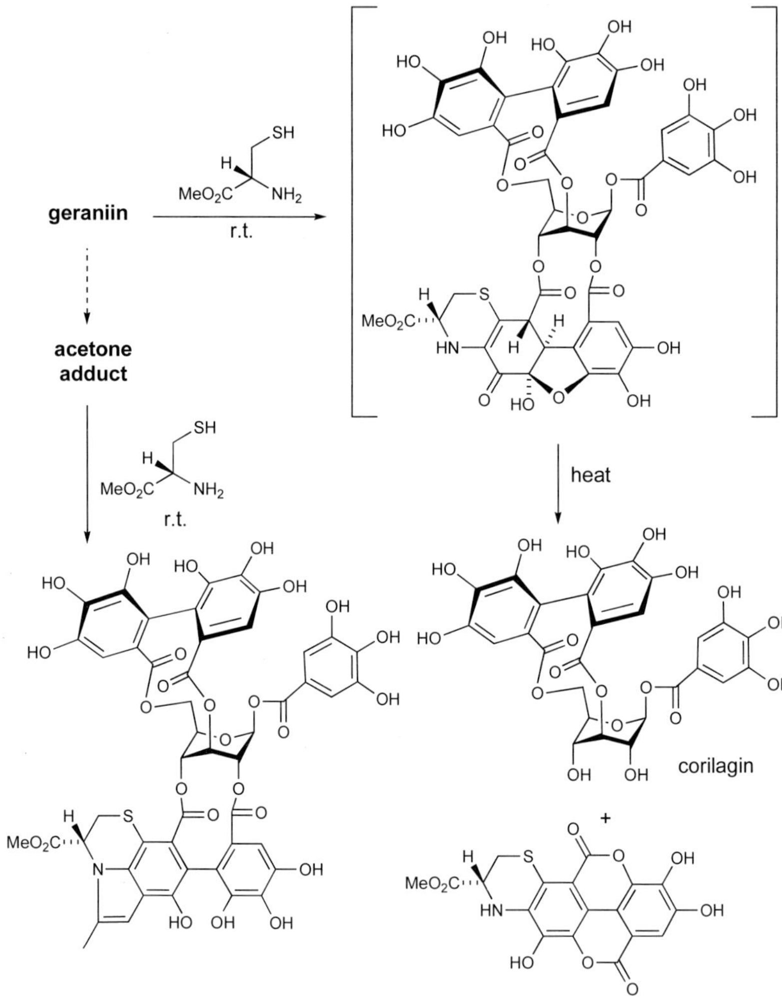

Fig. 4.17 Reaction of geraniin and its acetone adduct with L-cysteine methyl ester.

The DHHDP groups also react with L-cysteine methyl ester to give unstable condensation products, which readily decompose to give the corresponding hydrolysates in good yields. This is a convenient method to selectively hydrolyze DHHDP bisesters (Tanaka *et al.*, 1992b). In addition, L-cysteine reacts with DHHDP acetone adducts (Fig. 4.17) (Tanaka *et al.*, 1992a). The fact that dehydroellagitannins readily react with thiol compounds might have some biological significance, in particular for euphorbiaceous plants, which accumulate these ellagitannins.

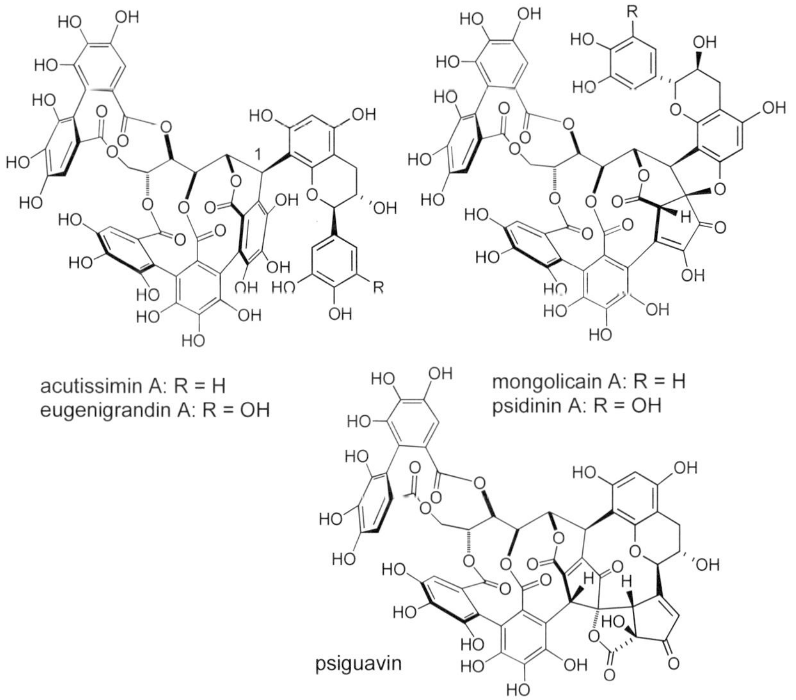

Fig. 4.18 Structures of some oxidized *C*-glycosidic ellagitannins.

4.3.5 Additional oxidation reactions of HHDP and related acyl groups

The oxidative metabolism of HHDP and related acyl groups is also observed in *C*-glycosidic ellagitannins. Typical examples can be found in the case of complex ellagitannins comprising flavan-3-ol units. Acutissimin A (Ishimaru *et al.*, 1987) and eugenigrandin A (Lin *et al.*, 1991) have (+)-catechin and (+)-gallocatechin units, respectively, at the vescalagin unit C-1 position, and these tannins are usually accompanied by their oxidized form, mongolicain A (Nonaka *et al.*, 1988) and psidinin A (Tanaka *et al.*, 1992c), respectively (Fig. 4.18). The cyclopentenone ring attached to the glucose C-1 position of these tannins is generated by oxidation of the pyrogallol ring, followed by a benzylic acid-type rearrangement (BAR) and subsequent decarboxylation and dehydrogenation (Fig. 4.19).

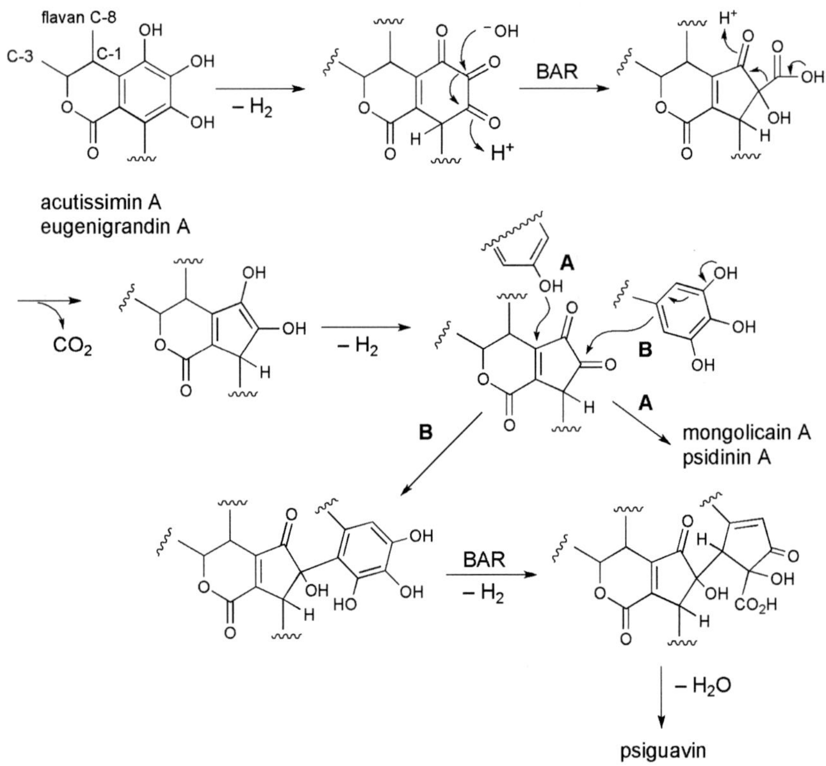

Fig. 4.19 Oxidative metabolism of *C*-glycosidic ellagitannins.

Furthermore, psiguavin isolated from guava (Tanaka *et al.*, 1992c) has an even more complex structure. The B-ring of the gallocatechin unit of eugenigrandin A is electron-richer and more highly susceptible to oxidation than the catechol B-ring of acutissimin A. After oxidation of the glucose C-1-linked pyrogallol ring, addition of the gallocatechin pyrogallol B-ring to the carbonyl group of the intermediate cyclopentenedione moiety (see route B in Fig. 4.19), followed by further oxidation of the B-ring, results in the generation of psiguavin.

4.4 Conclusion

All ellagitannins are biosynthesized via galloyl glucoses. Since most of the ellagitannins have a fully acylated glucose core, pentagalloylglucose (PGG) is a key intermediate in ellagitannin biosynthesis. If the tannins are produced for plant defense, then the use of pentagalloylglucose would seem ideal, because it strongly interacts with proteins and has a severely bitter and astringent taste. However, no plants were found to accumulate this particular tannin as a major polyphenol, even though there are many plants that accumulate proanthocyanidins and ellagitannins. The most inconvenient feature of pentagalloylglucose may be its poor water solubility. In addition, in *Paeonia* sp., a large amount of paeoniflorin (see Fig. 4.4), which bears a hydrophobic benzoyl group, was found to solubilize tannins by hydrophobic association. Furthermore, the oxidative metabolism of pentagalloylglucose into ellagitannins also increases water solubility, and this mechanism may enable many plants to accumulate tannins in higher concentration. The dehydroellagitannins are sometimes accumulated in large amounts in plant tissue. For example, the geraniin contents in the dried leaves of some *Geranium* and *Euphorbia* sp. were reported to range between 4 and 12% by weight of dry material (Okuda *et al.*, 1980a). The reason why geraniin is accumulated in such high concentrations is still unclear; however, it is strongly suggested that its characteristic structure and interesting reactivity, as described in this chapter, have some biological significance.

4.5 Bibliography

Ageta, M., Nonaka, G. and Nishioka, I. (1988). Tannins and related compounds. LXVII. Isolation and characterization of castanopsinins A-H, novel ellagitannins containing a triterpenoid glycoside core, from *Castanopsis cuspidata* var. *sieboldii* Nakai. (3), *Chem. Pharm. Bull.*, 36, pp. 1646–1663.

Barbehenn, R. V., Jones, C. P., Hagerman, A. E., Karonen, M. and Salminen J.-P. (2005). Ellagitannins have greater oxidative activities than condensed tannins and galloyl glucoses at high pH: potential impact on caterpillars, *J. Chem. Ecol.*, 32, pp. 2253–3367.

Baxter, N. J., Lilley, T. H., Haslam, E. and Williamson, M. P. (1997). Multiple interactions between polyphenols and a salivary proline-rich protein repeat result in complexation and precipitation, *Biochemistry*, 36, pp. 5566–5577.

Cai, Y., Gaffney, S. H., Lilley, T. H., Magnolato, D., Martin, R., Spencer, C. M. and Haslam, E. (1990). Polyphenol interactions. Part 4. Model studies with caffeine and cyclodextrins, *J. Chem. Soc. Perkin Trans. 2*, pp. 2197–2209.

Charlton, A. J., Baxter, N. J., Lilley, T. H., Haslam, E., McDonald, C. J. and Williamson, M. P. (1996). Tannin interactions with a full-length human salivary proline-rich protein display a stronger affinity than with single proline-rich repeats, *FEBS Letters*, 382, pp. 289–292.

Charlton, A. J., Haslam, E. and Williamson, M. P. (2002a). Multiple conformations of the proline-rich protein/epigallocatechin gallate complex determined by time-averaged nuclear Overhauser effects, *J. Amer. Chem. Soc.*, 124, pp. 9899–9905.

Charlton, A. J., Baxter, N. J., Khan, M. L., Moir, A. J. G., Haslam, E., Davies, A. P. and Williamson, M. P. (2002b). Polyphenol/peptide binding and precipitation, *J. Agric. Food Chem.*, 50, pp. 1593–1601.

Chen, H.-F, Tanaka, T., Nonaka, G., Fujioka, T. and Mihashi, K. (1993). Tannins and related compounds. Part 120. Hydrolyzable tannins based on a triterpenoid glycoside core, from *Castanopsis hystrix*, *Phytochemistry*, 32, pp. 1457–1460.

Gupta, R. K., Al-Shafi, S. M. K., Layden, K. and Haslam, E. (1982). The metabolism of gallic acid and hexahydroxydiphenic acid in plants. Part 2. Esters of (*S*)-hexahydroxydiphenic acid with D-glucopyranose ($^4C_1$), *J. Chem. Soc. Perkin Trans. 1*, pp. 2525–2534.

Haddock, E. A., Gupta, R. K. and Haslam, E. (1982). The metabolism of gallic acid and hexahydroxydiphenic acid in plants. Part 3. Esters of (*R*) and (*S*)-hexahydroxydiphenic acid and dehydrohexahydroxydiphenic acid with D-glucopyranose ($^1C_4$ and related conformations), *J. Chem. Soc. Perkin Trans. 1*, pp. 2535–2545.

Hara, Y. and Honda, M. (1990). The inhibition of α-amylase by tea polyphenols, *Agric. Biol. Chem.*, 54, pp. 1939–1945.

Hashimoto, F., Nonaka, G. and Nishioka, I. (1988). Tannins and related compounds. LXIX. Isolation and structure elucidation of B,B'-linked bisflavanoids, theasinensins D-G and oolongtheanin from oolong tea. (2), *Chem. Pharm. Bull.*, 36, pp 1676–1684.

Hashimoto, F., Nonaka, G. and Nishioka, I. (1992). Tannins and related compounds. CXIV. Structures of novel fermentation products, theogallinin, theaflavonin and desgalloyl theaflavonin from black tea, and changes of tea leaf polyphenols during fermentation, *Chem. Pharm. Bull.*, 40, pp. 1383–1389.

Haslam, E. and Cai, Y. (1994). Plant polyphenols (vegetable tannins): Gallic acid metabolism, *Nat. Prod. Rep.*, 11, pp. 41–66.

Haslam, E. (1996). Natural polyphenols (vegetable tannins) as drugs: possible modes of action, *J. Nat. Prod.*, 59, pp. 205–215.

Ishimaru, K., Nonaka, G. and Nishioka, I. (1987). Tannins and related compounds. LV. Isolation and characterization of acutissimins A and B, novel tannins from *Quercus* and *Castanea* species, *Chem. Pharm. Bull.*, 35, pp. 602–610.

Jiang, Z., Tanaka, T. and Kouno, I. (1995). Rhoipteleanins A and E, dimeric ellagitannins formed by intermolecular C-C oxidative coupling from *Rhoiptelea chiliantha*, *J. Chem. Soc., Chem. Commun.*, pp. 1467–1468.

Jiang, Z.-H., Hirose, Y., Iwata, H., Sakamoto, S., Tanaka, T. and Kouno, I. (2001). Caffeoyl, coumaroyl, galloyl, and hexahydroxydiphenoyl glucoses from *Balanophora japonica*, *Chem. Pharm. Bull.*, 49, pp. 887–892.

Lee, S. H., Tanaka, T., Nonaka, G. and Nishioka, I. (1990). Tannins and related compounds. XCV. Isolation and characterization of helioscopinins and helioscopins, four new hydrolyzable tannins from *Euphorbia helioscopia* L. (1), *Chem. Pharm. Bull.*, 38, pp. 1518–1523.

Lee, S.-H., Tanaka, T., Nonaka, G. and Nishioka, I. (1991). Tannins and related compounds. CV. Monomeric and dimeric hydrolyzable tannins having a dehydrohexahydroxydiphenoyl group, supinanin, euphorscopin, euphorhelin and jolkianin, from *Euphorbia* Species, *Chem. Pharm. Bull.*, 39, pp. 630–638.

Lee, S.-H., Tanaka, T., Nonaka, G. and Nishioka, I. (2004). Structure and biogenesis of jolkinin, a highly oxygenated ellagitannin from *Euphorbia jolkinii*, *J. Nat. Prod.*, 67, pp. 1018–1022.

Li, H., Tanaka, T., Zhang, Y.-J., Yang, C.-R., and Kouno, I. (2007a). Rubusaviins A-F, monomeric and oligomeric ellagitannins from Chinese sweet tea and their α-amylase inhibitory activity, *Chem. Pharm. Bull.*, 55, pp. 1325–1331.

Li, Y., Tanaka, T. and Kouno, I. (2007b). Oxidative coupling of the pyrogallol B-ring with a galloyl group during enzymatic oxidation of epigallocatechin 3-*O*-gallate, *Phytochemistry*, 68, pp. 1081–1088.

Lin, J. H., Ishimatsu, M., Tanaka, T., Nonaka, G. and Nishioka, I. (1990a). Tannins and Related Compounds. XCVI. Structures of macaranins and macarinins, new hydrolyzable tannins possessing macaranoyl and tergalloyl ester groups, from the Leaves of *Macaranga sinensis* (Baill) Muell.-Arg., *Chem. Pharm. Bull.*, 38, pp. 1844–1851.

Lin, J.-H., Nonaka, G. and Nishioka, I. (1990b). Tannins and related compounds. XCIV. Isolation and characterization of seven new hydrolysable tannins from the leaves of *Macaranga tanarius* (L.) Muell. et Arg., *Chem. Pharm. Bull.*, 38, pp. 1218–1223.

Lin, T.-C., Tanaka, T., Nonaka, G., Nishioka, I. and Yang, C.-R. (1991). Tannins and related compounds. CVIII. Isolation and characterization of novel complex tannins (flavano-ellagitannins), anogeissinin and anogeissusins A and B, from *Anogeissus acuminata* (Roxb ex DC.) Guill. et Perr. var. *lanceolata* Wall. ex Clarke, *Chem. Pharm. Bull.*, 39, pp. 1144–1147.

Luck, G., Liao, H., Murray, N. J., Grimmer, H. R., Warminski, E. E., Williamson, M. P., Lilley, T. H. and Haslam, E. (1994). Polyphenols, astringency and proline-rich proteins, *Phytochemistry*, 37, pp. 357–371.

Mehansho, H., Hagerman, A., Clements, S., Butler, L., Rogler, J. and Carlson, D. M. (1983). Modulation of proline-rich protein biosynthesis in rat parotid glants by sorghums with high tannin levels, *Proc. Natl. Acad. Sci. USA*, 80, pp. 3948–3952.

Mehansho, H., Ann, D. K., Butler, L. G., Rogler, J. and Carlson, D. M. (1987). Induction of proline-rich proteins in hamster salivary glands by isoproterenol treatment and an unusual growth inhibition by tannins, *J. Biol. Chem.*, 262, pp. 12344–12350.

Nakai, M., Fukui, Y., Asami, S., Toyoda-Ono, Y., Iwashita, T., Shibata, H., Mitsunaga, T., Hashimoto, F. and Kiso, Y. (2005). Inhibitory effects of oolong tea polyphenols on pancreatic lipase in vitro, *J. Agric. Food Chem.*, 53, pp. 4593–4598.

Niemetz, R. and Gross, G. G. (2005). Enzymology of gallotannin and ellagitannin biosynthesis, *Phytochemistry*, 66, pp. 2001–2011.

Nishimura, H., Nonaka, G. and Nishioka, I. (1984a). Tannins and related compounds. XX. Two new ellagitannins containing a *proto*-quercitol core from *Quercus stenophylla* Makino. (4), *Chem. Pharm. Bull.*, 32, pp. 1750–1753.

Nishimura, H., Nonaka, G. and Nishioka, I. (1984b). Tannins and related compounds. XVIII. A gallotannin and two ellagitannins containing a *p*-hydroxyphenethyl alcohol 1-O-β-D-glucoside (salidroside) core from *Quercus stenophylla* Makino. (2), *Chem. Pharm. Bull.*, 32, pp. 1735–1740.

Nishimura, H., Nonaka, G. and Nishioka, I. (1986). Tannins and related compounds. Part 44. *scyllo*-Quercitol gallates and hexahydroxydiphenoates from *Quercus stenophylla*, *Phytochemistry*, 25, pp. 2599–2604.

Nonaka, G., Harada, M. and Nishioka, I. (1980). Eugeniin, a new ellagitannin from cloves, *Chem. Pharm. Bull.*, 28, pp. 3685–3687.

Nonaka, G., Kawahara, O. and Nishioka, I. (1983). Tannins and related compounds. XV. A new class of dimeric flavan-3-ol gallates, theasinensins A and B, and proanthocyanidin gallates from green tea leaf. (1), *Chem. Pharm. Bull.*, 31, pp. 3906–3914.

Nonaka, G., Ishimaru, K., Mihashi, K., Iwase, Y., Ageta, M. and Nishioka, I. (1988). Tannins and related compounds. LXIII. Isolation and characterization of mongolicains A and B, novel tannins from *Quercus* and *Castanopsis* species, *Chem. Pharm. Bull.*, 36, pp. 857–869.

Nonaka, G., Mihashi, K. and Nishioka, I. (1990). Novel metabolites of hexahydroxydiphenic acid esters (ellagitannins) from *Carpinus japonica*, *J. Chem. Soc., Chem. Commun.*, pp. 790–791.

Nonaka, G., Akazawa, M. and Nishioka, I. (1992). Tannins and related compound. Part III. Two new ellagitannin metabolites, carpinusin and carpinusnin from *Carpinus laxiflora*, *Heterocycles*, 33, pp. 597–606.

Okuda, T., Mori, K. and Hatano, T. (1980a). Constitution of Geranium thunbergii Sieb. et Zucc. Part 10. The distribution of geraniin and mallotusinic acid in the order Geraniales, *Phytochemistry*, 19, pp. 547–551.

Okuda, T., Hatano, T., Nitta, H. and Fujii, R. (1980b). Hydrolyzable tannins having enantiomeric dehydrohexahydroxydiphenoyl group: revised structure of terchebin and structure of granatin B, *Tetrahedron Lett.*, 21, pp; 4361–4364.

Okuda, T., Yoshida, T. and Hatano, T. (1982). Constituents of *Geranium thunbergii* Sieb. et Zucc. Part 12. Hydrated stereostructure and the equilibration of geraniin, *J. Chem. Soc. Perkin Trans 1*, pp. 9–14.

Okuda, T., Yoshida, T. and Hatano, T. (2000). Correlation of oxidative transformations of hydrolyzable tannins and plant evolution, *Phytochemistry*, 55, pp. 513–529.

Okuda, T. (2005). Systematics and health effects of chemically distinct tannins in medicinal plants, *Phytochemistry*, 66, pp. 2012–2031.

Robbins, C. T., Hanley, T. A., Hagerman, A. E., Hjeljord, O., Baker, D. L., Schwartz, C. C. and Mautz, W. W. (1987). Role of tannins in defending plants against ruminants: reduction in protein availability, *Ecology*, 68, pp. 98–107.

Saijo, R., Nonaka, G. and Nishioka, I. (1989a). Tannins and related compounds. LXXXVII. Isolation and characterization of four new hydrolysable tannins from the leaves of *Mallotus repandus*, *Chem. Pharm. Bull.*, 37, pp. 2624–2630.

Saijo, R., Nonaka, G. and Nishioka, I. (1989b). Tannins and related compounds. LXXXIV. Isolation and characterization of five new hydrolysable tannins from the bark of *Mallotus japonicus*, *Chem. Pharm. Bull.*, 37, pp. 2063–2070.

Scalbert, A. (1991). Antimicrobial properties of tannins, *Phyrochemistry*, 30, pp. 3875–3883.

Schmidt, O. T. and Mayer, W. (1951). Natural tannins. V. Constitution of the "fission acid," $C_{14}H_{12}O_{11}$, obtained from chebulinic and chebulagic acid, *Justus Liebigs Ann. Chem.*, 571, pp. 1–15.

Schmidt, O. T. and Nieswandt, W. (1950). Natural tannins. III. Chebulagic acid, a crystalline ellagitannin from myrobalans, *Justus Liebigs Ann. Chem.*, 568, pp. 165–173.

Schmidt, O. T., Schanz, R., Eckert, R. and Wurmb, R. (1967a). Natural tannins. XXXIV. Brevilagin 1, *Justus Liebigs Ann. Chem.*, 706, pp. 131–153.

Schmidt, O. T. and Wieder, G. (1967b). Natural tannins. XXXIX. Synthesis of dehydrohexahydroxydiphenic acid, *Justus Liebigs Ann. Chem.*, 706, pp. 198–203.

Tanaka, T., Nonaka, G. and Nishioka, I. (1985). Tannins and related compounds. Part 28. Revision of the structures of sanguiins H-6, H-2, and H-3, and isolation and characterization of sanguiin H-11, a novel tetrameric hydrolyzable tannin, and seven related tannins, from *Sanguisorba officinalis*, *J. Chem. Research*, (S) pp. 176–177, (M) pp. 2001–2029.

Tanaka, T., Nonaka, G., Nishioka, I., Miyahara, K. and Kawasaki, T. (1986). Tannins and related compounds. Part 37. Isolation and structure elucidation of elaeocarpusin, a novel ellagitannin from *Elaeocarpus sylvestris* var. *ellipticus*, *J. Chem. Soc. Perkin Trans. 1*, pp. 369–376.

Tanaka, T., Nonaka, G. and Nishioka, I. (1990). Tannins and related compounds. C. Reaction of dehydrohexahydroxydiphenic acid esters with bases, and its application to the structure determination of pomegranate tannins, granatins A and B, *Chem. Pharm. Bull.*, 38, pp. 2424–2428.

Tanaka, T., Fujisaki, H., Nonaka, G. and Nishioka, I. (1992a). Tannins and related compounds. CXVIII. Structures, preparation, high-performance liquid chromatography and some reactions of dehydroellagitannin-acetone condensates, *Chem. Pharm. Bull.*, 40, pp. 2937–2944.

Tanaka, T., Fujisaki, H., Nonaka, G. and Nishioka, I. (1992b). Reaction of dehydroellagitannins with L-cysteine methyl ester, *Heterocycles*, 33, pp. 375–383.

Tanaka, T., Ishida, N., Ishimatsu, M., Nonaka, G. and Nishioka, I. (1992c). Tannins and related compounds. CXVI. Six new complex tannins, guajavins, psidinins and psiguavin from the bark of *Psidium guajava* L, *Chem. Pharm. Bull.*, 40, pp. 2092–2098.

Tanaka, T., Orii, Y., Nonaka, G. and Nishioka, I. (1993a). Tannins and related compounds. CXXIII. Chromone, acetophenone and phenylpropanoid glycosides and their galloyl and/or hexahydroxydiphenoyl esters from the leaves of *Syzygium aromaticum* Merr. et Perry, *Chem. Pharm. Bull.*, 41, pp. 1232–1237.

Tanaka, T., Tachibana, H., Nonaka, G., Nishioka, I., Hsu, F. L., Kohda, H. and Tanaka, O. (1993b). Tannins and related compounds. CXXII. New dimeric, trimeric and tetrameric ellagitannins, lambertianins A-D, from *Rubus lambertianus* Seringe, *Chem. Pharm. Bull.*, 41, pp. 1214–1220.

Tanaka, T., Kouno, I. and Nonaka, G. (1996). Glutathione-mediated conversion of the ellagitannin geraniin into chebulagic acid, *Chem. Pharm. Bull.*, 44, pp. 34–40.

Tanaka, T., Zhang, H., Jiang, Z. and Kouno, I. (1997a). Relationship between hydrophobicity and structure of hydrolysable tannins, and association of tannins with crude drug constituents in aqueous solution, *Chem. Pharm. Bull.*, 45, pp. 1891–1897.

Tanaka, T., Jiang, Z. and Kouno, I. (1997b). Structures and biogenesis of rhoipteleanins, ellagitannins formed by stereospecific intermolecular C-C oxidative coupling, isolated from *Rhoiptelea chiliantha*, *Chem. Pharm. Bull.*, 45, pp. 1915–1921.

Tanaka, T., Mine, C., Watarumi, S., Fujioka, T., Mihashi, K., Zhang, Y.-J. and Kouno, I. (2002). Accumulation of epigallocatechin quinone dimers during tea fermentation and formation of theasinensins, *J. Nat. Prod.*, 65, pp. 1582–1587.

Tanaka, T., Watarumi, S., Matsuo, Y., Kamei, M. and Kouno, I. (2003). Production of theasinensins A and D, epigallocatechin gallate dimers of black tea, by oxidation-reduction dismutation of dehydrotheasinensin A, *Tetrahedron*, 59, pp. 7939–7947.

Tanaka, T., Uehara, R., Nishida, K. and Kouno, I. (2005). Galloyl, caffeoyl and hexahydroxydiphenoyl esters of dihydrochalcone glucosides from Balanophora tobiracola, *Phytochemistry*, 66, pp. 675–681.

Yoshida, T., Fujii, R. and Okuda, T. (1980). Revised structures of chebulinic acid and chebulagic acid, *Chem. Pharm. Bull.*, 28, pp. 3713–3715.

Yoshida, T., Okuda, T., Koga, T. and Toh, N. (1982). Absolute configurations of chebulic, chebulinic and chebulagic acid, *Chem. Pharm. Bull.*, 30, pp. 2655–2658.

Yoshida, T., Amakura, Y., Liu, Y.-Z. and Okuda, T. (1994). Tannins and related polyphenols of Euphorbiaceous plants. XI. Three new hydrolyzable tannins and a polyphenol glucoside from *Euphorbia humifusa*, *Chem. Pharm. Bull.*, 42, pp. 1803–1807.

Zhang, Y.-J., Betsumiya, Y., Matsuo, A., Ueda, T., Tanaka, T. and Kouno, I. (2000). Interaction of Peptide Turn Structure with Tannins: Stereochemical Analysis by NMR Spectroscopy. In *Peptide Science 1999*, ed. by N. Fujii, The Japanese Peptide Society: Minoh, pp. 315–318.

Zhang, Y.-J., Abe, T., Tanaka, T., Yang, C.-R. and Kouno, I. (2001). Phyllanemblinins A-F, new ellagitannins from *Phyllanthus emblica*, *J. Nat. Prod.*, 64, pp. 1527–1532.

Zhang, Y.-J., Tanaka, T., Betsumiya, Y., Kusano, R., Matsuo, A., Ueda, T. and Kouno, I. (2002). Association of tannins and related polyphenols with the cyclic peptide gramicidin S, *Chem. Pharm. Bull.*, 50, pp. 258–262.

Zucker, W. V. (1983). Tannins: does structure determine function? An ecological perspective, *The American Naturalist*, 121, pp. 335–365.

Chapter 5

Strategies for the Synthesis of Ellagitannins

Karamali Khanbabaee

University of Paderborn, Faculty of Science, Department of Chemistry, Warburgerstr. 100, D-33098 Paderborn, Germany;
kkh@zitmail.uni-paderborn.de

As a result of recent insights in the molecular structures and medicinal uses of tannins, this class of natural products has enjoyed renewed scientific interest. This is reflected, on the one hand, in the enormously increased number of publications reporting on the isolation, characterisation and biological activity of tannins, and on the other hand, in the steadily growing number of papers reporting on tannin syntheses. In this chapter, the most recent advances and effective concepts in the total synthesis of ellagitannins are discussed. In nature, ellagitannins occur usually only as the (*R*)- or only as the (*S*)-configured atropisomer, with the corresponding opposite atropisomers seldom found. The configuration of the HexaHydroxyDiPhenoyl units (HHDPs) found in ellagitannins is usually determined by the linkage position of the HHDPs on their polyol core. Those ellagitannins that have their HHDPs linked to the 2,3- or 4,6-positions of D-glucopyranose have the (*S*)-configuration, whereas the corresponding 3,6-HHDP-bearing ellagitannins are (*R*)-configured. Very few exceptions are known in which both atropisomeric HHDPs co-exist (Schmidt *et al.*, 1965, Gupta *et al.*, 1982).

5.1 Introduction

With more than 500 structurally characterised members so far, ellagitannins constitute one of the most important classes of tannins, exhibiting a variety of interesting biological activities such as anti-oxidative, anticancer and antiviral activities. This and the fact that ellagitannins are usually not toxic for humans make them important and interesting compounds for pharmaceutical purposes. However, their broad application in pharmacy is hindered by the fact that they are hardly accessible in pure forms and in sufficient quantities from their natural sources. The first total synthesis of an ellagitannin was performed by Feldman in 1994.

The search for biologically active compounds is one of the most important aims in pharmaceutical research. The isolation of secondary metabolites from natural sources, especially from bacteria, fungi, and higher plants, has become increasingly important in the search for new lead compounds. In the search for biologically active compounds from plant extracts, a large number of tannins with various activities has been identified and characterised since the early 1980s (Haslam, 1996). Extensive biological evaluations have shown that numerous tannins possess antibacterial (Hada *et al.*, 1989, see also Chapter 2), antiviral (Fukuchi *et al.*, 1989, Nakashima *et al.*, 1992, Xie *et al.*, 1995) and anticancer properties (Miyamoto *et al.*, 1987, Yoshida *et al.*, 1989, 1990a/b, 1991a/b, Kashiwada *et al.*, 1992b, 1993). The observed high selectivity of the tannins is frequently caused by the inhibition of specific enzymes (Jankun, 1997). The natural products that inhibit specific enzymes have a great potential for the development of new pharmaceuticals, especially in AIDS and cancer therapy.

Many papers have appeared on the biosynthesis, isolation, and biological activity of tannins, especially ellagitannins, over the last twenty years (*e.g.*, Xie *et al.*, 1995, Yoshida *et al.*, 1982, 1984, 1985, 1986, 1989, 1990a/b, 1991a-d, 1992a/b, 1995, Nonaka *et al.*, 1980, 1984, 1989a-c, 1990, Tanaka *et al.*, 1986a/b, 1990, 1992a/b, 2001, Hatano *et al.*, 1988, 1989, 1990a-c, 1991, 1995, Lin *et al.*, 1990, Nishizawa *et al.*, 1982, 1983, Haddock *et al.*, 1982a/b, Kashiwada *et al.*, 1992a/b, 1993, Kadota *et al.*, 1990, Okuda *et al.*, 1982a-e, 1983a/b, El-Mekkawy *et al.*,

1995, Tsai *et al.*, 1992, Han *et al.*, 1995, Chen *et al.*, 1995, Morimoto *et al.*, 1986a/b, Saijo *et al.*, 1989). Access to pure ellagitannins by isolation from natural sources is often cumbersome and yields only small quantities of pure natural products (Okuda *et al.*, 1982a/b). For the organic chemist, it is therefore a great challenge to provide synthetic access to this type of natural product in order to provide sufficient quantities of pure biologically active compounds and to optimise the biological activity (*e.g.*, lowering cytotoxicity, improving absorption or improving selectivity) by derivatisation or structural modifications.

5.2 Strategies for the Synthesis of Ellagitannins

Up to now, only two different strategies have been developed for the synthesis of monomeric and dimeric ellagitannins (Fig. 5.1), *i.e.*, **A**: double esterification of a suitably protected hexahydroxydiphenic acid with a diol or a tetrol derivative of D-glucopyranose and **B**: esterification of a suitably protected gallic acid with a diol derivative of D-glucopyranose, followed by coupling of the galloyl residues. The coupling reaction of the 2,3- or 4,6-galloyl units occurred diastereoselectively, using different reagents, to form the (*S*)-HHDP unit. The stereoselectivity is induced by a template effect brought about by the chiral glucopyranose core (Feldman and Ensel, 1993, 1994a, Feldman *et al.*, 1994, Feldman and Sambandam, 1995, Feldman and Smith, 1996, Dai and Martin, 1998, Arisawa *et al.*, 1999). Method **A** offers an alternative way for the synthesis of ellagitannins using hexabenzyloxydiphenic acid as starting material.

Compared to method **B**, method **A** has several advantages. There is no need for for orthogonal protection of the phenolic hydroxyl groups of the hexahydroxydiphenic acid used. Consequently, there is no need for any deprotection of multiple protective groups. The coupling step in method **B**, which usually results in the formation of regioisomeric mixtures that are difficult to separate by chromatography, is not required in method **A**. Furthermore, method **A** can also be used for the synthesis of (*R*)-configured 2,3-HHDP ellagitannins (Itoh and Chika, 1995, Itoh *et al.*, 1996, Khanbabaee and Lötzerich, 1998).

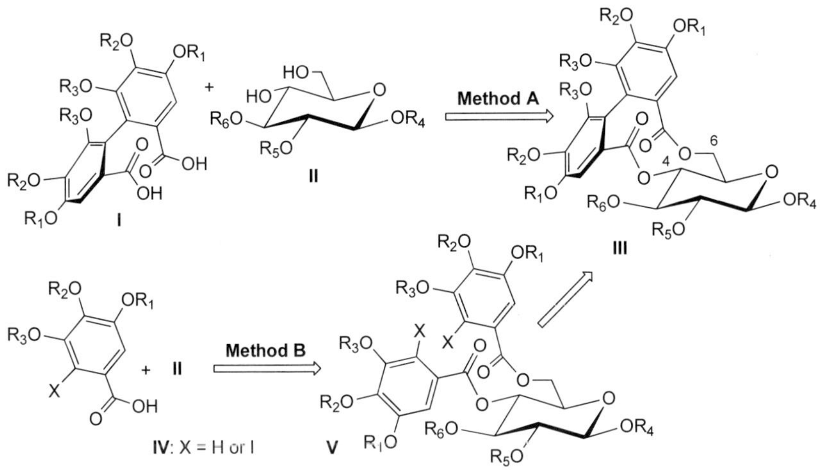

Fig. 5.1 Two strategies for the construction of the ellagitannin framework.

There are different ways to access the protected hexahydroxydiphenic (HHDP) acid that is a pre-requisite for method **A**. For example, as we shall discuss later in this chapter, the synthesis of the enantiopure (*S*)-hexamethoxydiphenic acid (*S*)-**50** has been accomplished by enantioselective reductive coupling reactions between chiral aryl components either in an intermolecular (Nelson and Meyers, 1994) or in an intramolecular fashion (Lipshutz *et al.*, 1993, 1994a/b). The subsequent esterification of the enantiopure (*S*)-**50** with a suitably functionalised diol derivative of D-glucopyranose opened the way for the synthesis of permethylated ellagitannins (see Fig. 5.10 and 5.11). A different route to the stereoselective synthesis of this type of permethylated ellagitannins is provided by application of the concept of kinetic racemate resolution through which, for example, the racemic mixture of hexamethoxydiphenoyl dichloride *rac*-**80** (see Fig. 5.16) serves to bisacylate a suitably functionalised diol derivative of D-glucopyranose (Itoh and Chika, 1995, Itoh *et al.*, 1996).

However, due to the dearth of suitable methods for the cleavage of the methyl ether groups of the permethylated precursor of natural ellagitannins, the total syntheses of the corresponding ellagitannins could not be successfully completed (Nelson and Meyers, 1994, Lipshutz *et al.*,

1993, 1994a/b, Itoh and Chika, 1995, Itoh *et al.*, 1996, Dai and Martin, 1998, Arisawa *et al.*, 1999, Ikeda *et al.*, 2004). Application of known methods for the cleavage of methyl ether groups, *e.g.*, using boron tribromide-etherate, resulted in product decomposition, as observed in the case of attempted syntheses of gallotannins (Khanbabaee and Lötzerich, 1997a). Therefore, the appropriate selection of suitable protecting groups became a decisive pre-requisite for a successful ellagitannin synthesis. In our own work, we chose benzyl protective groups for the phenolic hydroxyl functions, as these groups are stable under the reaction conditions of all the different steps of our synthetic sequence, and can be cleaved mildly at the end of the synthesis. Our strategy for the synthesis of ellagitannins is then also based on the concept of kinetic racemate resolution, thus using racemic hexabenzyloxydiphenic acid *rac*-**16** (see Fig. 5.4, Schmidt *et al.*, 1954) in a double esterification reaction with a suitable diol derivative of D-glucopyranose (*i.e.*, method **A**). This method enabled us to synthesise a series of natural and non-natural ellagitannins (Khanbabaee *et al.*, 1997, 1999, Khanbabaee and Lötzerich, 1997b/c, 1998, 1999, Khanbabaee and Großer, 2002, 2003).

5.2.1 2,3-HHDP-containing ellagitannins

5.2.1.1 *Total synthesis of sanguiin H-5*

The first total synthesis of the natural product sanguiin H-5 (**6**), also known as isostrictinin (Okuda, 1982e), was reported by Feldman and Sambandam in 1995 (Fig. 5.2). This is the first instance of a coupling of galloyl residues linked to the 2,3-positions of a D-glucosyl unit to form the 2,3-HHDP-containing intermediate **4** by means of $Pb(OAc)_4$ (*i.e.*, Wessely oxidation). Two problems had to be solved in order to synthesise a stereochemically pure product. Firstly, concerning the oxidative coupling to be achieved between the neighbouring C-2 and C-3 galloyl units on the D-glucosyl unit **3**, only the (*S*)-atropisomer was to be formed. Secondly, concerning the galloylation of the anomeric position of the D-glucosyl unit, only the β anomer was to be generated. In the first step of this total synthesis, the O-1-protected D-glucopyranose **1** is selectively acetalized with benzaldehyde at positions O-4 and O-6. The resulting acetal **2** is then esterified with 3,4-*O*-diphenylketal-protected gallic acid **8** and subsequently desilylated with *n*-Bu_4NF. The resulting diester **3** is then subjected to a fully diastereoselective biaryl coupling using $Pb(OAc)_4$.

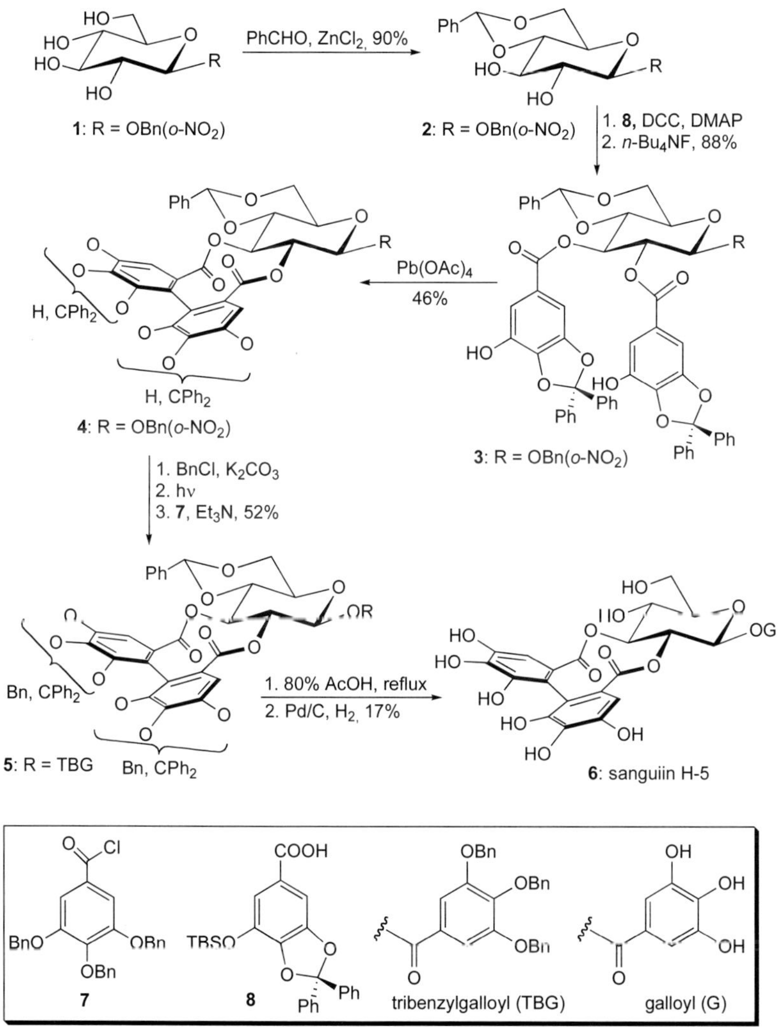

Fig. 5.2 Total synthesis of sanguiin H-5 (**6**).

The resulting product **4** consists of a mixture of regioisomeric diphenylketals. Benzylation of the remaining free phenolic hydroxyl functions afforded a fully protected system (not shown), which was then submitted to a photolytic cleavage to release the nitrobenzyl protective group at the anomeric center of the D-glucosyl unit. Acylation of this position with the tribenzylated galloyl chloride **7** in the presence of triethylamine (Et_3N) gave fully protected variants of the target. Acidic cleavage of the diphenylketals and the benzylidene acetal, followed by a final hydrogenolytic cleavage of the benzyl groups, furnished the natural product sanguiin H-5 (**6**, Fig. 5.2).

5.2.1.2 *Construction of a 2,3-bridged ellagitannin scaffold*

A copper-catalysed reductive Ullmann coupling of two methyl ether-protected iodinated galloyl moieties linked to the appropriate positions of a D-glucopyranosyl core system (*i.e.*, application of method **B**, see Fig. 5.1) was described by Dai and Martin (1998).

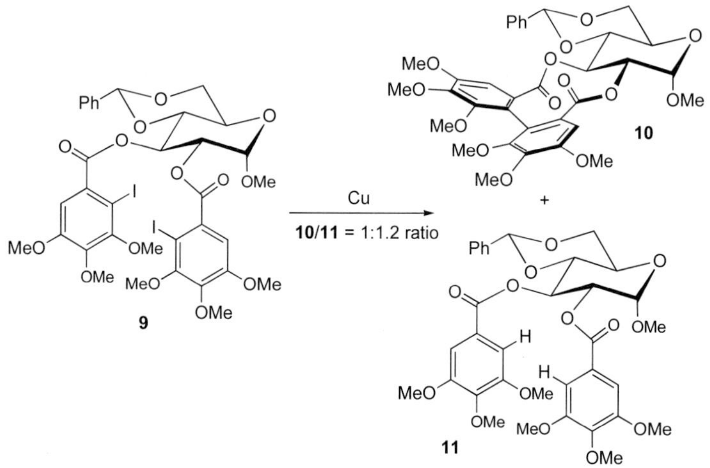

Fig. 5.3 Construction of a 2,3-HHDP-containing ellagitannin framework.

They could show that such coupling reaction conditions could be applied to substrates in which galloyl-derived units are linked to the 2,3-positions of a D-glucosyl unit (Fig. 5.3). However, the reductive coupling of dihalide **9** not only produced the desired 2,3-hexamethoxydiphenoyl-containing product **10**, but also the non-coupled and reduced product **11** in a **10**/**11** ratio of 1:1.2. As in the case of the Feldman coupling, this reaction was also (*S*)-atropodiastereoselective, as a consequence of conformational preferences in D-glucopyranose.

5.2.1.3 *Total syntheses of (S)-configured ellagitannins praecoxin B and pterocaryanin C, and (R)-configured artificial ellagitannins mahtabin A and pariin M*

The isolation and full characterisation of the 2,3-HHDP-bearing ellagitannin, praecoxin B (**15**), from the aerial parts of *Stachyurus praecox* SIEB. *et* ZUCC. (*Stachyuraceae*), a plant known for its high tannin content, was first reported in 1983, and then, in 1991 by Okuda and co-workers (Okuda *et al.*, 1983a, Hatano *et al.*, 1991). This ellagitannin was also found in the bark of *Mallotus japonicus* (THUNB.) MUELLER-ARG. (*Euphorbiaceae*) by Nishioka and co-workers (Saijo *et al.*, 1989) and in the leaves of *Tibouchina semidecandra* COGN by Okuda, Yoshida and co-workers (Yoshida *et al.*, 1991c). The tannin-containing extracts from these plants are used in Asian folk medicine as diuretics, for the treatment of stomach and duodenal tumours, and also as natural medicines against diarrhoea, dysentery, fluor genitalis and various skin conditions, and as astringents and haemostatic agents.

It is important to point out that the HHDP units of nearly all naturally occurring 2,3- and 4,6-HHDP-containing ellagitannins are (*S*)-configured (Khanbabaee and van Ree, 2001a). To our knowledge, up to now, only a handful unusual ellagitannins have been described as exceptions to this rule (Khanbabaee and van Ree, 2001b). For example, cercidinins A and B were reported as two unusual ellagitannins bearing a 2,3-linked (*R*)-configured HHDP unit (Nonaka *et al.*, 1989b). Structurally, cercidinins A and B would be the diastereoisomeric counterparts of pterocaryanin C (**14**) and praecoxin B (**15**), respectively. We synthesized all four diastereoisomers, compared their physical properties and hence found out that the structural determination of cercidinins A and B was wrong (Khanbabaee and Lötzerich, 1997b, 1998). We thus named the corresponding synthetic ellagitannins mahtabin A (**12**) instead of cercidinin A and pariin M (**13**) instead of cercidinin B. On the basis of our results, the chemical structure of cercidinin A has been re-examined and revised (Tanaka *et al.*, 2001).

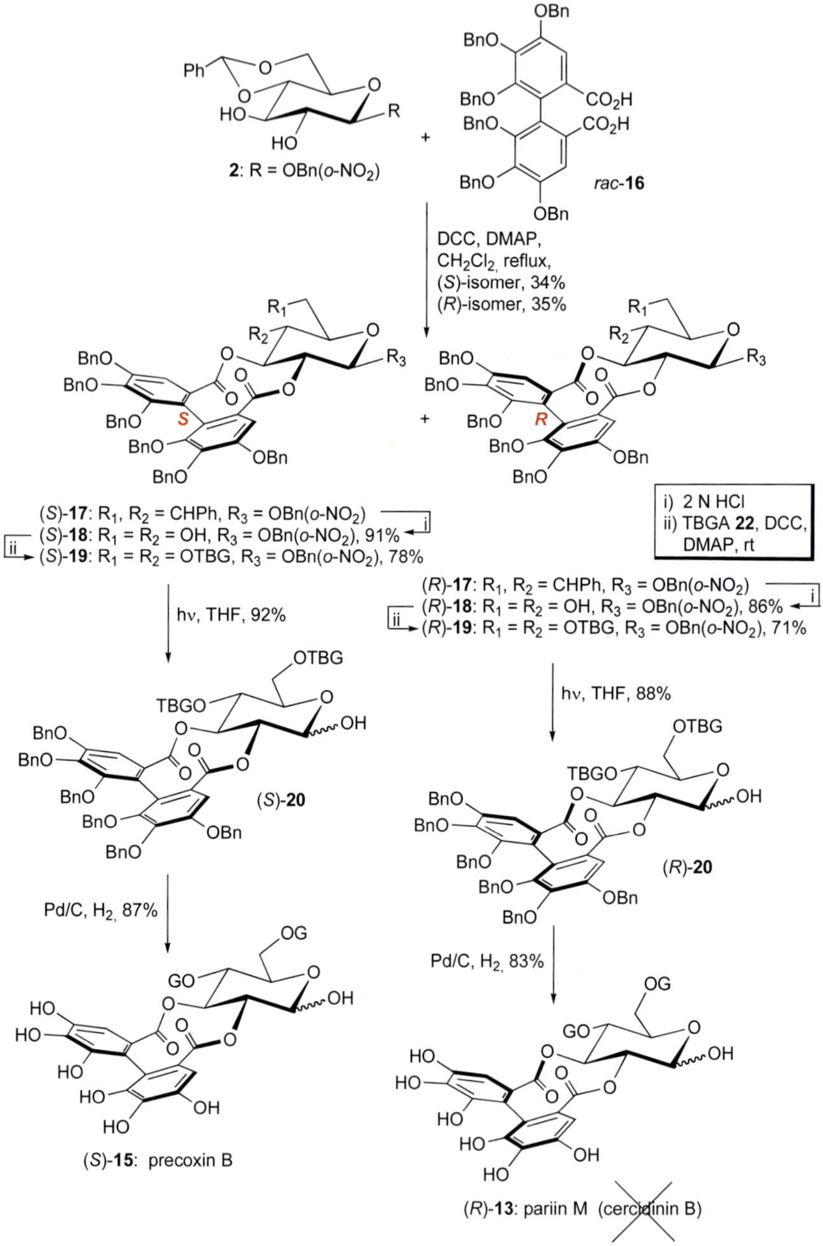

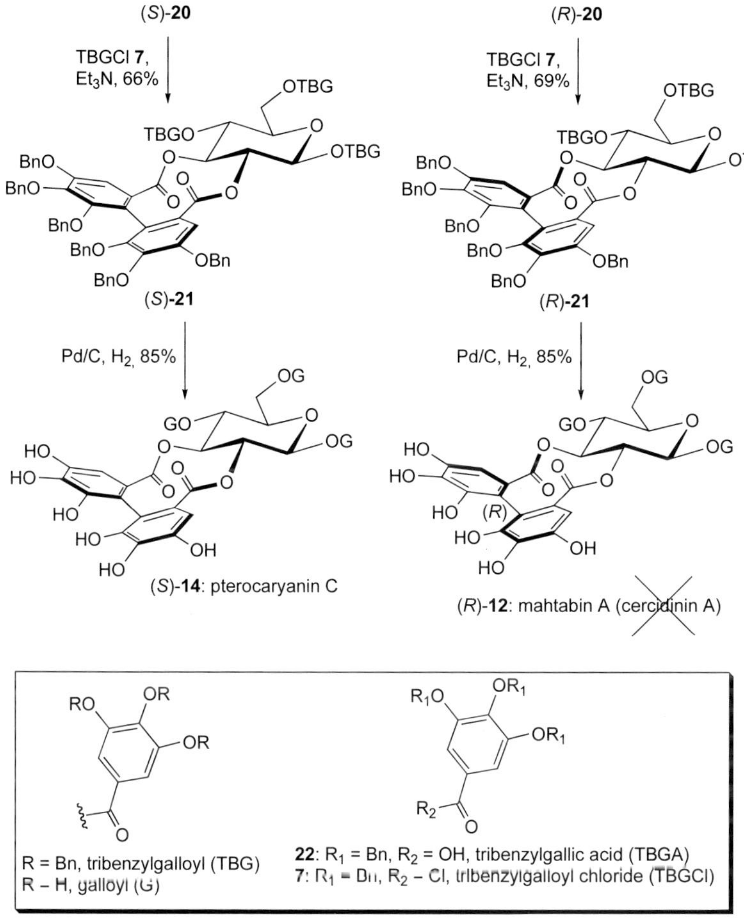

Fig. 5.4 Total synthesis of mahtabin A (**12**) and pariin M (**13**) and their diastereoisomeric counterparts, pterocaryanin C (**14**) and praecoxin B (**15**).

The total synthesis of all four diastereoisomeric ellagitannins **12-15** started from the 4,6-benzylidene acetal-protected D-glucopyranose derivative **2**, which was first acylated with racemic hexabenzyloxydiphenic acid *rac*-**16** to give the two diastereoisomers (*S*)-**17** and (*R*)-**17** (Fig. 5.4). The absolute configuration of the biaryl component in (*S*)-**17** and (*R*)-**17** could be determined after ester hydrolysis by using the Gassman method (Gassman and Schenk, 1977). The so obtained optically pure hexabenzyloxydiphenic acids (*S*)-**16** and

(*R*)-**16** could be assigned by comparison of their specific rotation with values published by Schmidt and Demmler (1952, 1954). A characteristic of both (*S*)-**16** and (*R*)-**16** is their highly stable configuration. Schmidt and Demmler (1952, 1954) also showed that even under the most extreme reaction conditions such as boiling in acetic acid or melting, no racemisation takes place.

In the next reaction step, the benzylidene acetal protecting groups in (*S*)-**17** and (*R*)-**17** were cleaved by acidic hydrolysis to yield the corresponding diols (*S*)-**18** and (*R*)-**18**. The subsequent acylation of these diols with 3,4,5-tri-*O*-benzylgallic acid (TBGA, **22**) via the Steglich method (Neises and Steglich, 1978, Höfle *et al.*, 1978) produced the fully protected intermediates (*S*)-**19** and (*R*)-**19**. Finally, the natural product praecoxin B (**15**) and pariin M (**13**) could be synthesised by removal of all protecting groups from (*S*)-**19** and (*R*)-**19**, respectively, in two additional steps via irradiation to cleave the photolabile *ortho*-nitrobenzyl group, thus furnishing α,β-anomeric mixtures of (*S*)-**20** and (*R*)-**20**, and via hydrogenolysis using palladium on activated charcoal and hydrogen to cleave all twelve benzyl ethers of the aromatic rings (Fig. 5.4). A small variation of this strategy enabled the synthesis of pterocaryanin C (**14**) and mahtabin A (**12**) from the α,β-anomeric mixtures of (*S*)-**20** and (*R*)-**20**, each consisting of 80% of α-anomer and 20% of β-anomer. Esterification of these anomeric mixtures with 3,4,5-tri-*O*-benzylgalloyl chloride (TBGCl, **7**) in the presence of Et_3N gave the corresponding β-anomers (*S*)-**21** and (*R*)-**21**, respectively. No corresponding α anomer could be detected by thin layer chromatography and NMR analysis of the reaction mixtures in each case. The total syntheses of pterocaryanin C (**14**) and mahtabin A (**12**) were then completed by hydrogenolysis of their respective precursors (*S*)-**21** and (*R*)-**21** using palladium on activated charcoal and hydrogen (Fig. 5.4).

The comparison of NMR data and specific rotations of the synthetic ellagitannins **12** and **13** (Khanbabaee and Lötzerich, 1998) with the data sets for cercidinin A and cercidinin B (Nonaka *et al.*, 1989b) clearly showed several differences. This is why we first presumed that the postulated chemical structures for cercidinins A and B were in fact not corresponding to those of the synthetic 1-*O*-galloyl-4,6-di-*O*-galloyl-2,3-*O*-(*R*)-hexahydroxydiphenoyl-β-D-glucopyranose (**12**) and 4,6-di-*O*-

galloyl-2,3-*O*-(*R*)-hexahydroxydiphenoyl-D-glucopyranose (**13**). Several physical and chemical analyses then provided the proofs of this assumption. The regiochemistry of every single reaction step was always unambiguously ensured by the chosen synthesis strategy. We could determine in two different ways that the HHDP unit at the 2,3-positions of the glucopyranose core had the (*R*)-configuration. As mentioned before, the basic hydrolysis of (*R*)-**17** with anhydrous KOH (Gassman and Schenk, 1977) yielded diphenic acid (*R*)-**16** with a negative specific rotation. Furthermore, CD spectroscopy was used to verify the absolute configurations (Berova *et al.*, 1994, Snatzke, 1981 and 1982). It is well known that CD spectroscopy is a suitable empirical method for the determination of the absolute stereochemistry of ellagitannins without the need for chemical degradation reactions (Okuda *et al.*, 1982c/d, Khanbabaee and Lötzerich, 1998).

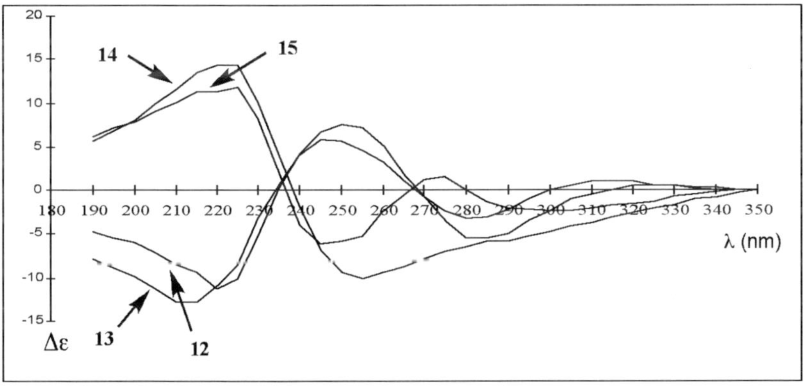

Fig. 5.5 CD Spectra of (*S*)-configured ellagitannins, pterocaryanin C (**14**) and praecoxin B (**15**), and (*R*)-configured 1-*O*-galloyl-4,6-di-*O*-galloyl-2,3-*O*-hexahydroxydiphenoyl-β-D-glucopyranose (mahtabin A, **12**) and 4,6-di-*O*-galloyl-2,3-*O*-hexahydroxydiphenoyl-D-glucopyranose (pariin M, **13**).

The Cotton effects near 220 and 250 nm can be correlated with the absolute configuration of the HHDP unit, using the sign of the Cotton effect between 220 and 235 nm as a diagnostic criterion. A positive value in this region unambiguously indicates the presence of a (*S*)-configured biaryl unit, whereas a negative value indicates the presence of the (*R*)-

configuration. This correlation is independent of the presence or absence of galloyl groups, the conformation of the D-glucopyranose core, and the linkage positions of the biaryl component (Okuda *et al.*, 1982c). As expected, the CD spectra of pterocaryanin C (**14**) and praecoxin B (**15**) showed a strong positive Cotton effect near 220 nm, while compounds **12** and **13** showed a strong negative effect at the same wavelength. Therefore, there can be no doubt concerning the absolute configuration of the biaryl components in all four substances **12-15** (Fig. 5.5).

The striking similarity of the ^{13}C NMR spectra of the compound pairs praecoxin B (**15**) and **13**, and pterocaryanin C (**14**) and **12**, is a further strong indication of the identical regiochemistry of these compound pairs (Khanbabaee and Lötzerich, 1997b, 1998). These facts brought us to the conclusion that the postulated structures for cercidinin A and cercidinin B should be revised. Since 1-*O*-galloyl-4,6-di-*O*-galloyl-2,3-*O*-(*R*)-hexahydroxydiphenoyl-β-D-glucopyranose (**12**) and 4,6-di-*O*-galloyl-2,3-*O*-(*R*)-hexahydroxydiphenoyl-D-glucopyranose (**13**) are novel compounds, we named **12** mahtabin A and **13** pariin M. On the basis of these results, the chemical structure of cercidinin A was re-examined and turned out to correspond to a novel ellagitannin structure possessing an unusual 3,4-(*R*)-HHDP moiety (Tanaka *et al.*, 2001).

5.2.2 4,6-HHDP-containing ellagitannins

5.2.2.1 *Total synthesis of strictinin*

Strictinin (**27**) was isolated from the dried leaves of *Psidium guajava* (Okuda *et al.*, 1982e). The structure of the isolated material was described as 1-*O*-galloyl-4,6-*O*-(*S*)-hexahydroxydiphenoyl-β-D-glucopyranose (Nonaka *et al.*, 1984). Strictinin (**27**) was found to possess anti-topoisomerase II activity with an IC_{100} value of 500 nM (Nonaka *et al.*, 1989c, Okuda *et al.*, 1983b), and also cytotoxic activity against PRMI-7951 melanoma cells with an ED_{50} value of 4.86 μg/ml, as shown in an *in vitro* study involving a large number of human cancer (Kashiwada *et al.*, 1992b).

For the development of a strategy for the synthesis of strictinin (**27**), two major points were to be addressed: the atropisomerism and the β-configuration of the anomeric center of the D-glucopyranose core. The two key building blocks needed for this total synthesis according to method **A** (see Fig. 5.1) were the *ortho*-nitrobenzylated 2,3-di-*O*-benzyl-β-D-glucopyranose **23**, available in four steps (Feldman and Sambandam, 1995, Zehavi, 1988, Zehavi and Patchornic, 1972, Zehavi *et al.*, 1972, Igarashi, 1977, Königs and Knorr, 1901, Rajasekharan Pillai, 1980, Gigg *et al.*, 1983, Kanai *et al.*, 1987, Czernecki *et al.*, 1976), and the racemic hexabenzyloxydiphenic acid *rac*-**16** (Schmidt *et al.*, 1954).

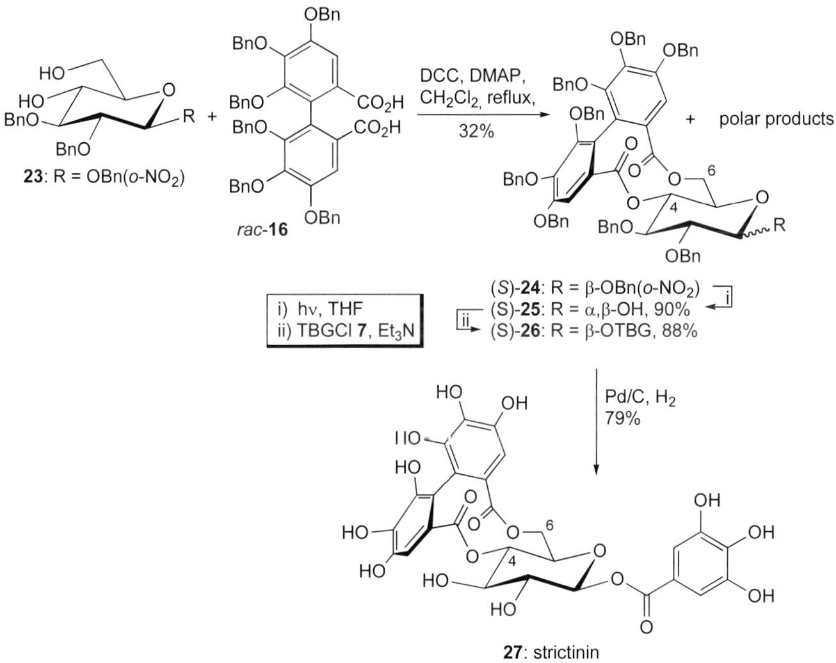

Fig. 5.6 Total synthesis of strictinin (**27**).

The total synthesis of strictinin (**27**) commenced with the double esterification of *rac*-**16** using diol **23** under Steglich conditions (Neises and Steglich, 1978, Höfle *et al.*, 1978) to furnish the (*S*)-configured esterification product **24** in a completely diastereoselective manner (Fig. 5.6). The (*R*)-hexabenzyloxydiphenic acid *(R)*-**16** present in the racemic starting material also reacted with the sugar derivative **23**, but in an open-chain manner to give polar products, which were easily separated by silica gel chromatography. The subsequent removal of the photolabile *ortho*-nitrobenzyl ether at the anomeric center of **24** furnished the O-1 deprotected D-glucopyranose derivative **25** as an anomer mixture (*i.e.*, α/β, 85:15). Acylation of **25** with protected galloyl chloride **7** yielded the required β-anomer **26**. Finally, the natural product strictinin (**27**) was obtained by removal of all protecting groups of **26** by hydrogenolysis under standard conditions (Fig. 5.6, Khanbabaee *et al.*, 1997).

5.2.2.2 *Total syntheses of gemin D and hippomanin A*

The isolation of gemin D (**37**) from the leaves of *Geum japonicum* (*Rosaceae*) and from the flowers of *Camellia japonica* (*Theaceae*) was first reported in 1982 (Yoshida *et al.*, 1982). In 1985, the Okuda group reported the complete structural determination of this ellagitannin (Yoshida *et al.*, 1985). The isolation of hippomanin A (**38**) from the aerial parts of *Hippomane mancinella* L. (N. O. *Euphorbiaceae*) had already been reported in 1974 (Rao, 1974), but again, the structural elucidation of this regioisomeric analogue of gemin D (**37**) was reported only three years later (Rao, 1977). Both gemin D (**37**) and hippomanin A (**38**) possess a (*S*)-configured HHDP unit linked to the 4,6-positions of D-glucopyranose. The only difference between gemin D (**37**) and hippomanin A (**38**) is the presence of a galloyl group at either the 3- or the 2-position of the glucosyl core of these ellagitannins. Gemin D (**37**) exhibits both anticancer (Miyamoto *et al.*, 1987) and anti-HIV (Vlietinck *et al.*, 1998) activities. The extracts of *Geum japonicum* and *Camellia japonica* have long been used in Japan and China as diuretics, astringent and haemostatic agents (Yoshida *et al.*, 1985).

Apart from the previously described protecting group strategies and diastereoselective double acylation of the 4,6-positions of the D-glucopyranose core, the selective monofunctionalisation of the 2-OH or 3-OH group of the sugar unit is decisive for a successful synthesis of both natural products. It is known that the C-2 hydroxyl function of a D-glucopyranose protected at the anomeric position is the most reactive secondary hydroxyl groups. This effect is often more pronounced in α-than in β-anomers (Garegg *et al.*, 1976, Iversen *et al.*, 1977). This phenomenon was to be exploited in our synthetic strategy.

The key starting material for the synthesis of both gemin D (**37**) and hippomanin A (**38**) is 1-*O*-benzyl-4,6-*O*-benzylidene-β-D-glucopyranose (**28**), which can be prepared by acetalisation of the commercially available 1-*O*-benzyl-β-D-glucopyranose using benzaldehyde and anhydrous $ZnCl_2$ (Petit and Sinay, 1978, Sen and Banerji, 1989).

Various alternative strategies exist for the selective monoalkylation of the hydroxyl group at C-2 of D-glucopyranose derivatives. For example, limiting the quantity of base or alkylating agent, as well as shortening the reaction time, can give the required selectivity (Ogawa *et al.*, 1978). Another method has been developed for the selective monobenzylation of the 2-OH group of 2,3-diol derivatives of D-glucopyranose protected at the 1,4,6-positions (Garegg *et al.*, 1976) by using tetra-*n*-butylammonium hydrogenosulfate (n-Bu_4NHSO_4) as a phase transfer catalyst. On the basis of these literature precedents, we performed the reaction of diol **28** with benzyl bromide (BnBr) in the presence of tetra-*n*-butylammonium iodide (n-Bu_4NI) as a phase transfer catalyst and obtained an inseparable mixture of both regioisomers **29** and **30** in 74% overall yield (Fig. 5.7).

The ratio of 2-*O*-benzylated product **29** to its 3-*O*-benzyl ether variant **30** was 3:1, as determined by NMR analysis. Due to the low selectivity of the monobenzylation reaction, the 1-*O*-benzylated 4,6-*O*-benzylidene-β-D-glucopyranose **28** thus provided access to both natural products **37** and **38**. The regioisomeric mixture **29**/**30** was then acylated with 3,4,5-tri-*O*-benzylgallic acid (**22**), under standard Steglich conditions, to afford the glucopyranose derivatives **31** and **32**, which were easily separated by silica gel chromatography. Both natural ellagitannins gemin D (**37**) and hippomanin A (**38**) were then synthesized in parallel. First, the benzylidene acetals were cleaved from **31** and **32** by acidic hydrolysis using 2 N HCl to yield the corresponding 4,6-diols **33** and **35**. This was followed by acylation with racemic hexabenzyloxydiphenic acid *rac*-**16**. These acylation reactions of the 4,6-positions of sugars **33** and **35** were both fully diastereoselective, as observed in the strictinin total synthesis (Khanbabaee *et al.*, 1997) and led to the perbenzylated precursors **34** and **36** in 68% and 66% yield, respectively. Gemin D (**37**, Yoshida *et al.*, 1982, 1985, Hatano *et al.*, 1988, Lee *et al.*, 1989) and hippomanin A (**38**, Rao, 1974, 1977, Hatano *et al.*, 1988, Lee *et al.*, 1989) were finally obtained by complete debenzylation under standard hydrogenolysis conditions (Fig. 5.7) (Khanbabaee *et al.*, 1999). The spectroscopic data of synthetic **37** agreed in all respects with those of natural **37** (Yoshida *et al.*, 1985). The complete NMR data of the natural product **38** were published in 1999 for the first time (Khanbabaee *et al.*, 1999).

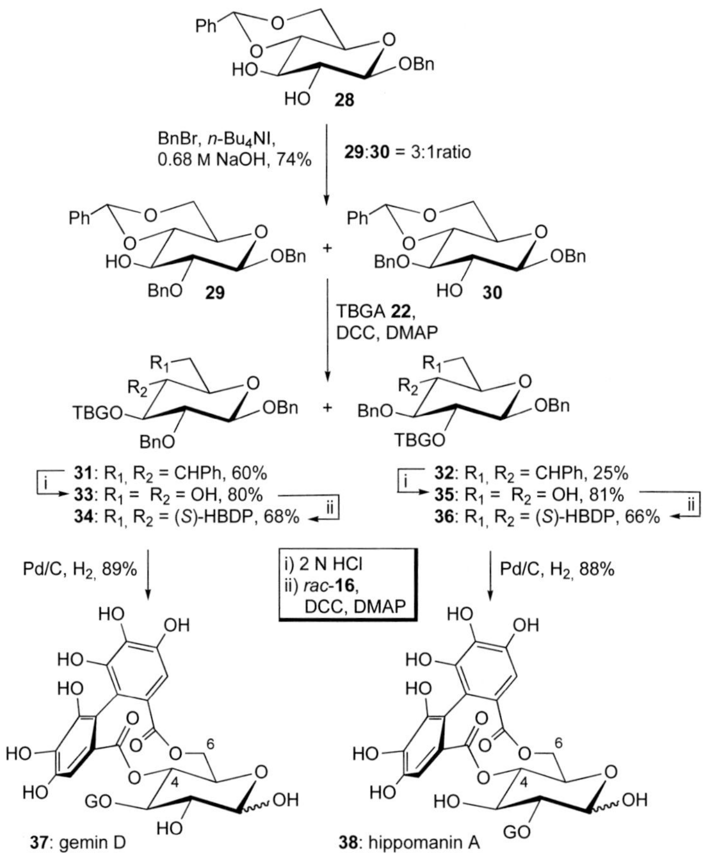

Fig. 5.7 Total synthesis of gemin D (**37**) and hippomannin A (**38**).

5.2.2.3 *Construction of a 4,6-bridged ellagitannin scaffold*

A copper-catalysed reductive Ullmann coupling of two methyl ether-protected iodinated galloyl moieties attached to the appropriate positions of a D-glucopyranosyl unit (see **39** in Fig. 5.8) produced the coupled (*S*)-configured product **40**, together with the non-coupled and reduced product **41**, in a 1:1 ratio. As in the case of the Feldman coupling, which also corresponds to an application of method **B** (see Fig. 5.1) to construct ellagitannin HHDP units, this reductive coupling reaction was also (*S*)-atropodiastereoselective, as a consequence of conformational preferences in D-glucopyranose (Dai and Martin, 1998).

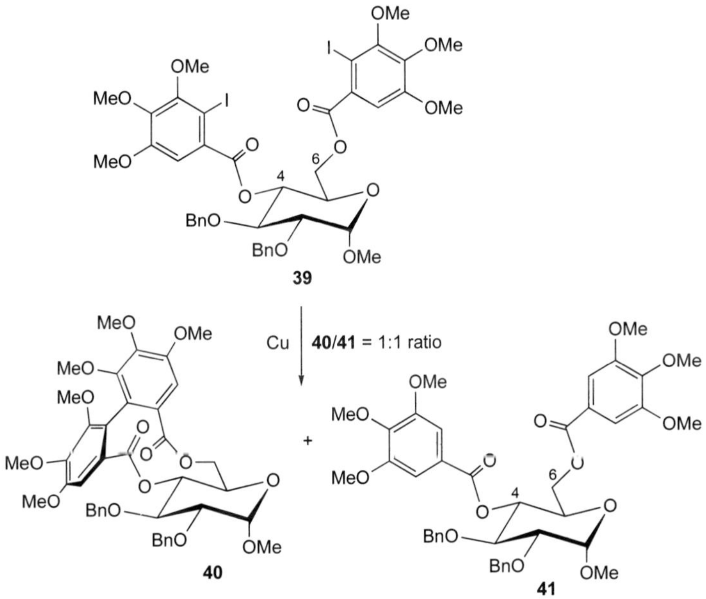

Fig. 5.8 Construction of the 4,6-HHDP-containing ellagitannin framework.

5.2.2.4 *Synthesis of O-perbenzyllagerstannin C*

The isolation of lagerstannin C (**47**) from the fruit and leaves of *Lagerstroemia speciosa* (L.) Pers. (L. *flosreginae* RETZ.) (*Lythraceae*) was first reported in 1992 along with that of some other lagerstannins featuring a D-gluconic acid unit as a special structural characteristic (Tanaka *et al.*, 1992b). It is well known that the tannin-rich extracts of fruit and leaves of *Lagerstroemia speciosa* have blood sugar lowering properties and are used to treat diabetes mellitus.

Apart from the regiochemical control issues fully addressed in the previously described total syntheses, a special aspect of the synthesis of lagerstannin C (**47**) was the construction of the D-gluconic acid unit by oxidative opening of the D-glucopyranose ring (Fig. 5.9). We thus planned to oxidise substrate **25** (see also Fig. 5.6) into the δ-lactone **42**, followed by opening of the sugar ring under weakly basic conditions (Kida *et al.*, 1990) in the aim of preparing the D-gluconic acid derivative **43**; strongly basic conditions might cleave the remaining ester groups in the molecule. Dimethylsulfoxide (DMSO) is an oxidant that has become very popular in carbohydrate chemistry, since its development in 1965 (Pfitzner and Moffatt, 1965a/b). In the presence of electrophilic activating agents such as dicyclohexylcarbodiimide (DCC, Albright and Goldman, 1965a/b, Pfitzner and Moffatt, 1965a/b), acetic anhydride (Ac_2O, Albright and Goldman, 1965a/b, 1967, Sowa and Thomas, 1966, Kuzuhara and Fletcher, 1967), phosphorus pentoxide (Onodera *et al.*, 1965) or oxalyl chloride (Mancuso and Swern, 1981), and a proton donor, primary and secondary hydroxyl functions are oxidized into a carbonyl group (Lehman, 1996). Other classical oxidants for primary and secondary alcohols are Cr(VI)-based reagents [CrO_3.2pyridine, pyridinium chlorochromate (PCC)] (Piancatelli and d'Auria, 1982, Hollenberg *et al.*, 1978 Bissember and Wightman, 1980). However, oxidation reactions of isolated secondary hydroxyl functions in carbohydrates are often relatively slow, with low conversion rates and modest yields (Czernecki *et al.*, 1985). Pyridinium dichromate (PDC) then became the reagent of choice (Herscovici and Antonakis, 1980, Herscovici *et al.*, 1982; Corey and Schmidt, 1979). Alcohols are converted nearly quantitatively to the corresponding carbonyl functions in an extremely fast and efficient reaction with PDC in the presence of freshly activated molecular sieve powder (3Å) and catalytic quantities of absolute acetic acid in dry dichloromethane. The addition of freshly activated molecular sieve powder (3Å) leads to the immediate adsorption of the chromate ester formed during the oxidation. It is thought that the participating bonds of the intermediate product are oriented in such a way by the adsorption that the C–H bond breaking at the alcohol carbon center is favoured, with a resultant drastic acceleration of the reaction rate. In fact, we found that oxidation of the free secondary hydroxyl group at the anomeric center of the D-glucopyranose derivative **25** was complete after 45 min, yielding smoothly the desired D-glucono-δ-lactone **42**. As the chosen oxidation method occurs under very mild conditions, and given the fact that reactions are complete in a very short time, it is possible to oxidise very sensitive substrates.

In contrast to what is observed for monosaccharides of similar configuration, the most stable cyclic form of D-gluconolactones is the five-membered ring (*i.e.*, γ-lactone or 1,4-lactone). The γ-lactone is indeed well suited to accommodate the structural requirements for its formation with respect to bond lengths and angles. In particular, the flattening of the structure caused by the sp^2-hybridation of the lactone carbon atom does not lead to conformational twisting, as it is the case for six-membered δ-lactones (or 1,5-lactones). In the latter case, this flattening causes the molecule to adopt a non-ideal half-chair or boat conformation (Lehmann, 1996). As a result, the six-membered δ-lactones have higher ring strain energies and hydrolyse faster than the corresponding five-membered γ-lactones (Brown *et al.*, 1989). In our case, the ring strain of the D-glucono-δ-lactone **42** is further increased by the presence of the bivalent and relatively rigid hexabenzyloxydiphenoyl unit, which opposes the twisting of the chair conformation. Due to the substitution pattern, an otherwise possible conversion to the more stable 1,4-lactone is in this case impossible. These considerations led us to believe that lactone **42** would be easily opened under mild conditions to D-gluconic acid derivative **43**. In fact, when the anomeric position of **25** was oxidized to give lactone **42**, we even observed its hydrolysis into **43** during silica gel chromatography (Fig. 5.9).

Subsequent attempts to selectively acylate the free hydroxyl function at C-5 of **43** were unsuccessful in spite of several variations of the reaction conditions. Neither the use of benzylated gallic acid **22** in the presence of DCC and DMAP, nor the use of galloyl chloride **7** in the presence of DMAP, led to the formation of the desired product. The trimethylsilyltriflate (TMSOTf)-catalysed acylation conditions reported by Procopiou *et al.* (1996, 1998) using benzylated gallic acid anhydride (Pacheco and Grouiller, 1965, Schmidt and Klinger, 1957) were also to no avail. However, once the carboxylic acid **43** was protected as its methyl ester **44** using a freshly prepared diazomethane solution, acylation with 3,4,5-tri-*O*-benzylgallic acid **22** under standard Steglich conditions (Neises and Steglich, 1978, Höfle *et al.*, 1978) then yielded the desired product **45** (Fig. 5.9).

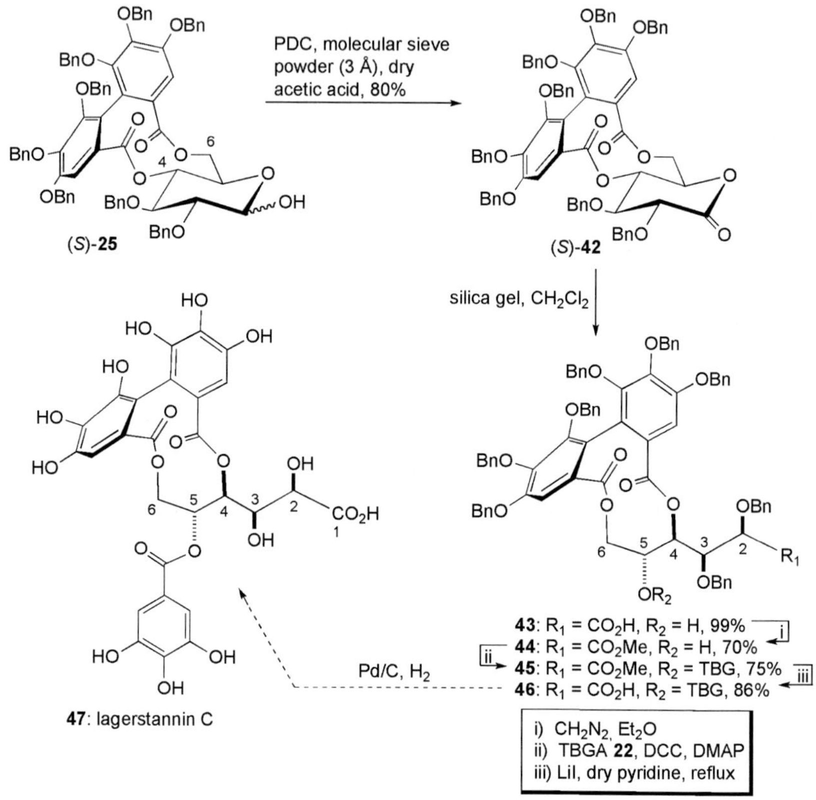

Fig. 5.9 Synthesis of undeca-*O*-benzyllagerstannin C (**46**).

It is known that methyl esters can be cleaved selectively under mild conditions using strong nucleophiles in dipolar, aprotic solvents (*i.e.*, carboxylate exchange by S_N2 dealkylation). Other carboxylic esters of more sterically demanding alcohols (even ethanol) that may be present in the molecule are not at all, or only very slowly, attacked under such conditions (Meinwald and Putzig, 1970, Dean, 1965, Borch *et al.*, 1972,

McMurry, 1976, Haslam, 1980). Cleavage of methyl ester **45** with LiI in absolute pyridine produced a well-defined, polar product **46**. Spraying the thin layer chromatogram with a bromocresol green solution immediately showed a yellow colour, which is indicative for the free carboxylic function in **46** (Khanbabaee and Lötzerich, 1999). The removal of the protective groups in the final step of this synthetic sequence with Pd/C-H_2 would lead to lagerstannin C (**47**) (Fig. 5.9).

5.2.2.5 *Tellimagrandin I and tellimagrandin II*

For the construction of the tellimagrandin framework, both the groups of Meyers and Lipshutz used the double esterification of enantiopure (*S*)-hexamethoxydiphenic acid (*S*)-**50** with suitable diol derivatives of D-glucopyranose to form the corresponding macrocyclic products (*i.e.*, method **A**, see Fig. 5.1). The two syntheses only differ in the way the enantiopure (*S*)-hexamethoxydiphenic acid (*S*)-**50** was constructed. Lipshutz used a cuprate-catalyzed biaryl coupling to form (*S*)-**50**, whereas Meyers used an Ullmann coupling of two identical aromatics containing oxazolines as chiral auxiliaries. Only the syntheses of the fully methyl ether-protected tellimagrandins **53** and **61** were described. Cleavage of the methyl ethers that would have completed the total syntheses to the respective natural products tellimagrandin I (**67**) and tellimagrandin II (**73**) was not executed (Lipshutz *et al.*, 1994b, Nelson and Meyers, 1994). Since methyl ethers cannot be easily removed (Quideau and Feldman 1996, Khanbabaee and Lötzerich, 1997a), benzyl ether-protected gallic acid derivatives could have constituted more appropriate coupling units, as benzyl ethers can be easily removed by hydrogenolysis at the final step of the synthesis. However, coupling of benzyl ether-protected gallic acid derivatives into the corresponding hexabenzyloxydiphenoyl unit has not yet been described; the possibility of such a coupling may be impaired by the higher steric demand of the benzyl groups relatively to that of smaller alkyl groups like methyl groups.

5.2.2.5.1 Synthesis of *O*-permethyltellimagrandin I

Nelson and Meyers (1994) were able to synthesise enantiopure (*S*)-hexamethoxydiphenic acid (*S*)-**50** using the asymmetric Ullmann coupling of two chiral (*S*)-bromooxazolines **48**. The diastereomerically pure biaryl compound (*S*,*S*,*S*)-**49** was obtained in 58% yield, and was then converted in three steps *via* an acid-catalysed cleavage of the oxazoline residues into (*S*)-**50** (Fig. 5.10).

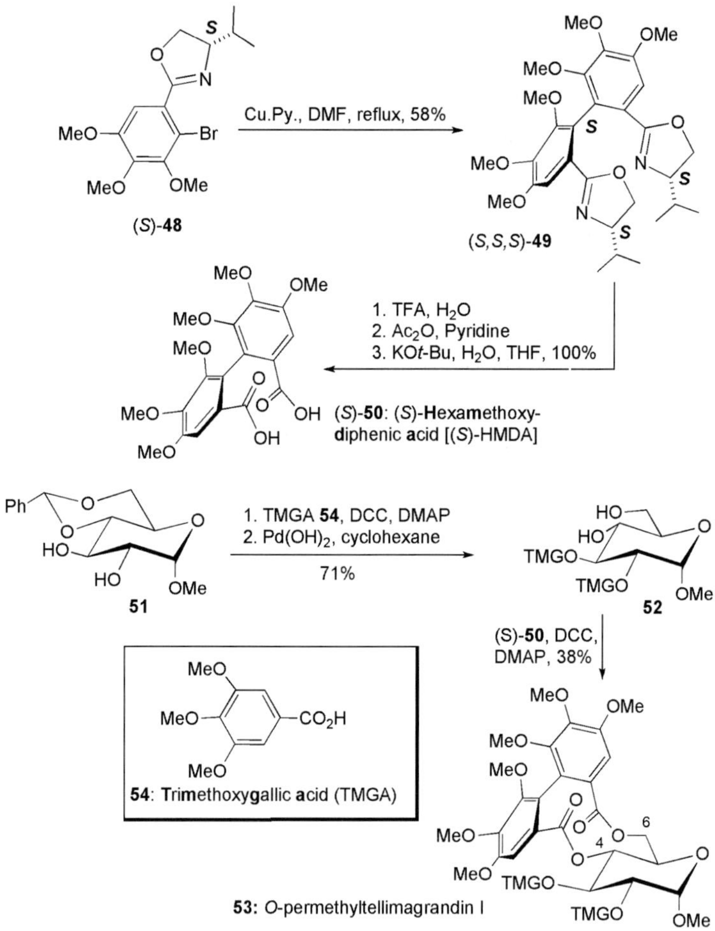

Fig. 5.10 Synthesis of *O*-permethyltellimagrandin I (**53**).

The second building block **52** was synthesised starting from commercially available benzylidene acetal **51** by acylation with 3,4,5-trimethylgallic acid **54**, followed by the removal of the benzylidene acetal with $Pd(OH)_2$ in cyclohexane. Esterification of (*S*)-**50** with the resulting D-glucopyranosyl derivative, methyl 2,3-di-*O*-(3,4,5-tri-*O*-methylgalloyl)-α-D-glucopyranose (**52**) yielded the permethylated tellimagrandin I (**53**, Fig. 5.10).

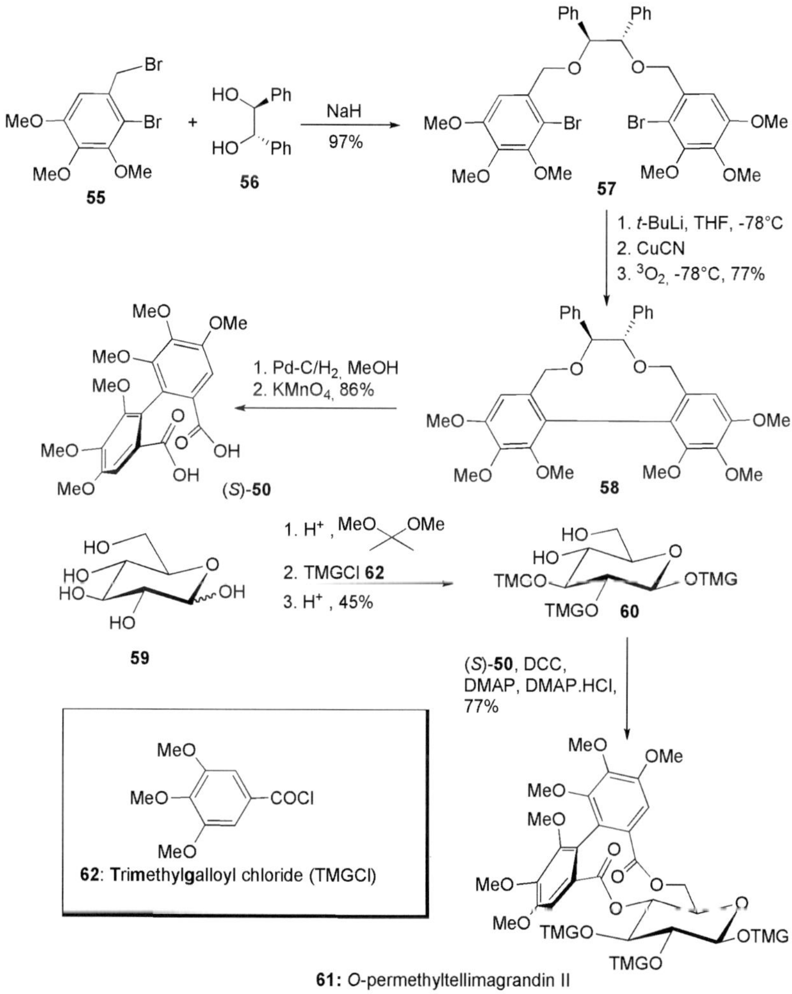

Fig. 5.11 Synthesis of *O*-permethyltellimagrandin II (**61**).

5.2.2.5.2 Synthesis of *O*-permethyltellimagrandin II

Also in 1994, Lipshutz *et al.* published the synthesis of *O*-permethyltellimagrandin II (**61**), the key step being this time a cyanocuprate-catalyzed diastereoselective biaryl coupling. Nucleophilic substitution of the benzylic bromide function of the starting dibromide **55** with enantiopure (*S*)-1,2-diphenylethane-1,2-diol (**56**) furnished the diether **57**. Reaction of **57** with *tert*-BuLi and CuCN formed a cyanocuprate intermediate, which was converted to the cyclic product **58** in 77% yield. Hydrogenolytic cleavage of the bisbenzylic ether bridge, followed by oxidation of the resulting benzylic alcohols, yielded the enantiomerically pure (*S*)-hexamethoxydiphenic acid (*S*)-**50** (Fig. 5.11).

The second building block used was the 4,6-diol derivative of D-glucopyranose **60**, which was prepared in three steps from D-glucopyranose **59**, *i.e.*, selective protection of its 4,6-hydroxyl groups, acylation with TMGCl **62** (Keck and Boden, 1985) and deprotection of the dimethoxyketal unit. The final step entails the esterification of (*S*)-**50** with **60** to furnish the permethylated tellimagrandin II (**61**, Fig. 5.11).

5.2.2.5.3 Total synthesis of tellimagrandin I

With the successful accomplishment of the total synthesis of tellimagrandin I (**67**, see Fig. 5.12), the Feldman group was the first to report a synthesis of a representative member of the ellagitannin class of natural products (Feldman *et al.*, 1994). As alluded to above (see method **B** in Fig. 5.1), their strategy for the stereoselective construction of an (*S*)-HHDP unit was based on a biomimetic oxidative biaryl coupling of neighbouring galloyl groups using lead tetraacetate (Quideau and Feldman, 1996).

In tellimagrandin I (**67**), the biaryl unit is linked to the 4,6-positions of its D-glucopyranosyl core; the galloyl residues at the 2- and 3-positions are not coupled to each other. In the first synthesis route, this regioselectivity was controlled as follows: starting from the D-glucopyranose derivative **28**, the anomeric center of which being protected as a benzyl ether and its O-4 and O-6 centers as a benzylidene acetal unit, the free hydroxyl functions at C-2 and C-3 were acylated with the tribenzylated gallic acid **22** to form the diester **63**. Hydrolysis of the benzylidene acetal unit gave **64**, the C-4 and C-6 hydroxyl functions of which were then acylated with the 3,4-*O*-diphenylketal-protected gallic acid derivative **8** (see Fig. 5.2). Desilylation using *n*-Bu_4NF afforded compound **65**, which was then submitted to the key $Pb(OAc)_4$-mediated diastereoselective coupling reaction to furnish the desired coupled product **66** as a mixture of diphenylketal regioisomers. Hydrogenolytic cleavage of all protecting groups finally furnished tellimagrandin I (**67**, see Fig. 5.12).

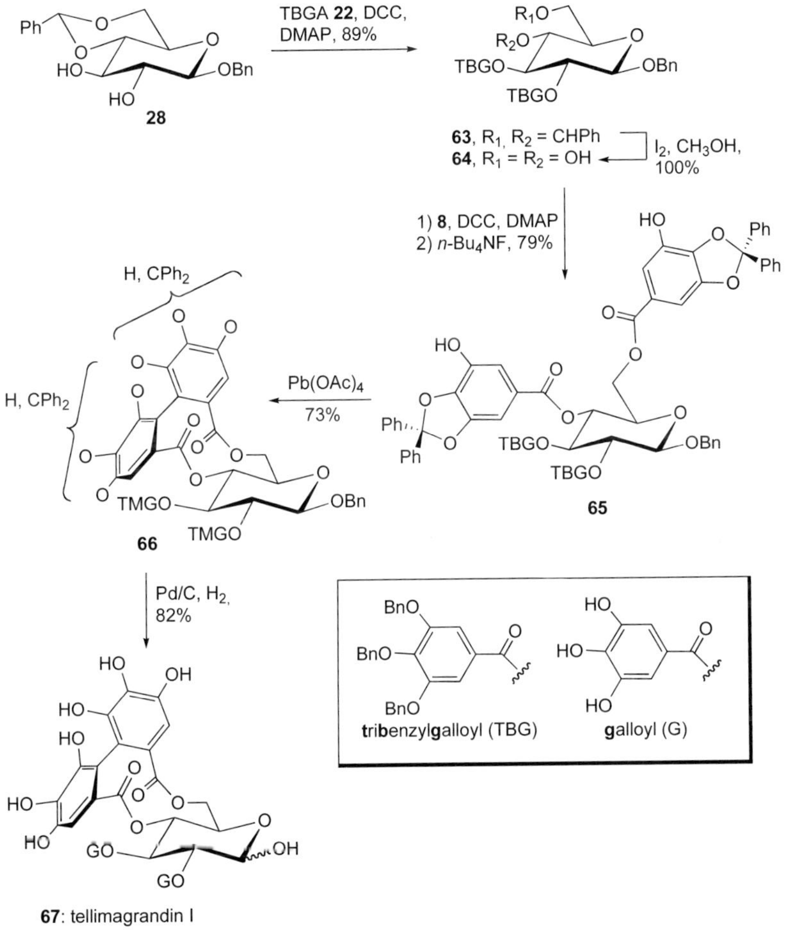

Fig. 5.12 First total synthesis of tellimagrandin I (**67**).

The second synthesis provided a much shorter route to the target molecule (**67**, Feldman *et al.*, 1994). In the first step, the O-1 benzylated D-glucopyranosyl derivative **68** is fully acylated with the acid **8** (see Fig. 5.2) and desilylated using n-Bu_4NF. The resulting D-glucopyranosyl derivative **69** is then subjected to the $Pb(OAc)_4$-mediated biaryl coupling reaction. An amazing regioselectivity was observed during this coupling reaction that took place only between the galloyl groups linked at the O-4 and O-6 positions. Thus, apparently, the galloyl group at the O-2 and O-3 positions could not adequately orient themselves relatively to each other to enable coupling or could not express a sufficient reactivity towards $Pb(OAc)_4$ to be oxidatively activated. Again, tellimagrandin I (**67**) was finally released by hydrogenolysis (Fig. 5.13).

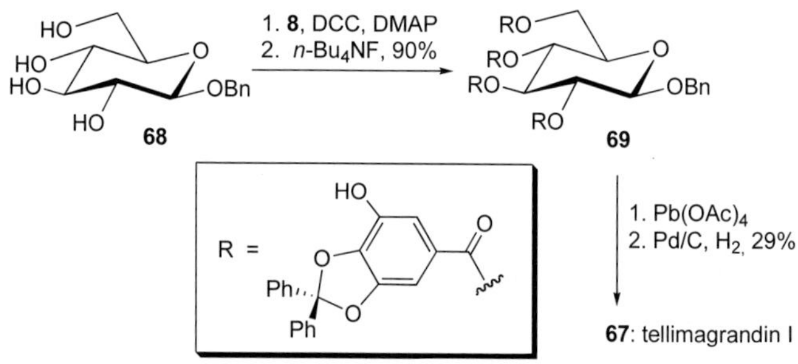

Fig. 5.13 Second total synthesis of tellimagrandin I (**67**).

5.2.2.5.4 Total synthesis of tellimagrandin II

The natural product tellimagrandin II (**73**, Feldman and Sahasrabudhe, 1999) differs from tellimagrandin I (**67**) by the presence of one additional galloyl unit at the anomeric center of the glucopyranosyl core. Thus, the synthetic sequence for tellimagrandin II (**73**) is mostly identical to that for the synthesis of tellimagrandin I (**67**). The only difference is that the anomeric center of the glucopyranosyl intermediate **70** is protected with the photolabile *ortho*-nitrobenzyl group for the synthesis of tellimagrandin II (**73**) instead of a benzyl group for the synthesis of tellimagrandin I (**67**). This enables the selective deprotection of the anomeric center, then allowing the acylation of that center with 3,4,5-tribenzylgalloyl chloride (TBGCl, **7**). Cleavage of the *ortho*-nitrobenzyl protecting group at the anomeric center of the D-glucopyranosyl unit of **70** was performed by UV irradiation, to give **71**, and was followed by a β-selective acylation at the thus released anomeric center. Reaction of **71** with TBGCl **7** in the presence of Et_3N produced the tellimagrandin II precursor **72**. Once again, hydrogenolytic debenzylation was applied at the final step of the synthesis to give rise to the natural product tellimagrandin II (**73**, Fig. 5.14).

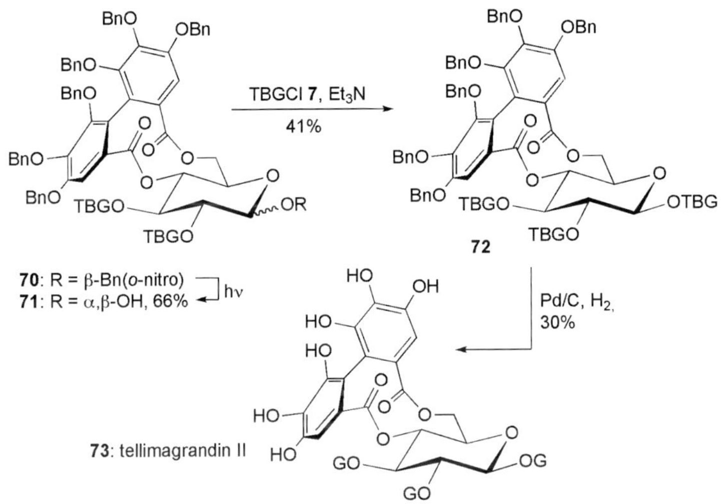

Fig. 5.14 Total synthesis of tellimagrandin II (**73**).

5.2.2.6 *Total synthesis of coriariin A, a dimeric ellagitannin*

This was the first synthesis of a dimeric ellagitannin, once again achieved by the Feldman group (Feldman *et al.*, 2000, Feldman and Lawlor, 2000). For this synthesis of coriariin A (**79**), the diaryl ether building block **75** was first esterified twice with the glucopyranosyl derivative **74**. After selective deprotection of the 2- and 3-positions of the glucopyranosyl units, the four liberated hydroxyl groups were acylated with the tribenzylated gallic acid **22** to form the dimeric intermediate **76**. The next step was the cleavage of the benzylidene acetal units at the 4,6-positions of the glucopyranosyl units. The four newly liberated hydroxyls were then acylated with the diphenylketal-protected gallic acid derivative **8**.

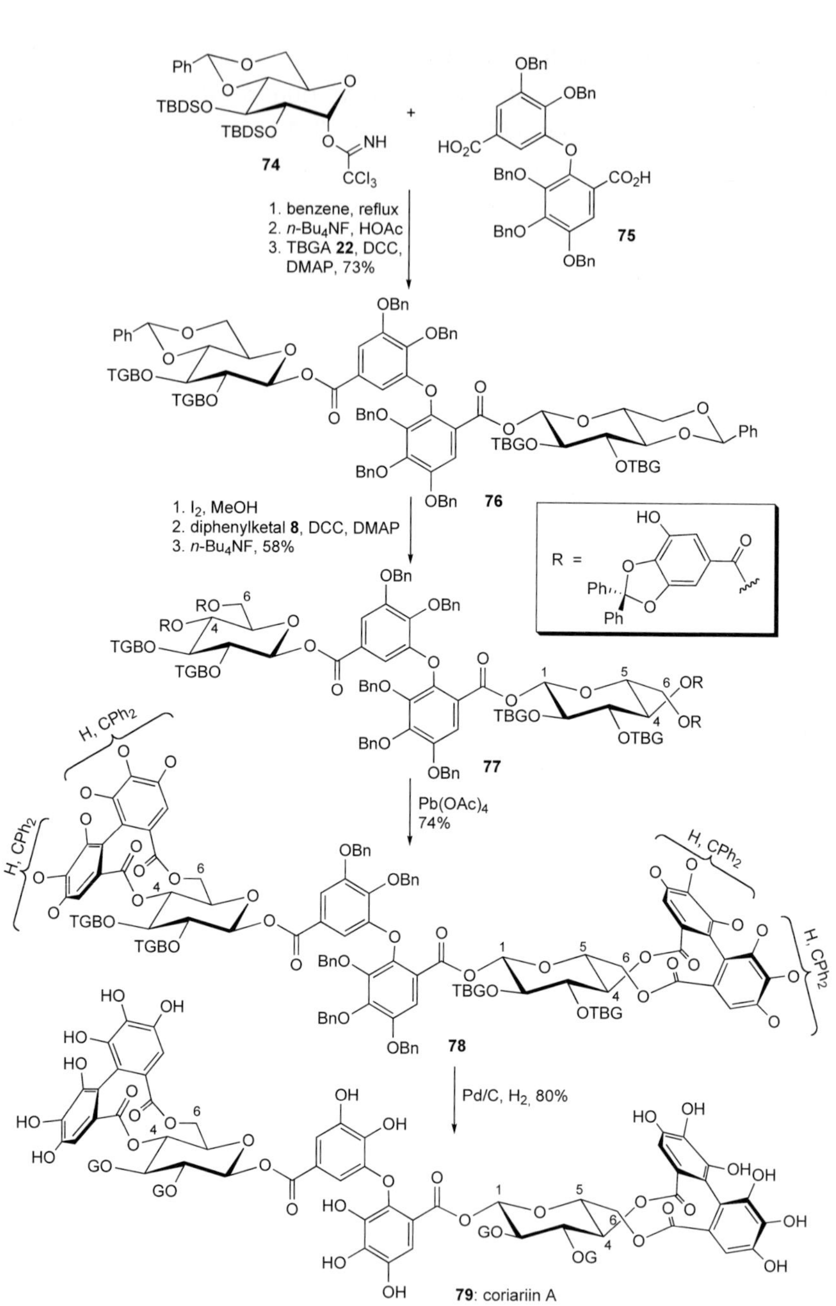

Fig. 5.15 Total synthesis of coriariin A (**79**).

The different protecting groups on the galloyl units of the resulting compound (not shown) permitted a stepwise deprotection sequence. First, the *t*-butyldimethylsilyl ether groups were cleaved using *n*-Bu_4NF to furnish **77**, in which the galloyl moieties at the 4- and 6-positions of both glucopyranosyl units were ready to be engaged in a double oxidative coupling step. This was again performed using $Pb(OAc)_4$, and the resulting regioisomeric mixture **78** was then fully debenzylated to furnish the dimer coriariin A (**79**, Fig. 5.15).

5.2.3 2,3,4,6-HHDP-containing ellagitannins

5.2.3.1 *Synthesis of O-permethyl-α-pedunculagin*

The chemical structure of pedunculagin (**87**) consists of two (*S*)-configured HHDP moieties that are located at the 2,3- and 4,6-positions of the D-glucopyranose core (Okuda *et al.*, 1983b). Pedunculagin (**87**) has been shown to inhibit the promising anticancer target enzyme DNA topoisomerase II *in vivo*, with an IC_{100} of 500 nM (Kashiwada *et al.*, 1993).

Itoh's strategy for the construction of ellagitannins is based on the concept of kinetic racemate resolution by esterification of racemic hexamethoxydiphenic acid dichloride *rac*-**80** with suitable D-glucopyranosyl derivatives (*i.e.*, see Fig. 5.1, method **A**). There is no need for atropodiastereoselective formation of the chiral biaryl axis, since the chiral axis is already present in the biaryl unit (Itoh and Chika, 1995, Itoh *et al.*, 1996). Their synthesis of trideca-*O*-methyl-α-pedunculagin (**84**, see Fig. 5.16) relied on the application of this kinetic resolution tactic (Itoh *et al.*, 1996). They also investigated the stereoselectivity outcome of the esterification reactions under different conditions, and found out that the diastereoisomeric ratio strongly varies depending upon the base used, for example, sodium hydride or dimethylaminopyridine (DMAP). The reaction of commercially available 1-*O*-methyl-4,6-*O*-benzylidene-α-D-glucopyranose (**51**) with *rac*-**80** in the presence of DMAP resulted in the formation of the two diastereomers **81** and **82**, which were then separated by silica gel. The benzylidene acetal group of the required (*S*)-diastereoisomer **82** was then cleaved under acid-catalyzed conditions and the resulting diol **83** was acylated with enantiopure (*S*)-hexamethoxydiphenoyl dichloride (*S*)-**80** to furnish trideca-*O*-methyl-α-pedunculagin (**84**, Fig. 5.16).

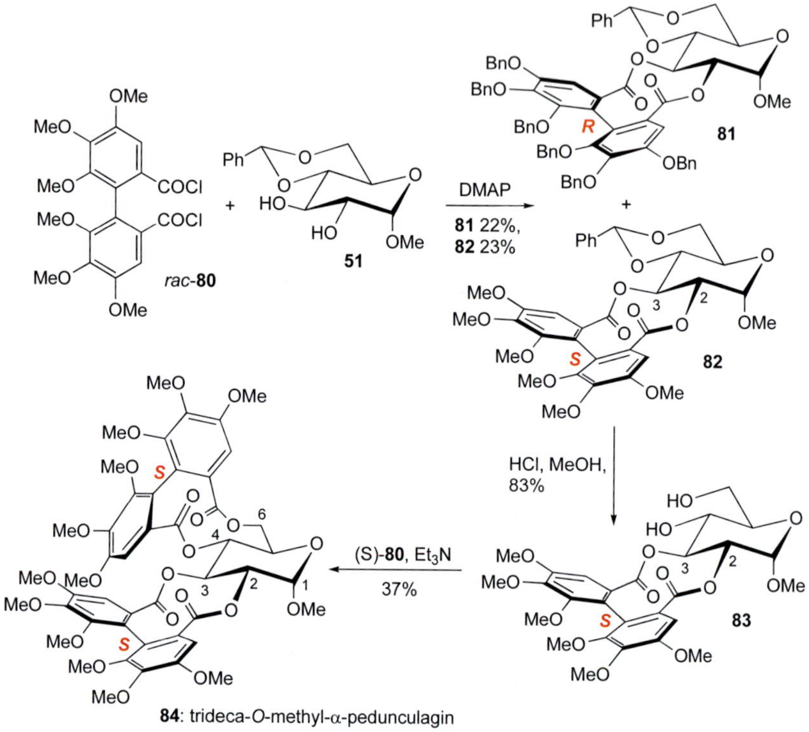

Fig. 5.16 Synthesis of trideca-*O*-methyl-α-pedunculagin (**84**).

5.2.3.2 *Total synthesis of pedunculagin via the biaryl coupling strategy*

Two different strategies have been developed for the total synthesis of pedunculagin (**87**). In the first strategy, the biaryl part of pedunculagin was constructed by a diastereoselective oxidative coupling of phenolic galloyl moieties attached to the D-glucopyranose core (Fig. 5.17, and see Fig. 5.1, method **B**) (Feldman and Smith, 1996). The second strategy relied on the double esterification approach (see Fig. 5.1, method **A**), through which an appropriately protected diphenic acid derivative is esterified with the sugar core (Fig. 5.18). For the total synthesis of pedunculagin (**87**) via method **B**, an important issue was the question of whether an existing chiral biaryl unit in the molecule would influence the stereochemical outcome of an oxidative coupling of neighbouring galloyl units. In fact, it was found that the sequence of the galloyl-galloyl coupling is decisive. If the 4,6-positioned galloyl units are coupled first, no subsequent coupling between the 2,3-positioned galloyl units is possible. Therefore, the C-2 and C-3 galloyl units must be coupled first and then the C-4 and C-6 galloyl units. Hence, the Feldman synthesis of **87** commenced with the acylation of the 2,3-positioned hydroxyl groups of the D-glucopyranosyl derivative **28** with the 3,4-*O*-diphenylketal gallic acid derivative **8** (see Fig. 5.2 and Fig. 5.17).

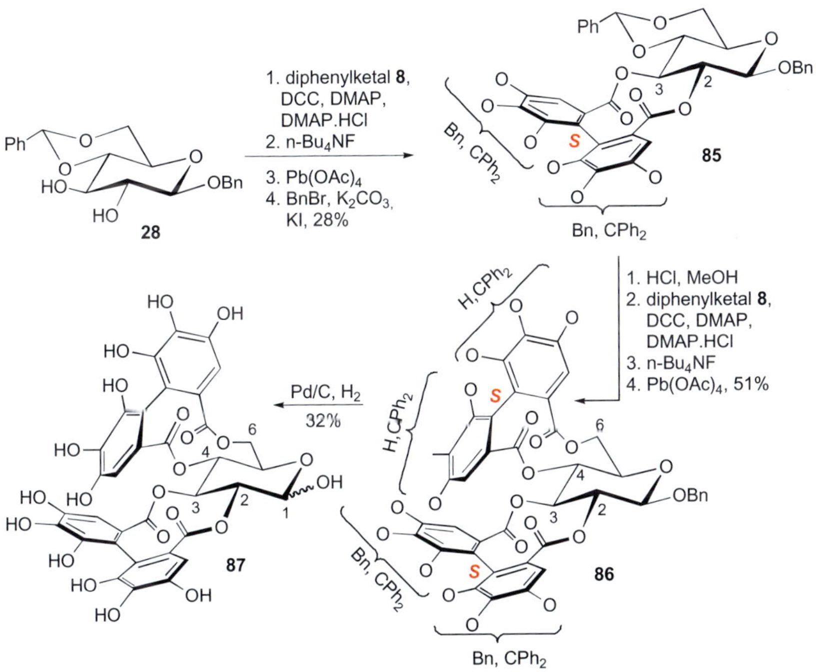

Fig. 5.17 Total synthesis of pedunculagin (**87**) via the biaryl coupling strategy.

Selective desilylation was followed by $Pb(OAc)_4$-mediated coupling of the two galloyl units at C-2 and C-3 of the D-glucopyranosyl core to give the 2,3-coupled product **85**. After removal of the benzylidene protecting group from the O-4 and O-6 centers, the released hydroxyl groups were then again acylated with the gallic acid derivative **8**. Subsequent desilylation was again followed by oxidative coupling of the two galloyl units at C-4 and C-6 to give the second (*S*)-configured biaryl unit, thus leading to the formation of the 2,3,4,6-coupled product **86**. Pedunculagin (**87**) is finally obtained in a 2.9% overall yield after hydrogenolytic cleavage of all benzyl protecting groups of the precursor **86** (Fig. 5.17, Feldman and Smith, 1996).

As mentioned before, the majority of naturally occurring 2,3- and 4,6-HHDP-containing ellagitannins displays the (*S*)-configuration. According to the postulate first proposed by Schmidt (Schmidt *et al.*, 1965) and later by Haslam (Gupta *et al.*, 1982), enzymes would not be responsible for such a strong diastereoselectivity in the biosynthesis of ellagitannins, but the chirality of the D-glucopyranosyl scaffold on which biaryl coupling occurs would induce this remarkable stereoselectivity. Even though the exact reaction sequence of galloyl-galloyl couplings is still unknown *in vivo* (Feldman and Smith, 1996), the stereoselectivity observed in Feldman's oxidative 2,3- and 4,6-phenolic coupling reaction is in agreement with ellagitannin stereochemistry, strongly favouring the (*S*)-configuration.

5.2.3.3 *Total synthesis of pedunculagin via the double esterification strategy*

For this total synthesis of pedunculagin (**87**), the 4,6-positions of the sugar derivative **88** were first protected as a benzylidene acetal and the resulting diol **2** was subsequently acylated with the enantiomerically pure (*S*)-configured hexabenzyloxydiphenic acid (*S*)-**16**. The benzylidene acetal protection was then hydrolytically removed to furnish diol **89** in a quantitative yield. This diol was then also acylated with (*S*)-**16** to give an unpolar product, which could be identified as the desired tetraester **90** by NMR analysis (Fig. 5.18). Finally, to complete the total synthesis of pedunculagin (**87**), the tetraester **90** was exposed to hydrogenolysis under standard conditions using H_2-Pd/C to cleave off all benzylic protective groups. From this reaction, the natural pedunculagin (**87**) could be isolated as a brownish solid after purification by preparative reversed phase thin layer chromatography (Khanbabaee and Großer, 2003).

Access to the key intermediate **90** was then re-examined in the aim of shortening the overall synthesis. In particular, the possibility of making it in one step by a direct acylation of tetrol **88** with (*S*)-**16** was explored. Remarkably, this reaction promoted under Steglich conditions using DCC and DMAP led to the formation of only one unpolar product, which could easily be identified as the tetraester **90** by comparing its characterization data with those of the same compound previously made *via* the stepwise strategy (Khanbabaee and Großer, 2003).

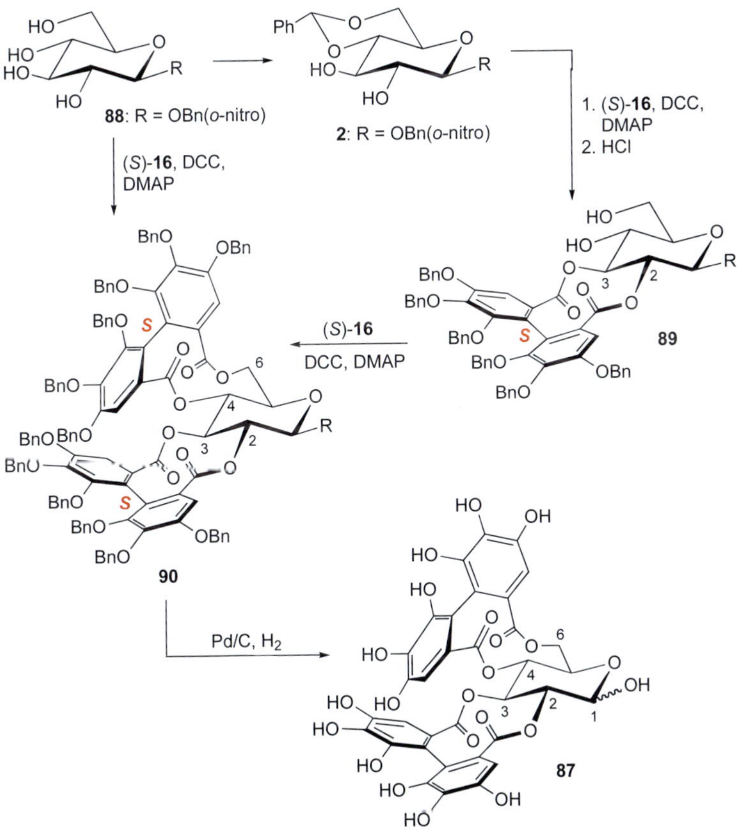

Fig. 5.18 Total synthesis of pedunculagin (**87**) via the double esterification strategy.

5.2.4 2,4-HHDP-containing ellagitannins

5.2.4.1 *Construction of a 2,4-bridged species*

From the synthetic point of view, the 2,4- and 3,6-HHDP-containing ellagitannins are much more challenging compounds than the 2,3- and 4,6-HHDP-containing ellagitannins, because their D-glucopyranose ring adopts the thermodynamically less stable $^1C_4$-conformation and their HHDP unit(s) the (*R*)-configuration. Up to now, there is no total synthesis of any ellagitannin of this type. However, the Feldman group recently reported the construction of the 2,4-HHDP-containing glucopyranose derivative **97**, starting from **91**, which possesses a 1,6-anhydro bridge that "freezes" the glucopyranose ring in the $^1C_4$-conformation (Feldman *et al.*, 2003). A two-fold acylation of **91** with the acid **8**, followed by desilylation gave the galloyl-derived bisester **92** (Fig. 5.19).

This material was then ready to be engaged in an oxidative intramolecular coupling reaction to form the key biaryl bond between the galloyl units linked to the 2- and 4-positions of the sugar core. As previously described in this Chapter, Feldman *et al.* had already demonstrated that such oxidative coupling reactions mediated by $Pb(OAc)_4$ (*i.e.*, Wessely oxidation) and performed on galloylated glucopyranose rings in their $^4C_1$-conformation are diastereoselective and give only the corresponding biaryl (*S*)-atropoisomers. Here, two questions then arose as to whether the coupling of 2,4-positioned galloyl moieties would be at all possible under Wessely conditions and whether the stereochemistry of the biaryl product will also be controlled in this case.

Feldman *et al.* observed that the $Pb(OAc)_4$-mediated oxidative coupling of **92** gave at least six distinct isomers consisting of the expected biaryl diphenylketal regioisomers and their corresponding atropisomers (see **93** in Fig. 5.19). Unexpectedly, the hydrogenolysis of this mixture gave the uncoupled bisgalloyl ester (not shown). Thus, it was assumed that the biaryl bond had been broken by the hydrogenolysis, maybe because of the relatively high strain energy of the HHDP-containing species **93**.

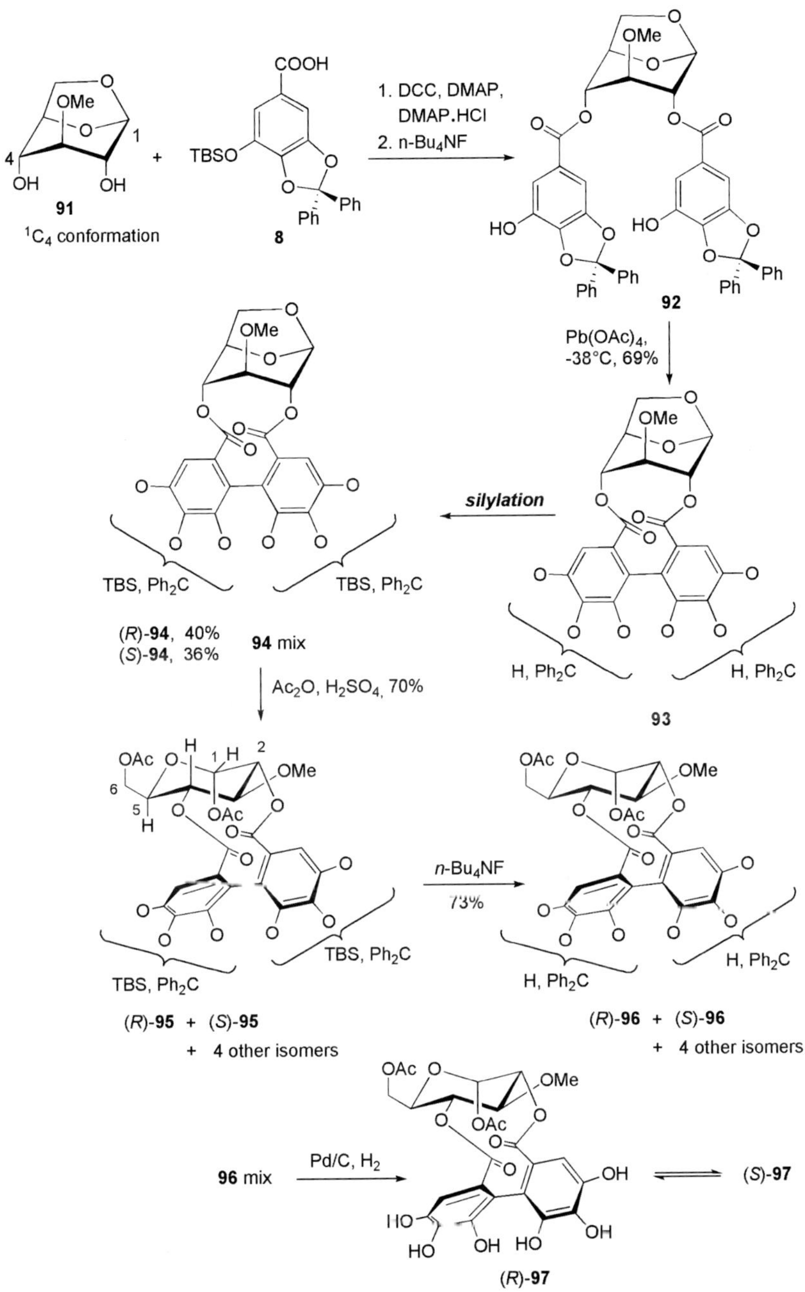

Fig. 5.19 Construction of a 2,4-HHDP-containing ellagitannin framework.

However, the silylation of the free phenolic hydroxyl groups within the mixture **93** led to another mixture consisting of four discrete isomers including atropisomer (*R*)-**94** (40% of the crude mixture of isomers) and (*S*)-**94** (36% of the crude mixture of isomers), which were isolated by column chromatography. The mixture **94** was then subjected to Ac_2O/H_2SO_4 to furnish a mixture of the atropisomers (*R*)-**95** and (*S*)-**95**, and four other undefined isomers. To verify these results, the separated compounds (*R*)-**94** and (S)-**94** were individually exposed to Ac_2O/H_2SO_4.

The exposure of (*R*)-**94** to Ac_2O/H_2SO_4 afforded a 1:1 mixture of the ring-opened diacetate (*R*)-**95** and a second but structurally unknown isomer, while the pure (*S*)-atropisomer (*S*)-**94** furnished a 1:1 mixture of the ring-opened diacetate (*S*)-**95** and a second also structurally unknown isomer. It is worth noting that these reactions largely proceeded with inversion of stereochemistry at C-1 to furnish the corresponding axial acetates (*R*)-**95** and (*S*)-**95**.

Cleavage of the silyl ether bonds of (*R*)-**95** and (*S*)-**95** by desylilation resulted in a mixture of at least 6 isomers, from which (*R*)-**96** and (*S*)-**96** could be isolated by chromatography. Feldman *et al.* observed that (*R*)-**96** and (*S*)-**96** equilibrate over the course of 48 h in acetone-d_6 solutions. Hydrogenolytic removal of the diphenylketals of (*R*)-**96** gave the single isomer (*R*)-**97** and that of (*S*)-**96** delivered the single isomer (*S*)-**97**, which also underwent facile atropisomerisation over 24 h (Feldman *et al.*, 2003).

5.2.5 3,6-HHDP-containing ellagitannins

5.2.5.1 *Synthesis of O-permethylcorilagin*

The first construction of the methyl ether-protected corilagin **108**, a 3,6-HHDP-containning ellagitannin featuring a glucopyranose ring in its $^1C_4$-conformation, has been achieved by the group of Yamada (Ikeda *et al.*, 2004). The synthesis of nonamethylcorilagin (**108**) started from phenyl 1-thio-β-D-glucopyranoside **98**, which possesses four free hydroxyl groups at C-2, C-3, C-4 and C-6 of the sugar in a $^4C_1$-conformation (Fig.

5.20). Treatment of **98** with methoxymethyl chloride (*i.e.*, MOMCl) led to the regioselective protection of the 3,6-hydroxyl groups and gave **99** in 64% yield. The free 2,4-hydroxyl groups of **99** were benzylated, and the resulted product (not shown) was converted to the α,β-anomeric mixture **100** by hydrolysis of the phenylthio group at the anomeric center of the D-glucopyranose ring. The challenging conversion of the stereochemistry of the D-glucopyranose ring from a $^4C_1$-conformation to a $^1C_4$-conformation and the establishment of the HHDP moiety in a correct absolute (*R*)-configuration were the main issues to be addressed in this synthesis work. To this aim, the authors used the following and quite elegant sequence: opening of the $^4C_1$-glucopyranose ring, installation of the methyl ether-protected galloyl residues, diastereoselective reductive Ullmann coupling of the galloyl residues to build the (*R*)-HHDP moiety (*i.e.*, method **B**, see Fig. 5.1), followed by closing of the sugar core into a glucopyranose ring having the $^1C_4$-conformation, and this key transformation was achieved by exploiting the template effect brought about by the (*R*)-HHDP moiety (Fig. 5.20).

Thus, the initial glucopyranose ring opening was accomplished *via* a Wittig olefination of **100** to generate **101**. Its free hydroxyl group at C-5 was protected as a *p*-methoxybenzyl (PMB) ether, and the MOM groups were then cleaved to afford the open-chain 3,6-diol **102**. Installation of two 2-iodo-3,4,5-trimethoxybenzoyl units onto **102** could be achieved by using 2-iodo-3,4,5-trimethoxybenzoic acid **103**, thus leading to the Ullmann coupling substrate **104**. The copper-mediated Ullmann coupling of **104** into the axially chiral biaryl intermediate **105** was examined at different concentrations, the highest yield of 48% for (*R*)-**105** being obtained when the final concentration of the product was 3 mM. The PMB protective group, as well as the double bond, were then cleaved under oxidative conditions. Remarkably, these conditions also allowed for the reconstruction of the glucopyranose ring, which adopted the desired $^1C_4$-conformation to furnish the advanced intermediate **106** in a good yield. Anomeric acylation of the intermediate **106** with the trimethylated galloyl chloride **62** under basic conditions led to the formation of a 4:6 mixture of the corresponding α- and β-isomers of **107**. These anomers were separated by chromatography, and hydrogenolytic debenzylation of the β-isomer **107** furnished nonamethylcorilagin (**108**).

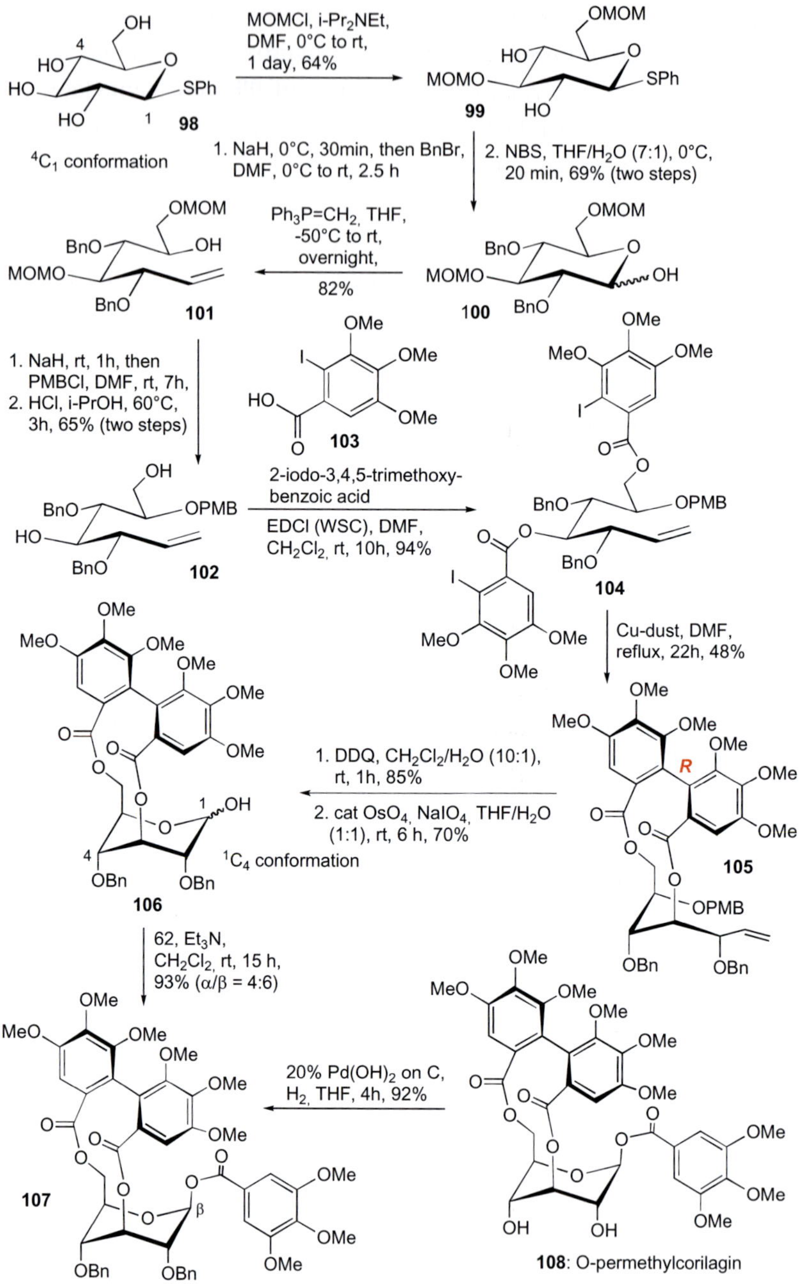

Fig. 5.20 Synthesis of *O*-permethylcorilagin (**108**).

A final cleavage of the methyl ether bonds would have led to the corresponding natural product corilagin, but in spite of several trials run under various reaction conditions, this ultimate transformation could not be cleanly done, notably because of an unavoidable competing cleavage of the anomeric galloyl ester function in nonamethylcorilagin (**108**).

5.3 Conclusion

Up to now, a large number of structurally characterized ellagitannins has been identified. Although these ellagitannins possess complex structures, all of them have at least one axially chiral hexahydroxydiphenoyl (HHDP) unit. Some ellagitannins possess interesting biological activities. However, it is not easy to isolate them in sufficient amount from natural sources for further studies. Although this class of natural products has been well established, not a single method had been developed for their total synthesis until 1994. Through the pioneer work of Feldman on the development of a method for the total synthesis of ellagitannins, some of them became accessible by total synthesis for the first time and the research activities on their synthesis consequently received a considerable boost. The characteristic step of this method is the diastereoselective oxidative galloyl coupling to the corresponding biaryl unit. Thereafter, a new method has been developed, in which a pre-synthesized biaryl compound is used either as its racemic mixture or in atropoisomerically pure form. This method allows the total synthesis of numerous ellagitannins in a few steps. The last developed method in this area concerned the construction of an ellagitannin precursor with a $^{1}C_{4}$-glucose unit. This method could constitute a successful route to the total synthesis of ellagitannins with a $^{1}C_{4}$-glucose unit in the future.

5.4 Bibliography

Abe, I., Kashiwagi, Y., Noguchi, H., Tanaka, T., Ikeshiro, Y. and Kashiwada, Y. (2001). Ellagitannins and hexahydroxydiphenoyl esters as inhibitors of vertebrate squalene epoxidase, *J. Nat. Prod.*, 64, pp. 1010–1014.

Albright, J. D. and Goldman, L. (1965a). Dimethyl sulfoxide-acid anhydride mixtures. New reagents for oxidation of alcohols, *J. Am. Chem. Soc.*, 87, pp. 4214–4216.

Albright, J. D. and Goldman, L. (1965b). Indole alkaloids. III. Oxidation of secondary alcohols to ketones, *J. Org. Chem.*, 30, pp. 1107–1110.

Albright, J. D. and Goldman, L. (1967). Dimethyl sulfoxide-acid anhydride mixtures for the oxidation of alcohols, *J. Am. Chem. Soc.*, 89, pp. 2416–2423.

Arisawa, M., Utsumi, S., Nakajima, M., Ramesh, N. G., Tohma, H. and Kita, Y. (1999). An unexpected intermolecular chiral biaryl coupling reaction induced by a hypervalent iodine(III) reagent: synthesis of potential precursor for ellagitannin, *Chem. Commun.*, pp. 469–470.

Berova, N., Woody, R. W. and Nakanishi, K. (1994). *Circular Dichroism – Principles and Applications*, VCH-Publishers: New York.

Bissember, B. B. and Wightman, R. H. (1980). The use of pyridinium chlorochromate for preparation of hexopyranosiduloses without epimerisation, *Carbohydr. Res.*, 81, pp. 187–191.

Borch, R. F., Grudzinskas, C. V., Peterson, D. A. and Weber, L. D. (1972). New synthesis of substituted 2(1H)-pyridones. Synthesis of a potential camptothecin intermediate, *J. Org. Chem.*, 37, pp. 1141–1145.

Brown, J. M., Conn, A. D., Pilcher, G., Leitão, M. L. P. and Meng-Yan, Y. (1989). On the strain energy of 5-ring and 6-ring lactones, *J. Chem. Soc., Chem. Commun.*, pp. 1817–1819.

Chen, L.-G., Yang, L.-L., Yen, K.-Y., Hatano, T., Yoshida, T. and Okuda, T. (1995). Tannins of euphorbiaceous plants. XIII. New hydrolyzable tannins having phloroglucinol residue from *glochidion rubrum* BLUME, *Chem. Pharm. Bull.*, 43, pp. 2088–2090.

Corey, E. J. and Schmidt, G. (1979). Useful procedures for the oxidation of alcohols involving pyridinium dichromate in approtic media, *Tetrahedron Lett.*, 20, pp. 399–402.

Czernecki, S., Georgoulis, C. and Provelenghiou, C. (1976). Nouvelle méthode de benzylation d'hydroxyles glucidiques encombrés, *Tetrahedron Lett.*, 17, pp. 3535–3536.

Czernecki, S., Georgoulis, C., Stevens, C. L., and Vijayakumaran, K. (1985). Pyridinium dichromate oxidation. Modifications enhancing its synthetic utility, *Tetrahedron Lett.*, 26, pp. 1699–1702.

Dai, D. and Martin, O. R. (1998). Studies relevant to ellagitannin chemistry: highly diastereoselective intramolecular biaryl coupling in bis(iodotrimethoxybenzoyl) hexopyranose derivatives, *J. Org. Chem.*, 63, pp. 7628–7633.

Dale, J. A. and Mosher, H. S. (1973). Nuclear magnetic resonance enantiomer regents. Configurational correlations via nuclear magnetic resonance chemical shifts of diastereomeric mandelate, *O*-methylmandelate, and α-methoxy-α-trifluoromethylphenylacetate (MTPA) esters, *J. Am. Chem. Soc.*, 95, pp. 512–519.

Dean, P. D. G. (1965). Halogenolysis of methyl glycyrrhetate with lithium iodide–dimethylformamide, *J. Chem. Soc.*, pp. 6655–6655.

El-Mekkawy, S., Meselhy, M. R., Kusumoto, I. T. and Kadota, S. (1995). Inhibitory effects of Egyptian folk medicines on human immunodeficiency virus (HIV) reverse transcriptase, *Chem. Pharm. Bull.*, 43, pp. 641–648.

Feldman, K. S. and Ensel, S. M. (1993). Ellagitannin chemistry. The first example of biomimetic diastereoselective oxidative coupling of a glucose-derived digalloyl substrate, *J. Am. Chem. Soc.*, 115, pp. 1162–1163.

Feldman, K. S., Ensel, S. M. and Minard, R. D. (1994). Ellagitannin chemistry. The first total chemical synthesis of an ellagitannin natural product, tellimagrandin I, *J. Am. Chem. Soc.*, 116, pp. 1742–1745.

Feldman, K. S. and Ensel, S. M. (1994). Ellagitannin chemistry. Preparative and mechanistic studies of the biomimetic oxidative coupling of galloyl esters, *J. Am. Chem. Soc.*, 116, pp. 3357–3366.

Feldman, K. S. and Sambandam, A. (1995). Ellagitannin chemistry. The first total chemical synthesis of an O(2),O(3)-galloyl-coupled ellagitannin, sanguiin H-5, *J. Org. Chem.*, 60, pp. 8171–8178.

Feldman, K. S. and Smith, R. S. (1996). Ellagitannin chemistry. First total synthesis of the 2,3- and 4,6-coupled ellagitannin pedunculagin, *J. Org. Chem.*, 61, pp. 2606–2612.

Feldman, K. S. and Sahasrabudhe, K. (1999). Ellagitannin chemistry. Syntheses of tellimagrandin II and a dehydrodigalloyl ether-containing dimeric gallotannin analogue of coriariin A, *J. Org. Chem.*, 64, pp. 209–216.

Feldman, K. S., Lawlor, M. D. and Sahasrabudhe, K. (2000). Ellagitannin chemistry. Evolution of a three-component coupling strategy for the synthesis of the dimeric ellagitannin coriariin A and a dimeric gallotannin analogue, *J. Org. Chem.*, 65, pp. 8011–8019.

Feldman, K. S. and Lawlor, M. D. (2000). Ellagitannin chemistry. The first total synthesis of a dimeric ellagitannin, coriariin A, *J. Am. Chem. Soc.*, 122, pp. 7396–7397.

Feldman, K. S., Iyer, M. R. and Liu, Y. (2003). Ellagitannin chemistry. Studies on the stability and reactivity of 2,4-HHDP-containing glucopyranose systems, *J. Org. Chem.*, 68, pp. 7433–7438.

Fukuchi, K., Sakagami, H., Okuda, T., Hatano, T., Tanuma, S., Kitajima, K., Inoue, Y., Inoue, S. and Ichikawa S. (1989). Inhibition of Herpes simplex virus infection by tannins and related compounds, *Antiviral Res.*, 11, pp. 285–297.

Garegg, P. J., Iversen, T. and Oscarson, S. (1976). Monobenzylation of diols using phase-transfer catalysis, *Carbohydr. Res.*, 50, pp. C12–C14.

Gassman, P. G. and Schenk, W. N. (1977). A general procedure for the base-promoted hydrolysis of hindered esters at ambient temperatures, *J. Org. Chem.*, 42, pp. 918–920.

Gigg, R., Payne, S. and Conant, R. (1983). The allyl group for protection in carbohydrate chemistry, Part 14. Synthesis of 2,3-di-*O*-methyl-4-*O*-(3,6-di-*O*-methyl-β-D-glucopyranosyl)-L-Rhamnopyranose (and its α-propyl glycoside): A haptenic portion of the major glycolipid from *Mycobacterium Leprae*, *J. Carbohydrate Chem.*, 2, pp. 207–223.

Gupta, R. K., Al-Shafi, S. M. K., Layden, K. and Haslam, E. (1982). The metabolism of gallic acid and hexahydroxydiphenic acid in plants. Part 2. Esters of (*S*)-hexahydroxydiphenic acid with D-glucopyranose ($^4C_1$), *J. Chem. Soc., Perkin Trans. 1*, pp. 2525–2534.

Hada, S., Kakiuchi, N., Hattori, M. and Namba, T. (1989). Identification of antibacterial principles against *Streptococcus* mutans and inhibitory principles against glucosyltransferase from the seed of *Areca catechu* L., *Phytotherapy Res.*, 3, pp. 140–144.

Haddock, E. A., Gupta, R. K., Al-Shafi, S. M. K. and Haslam, E. (1982a). The metabolism of gallic acid and hexahydroxydiphenic acid in plants. Part 1. Introduction. Naturally occurring galloyl esters, *J. Chem. Soc., Perkin Trans. 1*, pp. 2515–2524.

Haddock, E. A., Gupta, R. K. and Haslam, E. (1982b). The metabolism of gallic acid and hexahydroxydiphenic acid in plants. Part 3. Esters of (*R*)- and (*S*)-hexahydroxydiphenic acid and dehydrohexahydroxydiphenic acid with D-glucopyranose ($^{1}C_{4}$ and related conformations), *J. Chem. Soc., Perkin Trans 1*, pp. 2535–2545.

Han, L., Hatano, T., Okuda, T. and Yoshida, T. (1995). Tannins of *Stachyurus* species. III. Stachyuranins A, B and C, three new complex tannins from *Stachyurus praecox* leaves, *Chem. Pharm. Bull.*, 43, pp. 2109–2114.

Haslam, E. (1980). Recent developments in methods for the esterification and protection of the carboxyl group, *Tetrahedron*, 36, pp. 2409–2433.

Haslam, E. (1996). Natural Polyphenols (Vegetable Tannins) as Drugs: Possible Modes of Action, *J. Nat. Prod.*, 59, pp. 205–215.

Hatano, T., Yoshida, T., Shingu, T. and Okuda, T. (1988). ^{13}C Nuclear magnetic resonance spectra of hydrolyzable tannins. II. Tannins forming anomer mixtures, *Chem. Pharm. Bull.*, 36, pp. 2925–2933.

Hatano, T., Ogawa, N., Kira, R., Yasuhara, T. and Okuda, T. (1989). Tannins of cornaceous plants. I. Cornusiins A, B and C, dimeric monomeric and trimeric hydrolyzable tannins from *Cornus officinalis*, and orientation of valoneoyl group in related tannins, *Chem. Pharm. Bull.*, 37, pp. 2083–2090.

Hatano, T., Namba, O., Chen, L., Yasuhara, T., Yazaki, K., Yoshida, T. and Okuda, T. (1990a). Hydrolyzable tannins having a depsidone-forming valoneoyl group, *Heterocyles*, 31, pp. 1221–1224.

Hatano, T., Ogawa, N., Yasuhara, T. and Okuda, T. (1990b). Tannins of rosaceous plants. VIII. Hydrolyzable tannin monomers having a valoneoyl group from flower petals of *Rosa rugosa* THUNB, *Chem. Pharm. Bull.*, 38, pp. 3308–3313.

Hatano, T., Okonogi, A., Yazaki, K. and Okuda, T. (1990c). Trapanins A and B: oligomeric hydrolyzable tannins from *Trapa japonica* FLEROV, *Chem. Pharm. Bull.*, 38, pp. 2707–2711.

Hatano, T., Yazaki, K., Okonogi, A. and Okuda, T. (1991). Tannins of *Stachyurus* Species. II. Praecoxins A, B, C and D, four new hydrolyzable tannins from *Stachyurus praecox* leaves, *Chem. Pharm. Bull.*, 39, pp. 1689–1693.

Hatano, T., Han, L., Taniguchi, S., Okuda, T., Kiso, Y., Tanaka, T. and Yoshida, T. (1995). Camelliatannin D, a new inhibitor of bone resorption, from *Camellia Japonica*, *Chem. Pharm. Bull.*, 43, pp. 2033–2035.

Herscovici, J. and Antonakis, K. (1980). Molecular sieve-assisted oxidations: New methods for carbohydrate derivative oxidations, *J. Chem. Soc., Chem. Commun.*, pp. 561–562.

Herscovici, J., Egron, M.-J. and Antonakis, K. (1982). New oxidative systems for alcohols: Molecular sieves with chromium(VI) reagents, *J. Chem. Soc., Perkin Trans. 1*, pp. 1967–1973.

Höfle, G., Steglich, W. and Vorbrüggen, H. (1978). 4-Dialkylaminopyridine als hochwirksame Acylierungskatalysatoren, *Angew. Chem.*, 90, pp. 602–615.

Hollenberg, D. H., Klein, R. S. and Fox, J. J. (1978). Pyridinium chlorochromate for the oxidation of carbohydrates, *Carbohydr. Res.*, 67, pp. 491–494.

Igarashi, K. (1977). The Koenigs–Knorr reaction (Review), *Adv. Carbohydr. Chem. Biochem.*, 34 pp. 243–283.

Ikeda, Y., Nagao, K., Tanigakiuchi, K., Tokumaru, G., Tsuchiya, H. and Yamada, H. (2004). The first construction of a 3,6-bridged ellagitannin skeleton with $^{1}C_4$/B glucose core. Synthesis of nonamethylcorilagin, *Tetrahedron Lett.*, 45, pp. 487–489.

Itoh, T. and Chika, J.-i. (1995). New method of kinetic resolution of axially chiral biaryl compounds using a sugar template, *J. Org. Chem.*, 60, pp. 4968–4969.

Itoh, T., Chika, J.-i., Shirakami, S., Ito, H., Yoshida, T., Kubo, Y. and Uenishi, J.-i. (1996). Synthesis of trideca-*O*-methyl-α-pedunculagin. Diastereo-favoritism studies on intramolecular ester-cyclization of axially chiral diphenic acids with carbohydrate core, *J. Org. Chem.*, 61, pp. 3700–3705.

Iversen, T., Garegg, P. J. and Oscarson, S. (1977). Monotosylation of diols using phase-transfer catalysis, *Carbohydr. Res.*, 53, pp. C5–C7.

Jankun, J., Selman, S. H. and Swiercz, R. (1997). Why drinking green tea could prevent cancer, *Nature*, 387, p. 561.

Kadota, T., Takamori, Y., Nyein, K., Kikuchi, T., Tanaka, K. and Ekimoto, H. (1990). Constituents of the leaves of *Woodfordia fruticosa* KURZ. I. Isolation, structure, and proton and carbon-13 nuclear magnetic resonance signal assignments of woodfruticosin (Woodfordin C), an inhibitor of deoxyribonucleic acid topoisomerase II, *Chem. Pharm. Bull.*, 38, pp. 2687–2697.

Kanai, K., Sakamoto, I., Ogawa, S. and Suami, T. (1987). Synthesis on 1,4-diaminocyclitol antibiotics. III. Synthesis of 4-hydroxypurpurosamine B derivatives, *Bull. Chem. Soc. Jpn.*, 60, pp. 1529–1531.

Kashiwada, Y., Huang, L., Kilkuskie, R. E., Bodner, A. J. and Lee, K.-H. (1992a). New hexahydroxydiphenyl derivatives as potent inhibitors of HIV replication in H9 lymphocytes, *Bioorg. Med. Chem. Lett.*, 2, pp. 235–238.

Kashiwada, Y., Nonaka, G.-I., Nishioka, I., Chang, J.-J. and Lee, K.-H. (1992b). Antitumor agents. 129. Tannins and related compounds as selective cytotoxic agents, *J. Nat. Prod.*, 55, pp. 1033–1043.

Kashiwada, Y., Nonaka, G.-I., Nishioka, I., Jiann-Hung Lee, K., Bori, I., Fukushima, Y., Bastow, K. F. and Lee, K.-H. (1993). Tannins as potent inhibitors of DNA topoisomerase II in vitro, *J. Pharm. Sci.*, 82, pp. 487–492.

Keck, G. E. and Boden, E. P. (1985) Proton-transfer steps in Steglich esterification: a very practical new method for macrolactonization, *J. Org. Chem.*, 50, pp. 2394–2395.

Khanbabaee, K. and Lötzerich, K. (1997a). Efficient total synthesis of the natural products 2,3,4,6-tetra-*O*-galloyl-D-glucopyranose, 1,2,3,4,6-penta-*O*-galloyl-β-D-glucopyranose, and the unnatural 1,2,3,4,6-penta-*O*-galloyl-α-D-glucopyranose, *Tetrahedron*, 53, pp. 10725–10732.

Khanbabaee, K. and Lötzerich, K. (1997b). The first total synthesis of praecoxin B and pterocarinin C, two natural products of the tannin class, *Liebigs Ann. Chem./Recueil*, pp. 1571–1575. **NB:** the structure of the compound named "pterocarinin C" in this article corresponds to the structure of "pterocaryarin C", thus renamed in 1989, probably to avoid confusion with the *C*-glycosidic pterocarinins A and B (see Nonaka et *al.*, 1989a), by Nishioka and collaborators who first isolated and characterized this β-1,4,6-trigalloyl-2,3-(*S*)-hexahydroxydiphenoylglucopyranose compound (see Saijo *et al.*, 1989). Unfortunately, the use of the original name has persisted in the literature.

Khanbabaee, K., Schulz, C. and Lötzerich, K. (1997). Synthesis of enantiomerically pure strictinin using a stereoselective esterification reaction, *Tetrahedron Lett.*, 38, pp. 1367–1368.

Khanbabaee, K. and Lötzerich, K. (1998). Synthesis of enantiomerically pure unusual ellagitannins 1,4,6-tri-*O*-galloyl-2,3-(*R*)-hexahydroxydiphenoyl-β-D-glucopyranoside and 4,6-Di-*O*-galloyl-2,3-(*R*)-hexahydroxydiphenoyl-D-glucoside. The proposed chemical structures for cercidinin A and B must be revised, *J. Org. Chem.*, 63, pp. 8723–8728.

Khanbabaee, K. and Lötzerich, K. (1999). Studies relevant to ellagitannin chemistry: strain-energy induced opening of the D-Glucono-δ-lactone ring of an ellagitannin derivative and the synthesis of ellagitannin natural products with a D-gluconic acid moiety, *Eur. J. Org. Chem.*, pp. 3079–3083.

Khanbabaee, K., Lötzerich, K., Borges, M. and Großer, M. (1999). The first total syntheses of enantiomerically pure naturally occurring ellagitannins gemin D and its regioisomer hippomanin A, *J. Prakt. Chem.*, 341, pp. 159–166.

Khanbabaee, K. and van Ree, T. (2001a). Tannins: classification and definition, *Nat. Prod. Rep.*, 18, pp. 641–649.

Khanbabaee, K. and van Ree, T. (2001b). Strategies for the synthesis of ellagitannins, *Synthesis*, pp. 1585–1610.

Khanbabaee, K. and Großer, M. (2003). An efficient total synthesis of pedunculagin by using a twofold intramolecular double esterification strategy, *Eur. J. Org. Chem.*, pp. 2128–2131.

Kida, T., Masuyama, A. and Okahara, M. (1990). Preparation of trihydroxycarboxylates bearing a long-chain alkyl acetal group from glucono-1,5-lactone, *Tetrahedron Lett.*, 31, pp. 5939–5942.

Königs, W. and Knorr, E. (1901). Ueber einige Derivate des Traubenzuckers und der Galactose, *Chem. Ber.*, 34, pp. 957–981.

Kuzuhara, H. and Fletcher, H. G. (1967). Syntheses with partially benzylated sugars. VIII. Substitution at carbon-5 in aldose. The synthesis of 5-*O*-methyl-D-glucofuranose derivatives, *J. Org. Chem.*, 32, pp. 2531–2534.

Lee, S. H., Tanaka, T., Nonaka, G. and Nishioka, I. (1989). Sedoheptulose digallate from *Cornus officinalis*, *Phytochemistry*, 28, pp. 3469–3472.

Lehmann, J. (1996). *Kohlenhydrate – Chemie und Biologie*, second edition, Georg Thieme-Verlag: Stuttgart.

Lin, J.-H., Nonaka, G.-I. and Nishioka, I. (1990). Tannins and related compounds. XCIV. Isolation and characterization of seven new hydrolyzable tannins from the leaves of *Macaranga tanarius* (L.) MUELL. et ARG., *Chem. Pharm. Bull.*, 38, pp. 1218–1223.

Lipshutz, B. H., Siegmann, K., Garcia, E. and Kayser, F. (1993). Synthesis of unsymmetrical biaryls via kinetic higher order cyanocuprates: scope, limitations, and spectroscopic insights, *J. Am. Chem. Soc.*, 115, pp. 9276–9282.

Lipshutz, B. H., Kayser, F. and Liu, Z. (1994a). Asymmetrische Synthese von Biarylen durch intramolekulare oxidative Kupplung von Cyanocuprat-Zwischenstufen, *Angew. Chem.*, 106, pp. 1962–1964.

Lipshutz, B. H., Liu, Z.-P. and Kayser, F. (1994b). Cyanocuprate-mediated intramolecular biaryl couplings applied to an ellagitannin. Synthesis of (+)-*O*-permethyltellimagrandin II, *Tetrahedron Lett.*, 35, pp. 5567–5570.

Mancuso, A. J. and Swern, D. (1981). Activated dimethyl sulfoxide: Useful reagents for synthesis, *Synthesis*, pp. 165–184.

McMurry, J. (1976). Übersicht über Ester-Desalkylierungen, In *Organic Reactions*, vol. 24, Wiley: New York, pp. 187–224.

Meinwald, J. and Putzig, D. E. (1970). Bridgehead substitution vs. ring contraction in the deamination of 1-aminobicyclohept-5-en-2-ol, *J. Org. Chem.*, 35, pp. 1891–1894.

Miyamoto, K., Kishi, N., Koshiura, R., Yoshida, T., Hatano, T. and Okuda, T. (1987). Relationship between the structure and the antitumor activities of tannins, *Chem. Pharm. Bull.*, 35, pp. 814–822.

Morimoto, S., Nonaka, G.-I. and Nishioka, I. (1986a). Tannins and related compounds. XXXVIII. Isolation and characterization of flavan-3-ol glucosides and procyanidin oligomers from *Cassia* bark: *Cinnamomum cassia* BLUME, *Chem. Pharm. Bull.*, 34, pp. 633–642.

Morimoto, S., Nonaka, G.-I. and Nishioka, I. (1986b). Tannins and related compounds. XXXIX. Procyanidin *C*-glucosides and an acylated flavan-3-ol glucoside from the barks of *Cinnamomum cassia* BLUME and C. *obtusifolium* NEES, *Chem. Pharm. Bull.*, 34, pp. 643–649.

Nakashima, H., Murakami, T., Yamamoto, N., Sakagami, H., Tanuma, S., Hatano, T., Yoshida, T. and Okuda, T. (1992). Inhibition of human immunodeficiency viral replication by tannins and related compounds, *Antiviral Res.*, 18, pp. 91–103.

Neises, B. and Steglich, W. (1978). Einfaches Verfahren zur Veresterung von Carbonsäuren, *Angew. Chem.*, 90, pp. 556–557.

Nelson, T. D. and Meyers, A. I. (1994). A rapid total synthesis of an ellagitannin, *J. Org. Chem.*, 59, pp. 2577–2580.

Nishizawa, M. and Yamagishi, T. (1982). Tannins and related compounds. Part 5. Isolation and characterization of polygalloylglucoses from Chinese gallotannins, *J. Chem. Soc., Perkin Trans 1*, pp. 2963–2968.

Nishizawa, M. and Yamagishi, T. (1983). Tannins and related compounds. Part 9. Isolation and characterization of polygalloylglucoses from Turkish galls (*Quercus infectoria*), *J. Chem. Soc. Perkin Trans. 1*, pp. 961–965.

Nonaka, G.-I., Harada, M. and Nishioka, I. (1980). Eugeniin, a new ellagitannin from cloves, *Chem. Pharm. Bull.*, 28, pp. 685–687.

Nonaka, G.-I., Sakai, R. and Nishioka, I. (1984). Hydrolysable tannins and proanthocyanidins from green tea, *Phytochemistry*, 23, pp. 1753–1755.

Nonaka, G.-I., Ishimaru, K., Azuma, R., Ishimatsu, M. and Nishioka, I. (1989a). Tannins and related compounds. LXXXV. Structures of novel *C*-glycosidic ellagitannins, grandinin and pterocarinins A and B, *Chem. Pharm. Bull.*, 37, pp. 2071–2077.

Nonaka, G.-I., Ishimatsu, M., Ageta, M. and Nishioka, I. (1989b). Tannins and related compounds. LXXVI. Isolation and characterization of cercidinins A and B and cuspinin, unusual 2,3-(*R*)-hexahydroxydiphenoyl glucoses from *Cercidiphyllum japonicum* and *Castanopsis cuspidata* var. *sieboldii*, *Chem. Pharm. Bull.*, 37, pp. 50–53.

Nonaka, G.-I., Nakayama, S. and Nishioka, I. (1989c). Tannins and related compounds. LXXXIII. Isolation and structures of hydrolyzable tannins, phillyraeoidins A-E from *Quercus phillyraeoides*, *Chem. Pharm. Bull.*, 37, pp. 2030–2036.

Nonaka, G.-I., Sakai, T., Tanaka, T., Mihashi, K. and Nishioka, I. (1990). Tannins and related compounds. XCVII. Structure revision of *C*-glycosidic ellagitannins, castalagin, vescalagin, casuarinin and stachyurin, and related hydrolyzable tannins, *Chem. Pharm. Bull.*, 38, pp. 2151–2156.

Ogawa, T., Katano, K. and Matsui, M. (1978). Regio- and stereo-controlled synthesis of core oligosaccharides of glycopeptides, *Carbohydr. Res.*, 64, pp. C3–C9.

Okuda, T., Hatano, T., Yazaki, K. and Ogawa, N. (1982a). Rugosin A, B, C and praecoxin A, tannins having a valoneoyl group, *Chem. Pharm. Bull.*, 30, pp. 4230–4233.

Okuda, T., Hatano, T. and Ogawa, N. (1982b). Rugosin D, E, F and G, dimeric and trimeric hydrolyzable tannins, *Chem. Pharm. Bull.*, 30, pp. 4234–4236.

Okuda, T., Yoshida, T., Hatano, T., Koga, T., Toh, N. and Kuriyama, K. (1982c). Circular dichroism of hydrolysable tannins. I. Ellagitannins and gallotannins, *Tetrahedron Lett.*, 23, pp. 3937–3940.

Okuda, T., Yoshida, T., Hatano, T., Koga, T., Toh, N. and Kariyama, K. (1982d). Circular dichroism of hydrolysable tannins. II. Dehydroellagitannins, *Tetrahedron Lett.*, 23, pp. 3941–3944.

Okuda, T., Yoshida, T., Hatano, T., Yazaki, K. and Shida, M. (1982e). Ellagitannins of the *Casuarinaceae*, *Stachyuraceae* and *Myrtaceae*, *Phytochemistry*, 21, pp. 2871–2874.

Okuda, T., Hatano, T. and Yazaki, K. (1983a). Praecoxin B, C, D and E, novel ellagitannins from *Stachyurus Praecox*, *Chem. Pharm. Bull.*, 31, pp. 333–336.

Okuda, T., Yoshida, T., Ashida, M. and Yazaki, K. (1983b). Tannins of *Casuarina* and *Stachyurus* species. Part 1. Structures of pendunculagin, casuarictin, strictinin, casuarinin, casuariin, and stachyurin, *J. Chem. Soc., Perkin Trans. 1*, pp. 1765–1772.

Onodera, K., Hirano, S. and Kashimura, N. (1965). Oxidation of carbohydrates with dimethyl sulfoxide containing phosphorus pentoxide, *J. Am. Chem. Soc.*, 87, pp. 4651–4652.

Pacheco, H. H. and Grouiller, A. (1965). Chimie des composés flavoniques. Synthèse de mono et di-O-méthyl flavonols. Etudes de leurs propriétés spectrales dans l'ultraviolet et l'infrarouge, *Bull. Soc. Chim. Fr.*, pp. 779–785.

Petit, J.-M. and Sinay, P. (1978). Synthèse du 4-O-D-glucose et de son isomère (R), *Carbohydr. Res.*, 64, pp. 9–16.

Pfitzner, K. E. and Moffatt, J. G. (1965a). Sulfoxide-carbodiimide reactions. I. A facile oxidation of alcohols, *J. Am. Chem. Soc.*, 87, pp. 5661–5670.

Pfitzner, K. E. and Moffatt, J. G. (1965b). Sulfoxide-carbodiimide reactions. II. Scope of the oxidation reaction, *J. Am. Chem. Soc.*, 87, pp. 5670–5678.

Piancatelli, G. and D'Auria, S. M. (1982). Pyridinium chlorochromate: a versatile oxidant in organic synthesis, *Synthesis*, pp. 245–258.

Procopiou, P. A., Baugh, S. P. D., Flack, S. S. and Inglis, G. G. A. (1996). An extremely fast and efficient acylation reaction of alcohols with acid anhydrides in the presence of trimethylsilyl trifluoromethanesulfonate as catalyst, *J. Chem. Soc., Chem. Commun.*, pp. 2625–2626.

Procopiou, P. A., Baugh, S. P. D., Flack, S. S. and Inglis, G. G. A. (1998). An extremely powerful acylation reaction of alcohols with acid anhydrides catalyzed by trimethylsilyl trifluoromethanesulfonate, *J. Org. Chem.*, 63, pp. 2342–2347.

Quideau, S. and Feldman, K. S. (1996). Ellagitannin Chemistry, *Chem. Rev.*, 96, pp. 475–504.

Rajasekharan Pillai, V. N. (1980). Photoremovable protecting groups in organic synthesis, *Synthesis*, pp. 1–26.

Rao, K. V. (1974). Toxic principles of *Hippomane mancinella*, *Planta Medica*, 25, pp. 166–171.

Rao, K. V. (1977). Toxic principles of *Hippomane mancinella*. II. Structure of Hippomanin A, *Lloydia*, 40, pp. 169–172.

Saijo, R., Nonaka, G.-I. and Nishioka, I. (1989). Tannins and related compounds. LXXXIV. Isolation and characterization of five new hydrolyzable tannins from the bark of *Mallotus japonicus*, *Chem. Pharm. Bull.*, 37, pp. 2063–2070.

Schmidt, O. Th. and Demmler, K. (1952). Optisch aktive 2,3,4,2',3',4'-Hexamethoxy-diphenyl-dicarbonsäure-6,6'. XIII. Mitteilung über natürliche Gerbstoffe, *Liebigs Ann. Chem*, 576, pp. 85–93.

Schmidt, O. Th. and Demmler, K. (1954). Racemische und optisch aktive 2,3,4,2',3',4'-Hexaoxydiphenyl-6,6'-dicarbonsäure. XVII. Mitteilung über natürliche Gerbstoffe. *Liebigs Ann. Chem.*, 586, pp. 179–193.

Schmidt, O. Th., Voigt, H., Puff, W. and Köster, R. (1954). Benzyläther der Ellagsäure und Hexaoxy-diphensäure. XVI. Mitteilung über natürliche Gerbstoffe, *Liebigs Ann. Chem.*, 586, pp. 165–178.

Schmidt, O. Th. and Klinger, G. (1957). Über natürliche Gerbstoffe. XXVIII. Synthese der 1.3.6-Trigalloyl-Glucose, *Liebigs Ann. Chem.*, 609, pp.199–208.

Schmidt, O. Th., Würtele, L. and Harréus, A. (1965). Über natürliche Gerbstoffe, XXXIII. Pedunculagin, eine 2.3;4.6-Di-[(–)-hexahydroxy-diphenoyl]-glucose aus Knoppern, *Liebigs Ann. Chem.*, 690, pp. 150–162.

Sen, A. K. and Banerji, N. (1989). Synthesis of 4-*O*-beta-D-xylopyranosyl-D-glucopyranoses and 6-*O*-beta-D-xylopyranosyl-D-glucopyranoses and their protein conjugates, *Ind. J. Chem.*, 28B, pp. 818–823.

Snatzke, G. (1981). Chiroptische Methoden in der Stereochemie, Teil I, *Chemie in unserer Zeit*, 15, pp. 78–87.

Snatzke, G. (1982). Chiroptische Methoden in der Stereochemie, Teil II, *Chemie in unserer Zeit*, 16, pp. 160–168.

Sowa, W. and Thomas, G. H. S. (1966). Oxidation of 1,2:5,6-di-*O*-isopropylidene-D-glucose by dimethyl sulfoxide-acetic anhydride, *Can. J. Chem.*, 44, pp. 836–838.

Tanaka, T., Nonaka, G.-I. and Nishioka, I. (1986a). Tannins and related compounds. XL. Revision of the structures of punicalin and punicalagin, and isolation and characterization of 2-O-galloylpunicalin from the bark of *Punica granatum* L., *Chem. Pharm. Bull.*, 34, pp. 650–655.

Tanaka, T., Nonaka, G.-I. and Nishioka, I. (1986b). Tannins and related compounds. XLI. Isolation and characterization of novel ellagitannins, punicacorteins A, B, C, and D, and punigluconin from the bark of *Punica granatum* L., *Chem. Pharm. Bull.*, 34, pp. 656–663.

Tanaka, T., Nonaka, G.-I. and Nishioka, I. (1990). Tannins and related compounds. C. Reaction of dehydrohexahydroxydiphenic acid esters with bases, and its application to the structure determination of pomegranate tannins, granatins A and B, *Chem. Pharm. Bull.*, 38, pp. 2424–2428.

Tanaka, T., Fujisaki, H., Nonaka, G.-I. and Nishioka, I. (1992a). Tannins and related compounds. CXVIII. Structures, preparation, high-performance liquid chromatography and some reactions of dehydroellagitannin-acetone condensates, *Chem. Pharm. Bull.*, 40, pp. 2937–2944.

Tanaka, T., Tong, H.-H., Xu, Y.-M., Ishimaru, K., Nonaka, G.-I. and Nishioka, I. (1992b). Tannins and related compounds. CXVII. Isolation and characterization of three new ellagitannins, lagerstannins A, B and C, having a gluconic acid core, from *Lagerstroemia speciosa* (L.) PERS, *Chem. Pharm. Bull.*, 40, pp. 2975–2980.

Tanaka, T., Nonaka, G.-I., Ishimatsu, M., Nishioka, I. and Kouno, I. (2001). Revised structure of cercidinin A, a novel ellagitannin having (*R*)-hexahydroxydiphenoyl esters at the 3,4-positions of glucopyranose, *Chem. Pharm. Bull.*, 49, pp. 486–487.

Tsai, Y.-J., Aoki, T., Maruta, H., Abe, H., Sakagami, H., Hatano, T., Okuda, T. and Tanuma, S.-I. (1992). Mouse mammary tumor virus gene expression is suppressed by oligomeric ellagitannins, novel inhibitors of poly(ADP-ribose) glycohydrolase, *J. Biol. Chem.*, 267, pp. 14436–14442.

Vlietinck, A. J., De Bruyne, T., Apers, S. and Pieters, L. A. (1998). Plant-derived leading compounds for chemotherapy of human immunodeficiency virus (HIV) infection, *Planta Medica*, 64, pp. 97–109.

Xie, L., Xie, J.-X., Kashiwada, Y., Cosentino, L. M., Liu, S.-H., Pai, R. B., Cheng, Y.-C. and Lee, K.-H. (1995). Anti-AIDS (Acquired Immune Deficiency Syndrome) agents. 17. New brominated hexahydroxybiphenyl derivatives as potent anti-HIV agents, *J. Med. Chem.*, 38, pp. 3003–3008.

Yoshida, T., Maruyama, Y., Okuda, T., Memon, M. U. and Shingu, T. (1982). Gemin B and C, dimeric ellagitannins from *Geum Japonicum*, *Chem. Pharm. Bull.*, 30, pp. 4245–4248.

Yoshida, T., Hatano, T., Okuda, T., Memon, M. U., Shingu, T. and Inoue, K. (1984). Spectral and chromatographic analyses of tannins. I. ^{13}C Nuclear magnetic resonance spectra of hydrolyzable tannins, *Chem. Pharm. Bull.*, 32, pp. 1790–1799.

Yoshida, T., Maruyama, Y., Memon, M. U., Shingu, T. and Okuda, T. (1985). Gemins D, E and F, ellagitannins from *Geum japonicum*, *Phythochemistry*, 24, pp. 1041–1046.

Yoshida, T., Ikeda, Y., Ohbayashi, H., Ishihara, K., Ohwashi, W., Shingu, T. and Okuda, T. (1986). Dimeric ellagitannins in plants of *Melastomataceae*, *Chem. Pharm. Bull.*, 34, pp. 2676–2679.

Yoshida, T., Chou, T., Haba, K., Okano, Y., Shingu, T., Miyamoto, K.-I., Koshiura, R. and Okuda, T. (1989). Camelliin B and nobotanin I, macrocyclic ellagitannin dimers and related dimers, and their antitumor activity, *Chem. Pharm. Bull.*, 37, pp. 3174–3176.

Yoshida, T., Chou, T., Nitta, A., Miyamoto, K.-I., Koshiura, R. and Okuda, T. (1990a). Woodfordin C, a macro-ring hydrolyzable tannin dimer with antitumor activity, and accompanying dimers from *Woodfordia fruticosa* flowers, *Chem. Pharm. Bull.*, 38, pp. 1211–1217.

Yoshida, T., Chou, T., Maruyama, Y. and Okuda, T. (1990b). Tannins of theaceous plants. II. Camelliins A and B, two new dimeric hydrolyzable tannins from flower buds of *Camellia japonica* L. and *Camellia sasanqua* THUNB, *Chem. Pharm. Bull.*, 38, pp. 2681–2686.

Yoshida, T., Chou, T., Matsuda, M., Yasuhara, T., Yazaki, K., Hatano, T., Nitta, A. and Okuda, T. (1991a). Woodfordin D and oenothein A, trimeric hydrolyzable tannins of macro-ring structure with anti-tumor activity, *Chem. Pharm. Bull.*, 39, pp. 1157–1162.

Yoshida, T., Chou, T., Nitta, A. and Okuda, T. (1991b). Tannins and related polyphenols of theaceous plants. IV. Monomeric and dimeric hydrolyzable tannins having a dilactonized valoneoyl group from *Schima wallichii* KORTH, *Chem. Pharm. Bull.*, 39, pp. 2247–2251.

Yoshida, T., Ohbayashi, H., Ishihara, K., Ohwashi, W., Haba, K., Okano, Y., Shingu, T. and Okuda, T. (1991c). Tannins and related polyphenols of melastomataceous plants. I. Hydrolyzable tannins from *Tibouchina semidecandra* COGN, *Chem. Pharm. Bull.*, 39, pp. 2233–2240.

Yoshida, T., Ohwashi, W., Haba, K., Ohbayashi, H., Ishihara, K., Okano, Y., Shingu, T. and Okuda, T. (1991d). Tannins and related polyphenols of melastomataceous plants. II. Nobotanins B, C and E, hydrolyzable tannin dimer and trimers from *Tibouchina semidecandra* COGN, *Chem. Pharm. Bull.*, 39, pp. 2264–2270.

Yoshida, T., Haba, K., Nakata, F., Okano, Y., Shingu, T. and Okuda, T. (1992a). Tannins and related polyphenols of melastomataceous plants. III. Nobotanins G, H and I, dimeric hydrolyzable tannins from *Heterocentron roseum*, *Chem. Pharm. Bull.*, 40, pp. 66–71.

Yoshida, T., Hatano, T., Kuwajima, T. and Okuda, T. (1992b). Oligomeric hydrolyzable tannins – Their ^{1}H NMR spectra and partial degradation, *Heterocycles*, 33, pp. 463–482.

Yoshida, T., Namba, O., Lu, C.-F., Yang, L.-L., Yew, K.-Y. and Okuda, T. (1992c). Tannins of euphorbiaceous plants. X. Antidesmin A, a new dimeric hydrolyzable tannin from *Antidesma pentandrum* var. *barbatum*, *Chem. Pharm. Bull.*, 40, pp. 338–342.

Yoshida, T., Haba, K., Arata, R., Nakata, F., Shingu, T. and Okuda, T. (1995). Tannins and related polyphenols of melastomataceous plants. VII. Nobotanins J and K, trimeric and tetrameric hydrolyzable tannins from *Heterocentron roseum*, *Chem. Pharm. Bull.*, 43, pp. 1101–1106.

Zehavi, U. (1988). Perspectives of oligosaccharide and glycosphingolipid synthesis on polymer supports, *Adv. Carbohydr. Chem. Biochem.*, 46, pp. 179–204.

Zehavi, U., Amit, B. and Patchornik, A. (1972). Light-sensitive glycosides. I. 6-Nitroveratryl-β-D-glucopyranoside and 2-nitrobenzyl-β-D-glucopyranoside, *J. Org. Chem.*, 37, pp. 2281–2285.

Zehavi, U. and Patchornik, A. (1972). Light-sensitive glycosides. II. 2-Nitrobenzyl 6-deoxy-α-L-mannopyranoside and 2-nitrobenzyl 6-deoxy-β-L-galactopyranoside, *J. Org. Chem.*, 37, pp. 2285–2288.

Chapter 6

Immunomodulatory Ellagitannin Chemistry

Adiel Coca, Ken S. Feldman* and Michael D. Lawlor

*Department of Chemistry, The Pennsylvania State University, University Park, Pennsylvania 16802 USA; *corresponding author:* ksf@chem.psu.edu

6.1 Immune Function: LPS, lipid A and TNFα

Interest in the active components of the polyphenol-rich medicinal plants used for years in China and Japan to treat a variety of inflammatory diseases has led to the isolation of numerous novel compounds. Among these species, several dimeric and macrocyclic members of the ellagitannin family of secondary plant metabolites have demonstrated the surprising ability to induce substantial tumor remission in mice inoculated with sarcoma-180, MM2 mammary carcinoma, MH134 hepatoma, and Meth-A fibrosarcoma tumor cells. Careful studies of the biological mechanism-of-action of these active ellagitannins have prompted speculation that these natural products act as immunomodulators through inducible cytokine mediators such as interleukin-1 beta (IL-1β) and tumor necrosis factor alpha (TNFα).

The antitumor chemotherapeutic window is small however, as the putative tumor-lethal cytokine TNFα is implicated as a causative agent in an assortment of diseases including most notably bacterial sepsis. Sepsis, which causes over 200,000 deaths per year, is characterized by a systemic pro-inflammatory response to the lipid A component of lipopolysaccharide (LPS) in Gram-negative bacteria and to lipoteichoic acid, peptidoglycan, and exotoxins of Gram-positive bacteria. In addition, chronic over-expression of TNFα has been linked to inflammatory diseases such as cachexia, diabetes, rheumatoid arthritis, and Crohn's disease.

The antitumor properties of TNFα include direct necrosis of solid tumors *in vitro* and disruption of the tumor vasculature *in vivo*. In fact, one of the earliest successful human cancer treatments involved administering *E. coli* extracts containing LPS to patients, presumably a regimen that involved stimulation of the secretion of non-lethal doses of TNFα *in vivo*. Similar TNFα secretion dose-response curves upon exposure of human peripheral blood mononuclear cells (hPBMC's) to either some specific ellagitannins or to LPS have raised the question of whether the tannins may be achieving their documented tumoricidal action through the well-studied LPS/lipid A cellular receptor system. Consequently, the potential for ellagitannins to act as lipid A *agonists* (*i.e.*, mimicking LPS/lipid A activity), and the prospects for redesigning the ellagitannin structure to develop a lipid A *antagonist* (*i.e.*, counteracting LPS/lipid A activity) are under study.

6.1.1 Lipid A

The outer surface of the outer membrane of the cell wall of Gram-negative bacteria is composed primarily of the amphipathic glycolipid lipopolysaccharide (Nikaido and Vaara, 1985). LPS (also known as endotoxin and enteric LPS) is the major surface-associated antigen for several Gram-negative bacteria and is involved in several pathological activities associated with the immune response of the human host (Meredith *et al.*, 2006). A single *E. coli* cell contains around 2 to 4 million LPS molecules. They are highly heat-stable, as a temperature of about 180 °C is needed to inactivate endotoxins.

6.1.1.1 *Lipid A structure and chemistry*

The main constituent and the agent responsible for the biological response to LPS is a lipid component, lipid A. LPS also contains two other structural domains: the oligosaccharide core, and the O-antigen in enteric bacteria. The O-antigen is a long heteropolysaccharide chain that is strain-specific and is usually composed of identical repeating 20 to 40 oligosaccharide units of three sugars with a branching sugar at the first and third saccharide of each unit. The oligosaccharide core is further divided into the inner and outer core. The inner core is made up of two or more 2-keto-3-deoxyoctonic acid (KDO) units and two or three L-glycero-D-mannose-heptose saccharides. The outer core is usually composed of about three sugars with one or more appended as side chains. The lipid A region anchors LPS to the cell wall, and so lysis of the bacteria is necessary for exposure of lipid A. Lipid A, whose structure is highly conserved among different species of Gram-negative bacteria, is made up of a β-D-glucosaminyl-(1-6)-D-glucosamine disaccharide (Fig. 6.1) (Yin *et al.*, 2003, Dixon and Darveau, 2005).

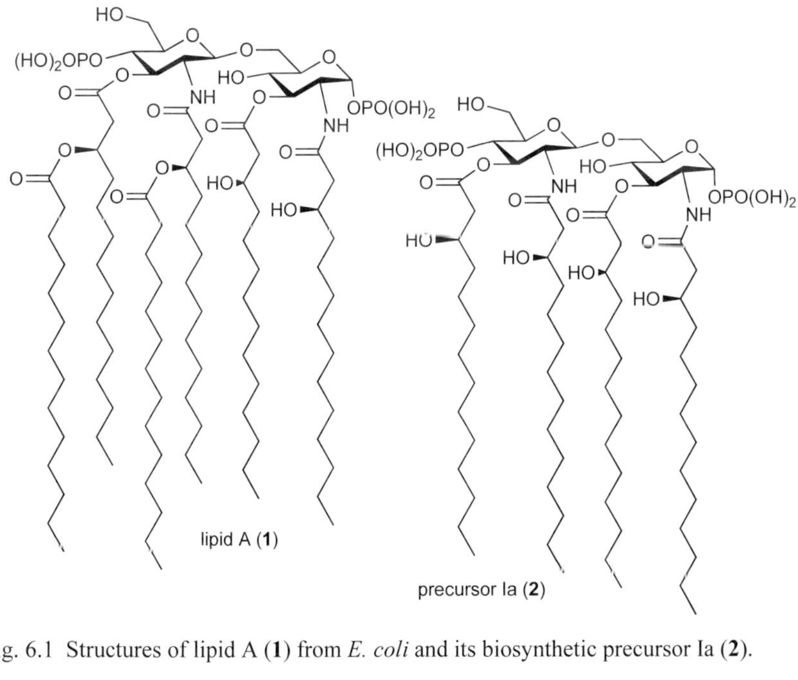

Fig. 6.1 Structures of lipid A (**1**) from *E. coli* and its biosynthetic precursor Ia (**2**).

The 1 and 4' positions are mono- or diphosphorylated, and the disaccharide can be acylated by up to four 3-hydroxy-containing or nonhydroxylated acyl residues as ester or amide linkages at specific positions (C2, C3, C2', C3'). Additional fatty acids can be attached (secondary substitution) in cases where there is a hydroxyl group on the hydrocarbon moiety. Despite the general structural conservation of lipid A seen across species, the amount and length of the fatty acid chains can vary greatly between different bacterial strains. Even LPS samples derived from the same species of bacteria can vary in structure (Lehmann and Rupprecht, 1977, Chang and Nowotny, 1975). Lipid A (**1**, Fig. 6.1) from *Escherichia coli*, which carries five fatty acids that are fourteen carbons in length and one which is twelve carbons in length, is exemplary of this structural class (Reitschel *et al.*, 1996).

The structural details of each lipid A isoform determines whether the compound will cause a host-mediated response or not. For example, lengthening the acyl chains leads to a reduction in cytotoxicity, as evidenced by the lipid A's of *Chlamydia psittaci*, *Bacteroides fragilis*, and *Legionella pneumophilia* (Heine *et al.*, 2003). Those lipid A structures featuring either a monosaccharide core or a dimeric structure with only four acyl side chains are significantly less toxic (Demchenko *et al.*, 2003). Pentacylated forms of lipid A are less active than the natural hexaacylated forms, whereas bisacylated and tetraacylated derivatives as seen in the antagonistic biosynthetic precursor Ia (**2**) (also known as lipid IVa or compd. 406) are completely inactive (Liu *et al.*, 1997, Flad *et al.*, 1993). Interestingly, synthetic *E. coli* lipid A is less potent than the natural material. The chirality of the 3-hydroxyacyl chains apparently plays no role on the biological activity of the substance as both *R*- and *S*-configurations induce similar responses. The incorporation of at least one phosphate group is a requirement for bioactivity. These residues apparently are needed for recognition by the cell-surface receptor(s) for LPS (Demchenko *et al.*, 2003).

6.1.1.2 *Role in disease*

LPS via the lipid A component is one of the most potent immunostimulatory substances known (as low as ng to pg/mL range)

(Rietschel and Westphal, 1999). LPS released in a human host due to bacterial growth or bacterial lysis interacts with a pattern-recognition receptor complex present in many host cells including monocytes and macrophages. This recognition event activates these cells to release several pro-inflammatory cytokines, members of a large class of small intercellular signaling molecules consisting of more than 160 polypeptides that orchestrate immune responses to invading pathogens by allowing cells to communicate with one another. Among these cytokines are tumor necrosis factor alpha (TNFα), interleukin-1 (IL-1), and interleukin-6 (IL-6). The cytokines then activate other phagocytic cells and up-regulate the innate immune system in an attempt to clear the invading microbe. This systemic inflammatory response to LPS leads to sepsis. Sepsis is characterized by fever, mental confusion, diarrhea, and hypotension. Severe sepsis, commonly referred to as *septic shock syndrome*, is a very dangerous disease that can lead to organ damage and even failure, and eventually death. An LPS concentration of less than 1 ng/mL is sufficient to cause the symptoms of sepsis, but more than 100 ng/mL is required to produce septic shock (Parrillo, 1993). After septic shock onset, patients also can succumb to secondary infections caused by opportunistic pathogens. That is, following the initial period of hyperimmune response, several anti-inflammatory cytokines are released to maintain homeostasis and as a result patients may become temporarily immunodepressed and susceptible to other pathogens.

6.1.1.3 *Impact on human health*

Despite great efforts, sepsis continues to be a severe health problem with mortality rates ranging between 20-40% overall, and 40-60% for septic shock cases (Friedman *et al.*, 1998, Wheeler and Bernard, 1999). Although the mortality rate has been decreasing over the last few decades, it is still the tenth leading cause of death in the United States. There were 660,000 cases (240 per 100,000 population) of sepsis reported in 2000 according to a joint study by the Center for Disease Control and Prevention and Emory University School of Medicine (Martin *et al.*, 2003). It has been an especially difficult problem in men (more common than in women by a factor of 1.28) and nonwhites (mean relative risk of 1.90 compared to whites). Sepsis is most dangerous in black men, the elderly, immuno-compromised individuals, and in critically ill patients.

Before 1987, Gram-negative bacterial infections were the principal cause of sepsis, but in more recent years, this role has been overtaken by Gram-positive bacteria. In 2000, Gram-positive bacteria accounted for 52.1 percent, Gram-negative bacteria 37.6 percent, polymicrobial origins 4.7 percent, fungi 4.6 percent, and anaerobes 1.0 percent of all cases (Martin *et al.*, 2003). The organs that failed the most frequently were the lungs (18%), followed by the kidneys (15%), and then cardiovascular (7%), hematological (6%), metabolic (4%), and neurological failure (2%). At present, there is no effective treatment for severe sepsis (Shiozaki *et al.*, 2006).

6.1.2 Lipoteichoic acid, peptidoglycan, and exotoxins

Unlike Gram-negative bacteria, Gram-positive bacteria have a number of active components that promote an immunological response, including lipoteichoic acid (LTA), peptidoglycan (PGN), and exotoxins. Compared to LPS, much less is known about these components of Gram-positive bacteria. LTA and PGNs, similarly to LPS, induce the release of TNFα, interleukin-1 (IL-1), IL-6, INF-γ, and nitric oxide from macrophages, monocytes, and several other cells (Bhakdi *et al.*, 1991, Keller *et al.*, 1992, Kengatharan *et al.*, 1996, Wang *et al.*, 2000). LTA (**3**), which resembles LPS in many aspects, is a glycolipid made up of a diacylglycerol lipid portion and an unbranched chain of repeating units of polyglycerophosphate that is bound to a hexapyranosyl sugar (Fig. 6.2). As with LPS, differences in the acyl chain length are observed among different bacterial species. The immunological potency of LTA is highly dependent on the bacterial species. PGN consists of a tetrapeptide portion of L-alanine, D-glutamine, L-lysine, and D-alanine, as well as a glycan portion containing repeating disaccharide *N*-acetylmuramic acid-(β1-4)-*N*-acetylglucosamine units (between 5 and 30). The short peptide is cross-linked to the peptide portion of another chain by a pentaglycine group to provide the molecular network that makes up the cell wall. The exotoxins are secreted peptides that may act as potent T cell activators and lead to uncontrolled release of proinflammatory cytokines.

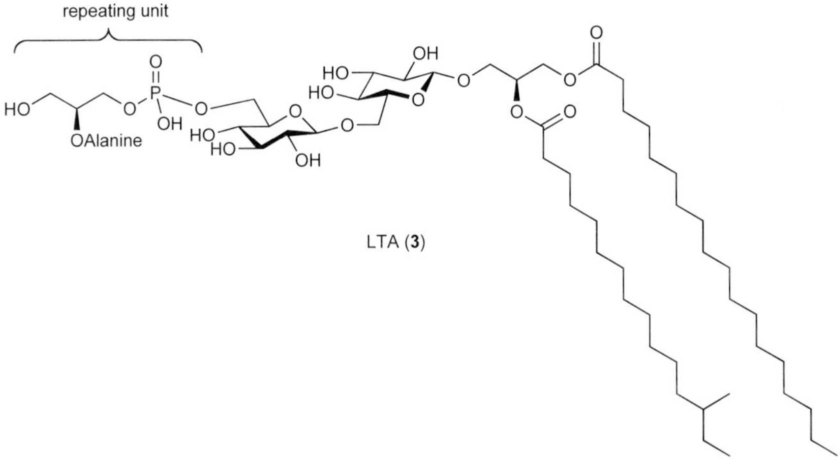

Fig. 6.2 Structure of LTA from *Staphylococcus aureus*.

6.1.3 TNFα

Tumor necrosis factor alpha (TNFα), also known as cachectin, is a pleiotropic cytokine with broad biological capabilities. Like other cytokines, TNFα has several advantageous properties including accelerating wound healing, cell growth modulation, cellular differentiation, and endogenous antimicrobial activity through activation of the immune system. TNFα was initially identified as an anticancer agent in 1975 when it was observed to destroy solid murine tumors *in vivo* and was cytotoxic against several tumor cell lines (Carswell *et al.*, 1975, Helson *et al.*, 1975). Since these seminal reports detailing TNFα's astonishing antitumor properties, its other (and major) role as a pro-inflammatory mediator implicated in the host of diseases discussed above has come to light. Many stimuli can activate the synthesis of TNFα including TNFα itself, other cytokines (such as IL-1, INF-γ, IL-2, platelet activating factor, and granulocyte macrophage-CSF), phorbol ester, IgE, reactive oxygen metabolites, X-rays, mycobacterial proteins, cyanobacteria, zymosan (an insoluble carbohydrate from the cell wall of yeast), malarial parasite antigens, viruses (Sendai, *Herpes simplex*, and Epstein-Barr), and, most importantly, LPS (Sidhu and Bollon, 1993). Several known agents are able to down-regulate TNFα production. These species include cytokines (IL-4, IL-10, and transforming growth factor-β), prostaglandin E_2, Iloprost (a prostacycline analog), histamine, antioxidants (such as *N*-acetylcystine and glutathione), glucocorticoids (including dexamethasone, cortisol, and danazol), lipoxygenase inhibitors, and certain drugs (such as pentoxifylline, thalidomide, ambroxol, and cyclosporin A).

6.1.3.1 *TNFα generated from LPS exposure: mediator of septic shock*

The link between TNFα and LPS goes back more than 100 years. At the conclusion of the 19th century, the New York surgeon William Coley reported that inoperable tumors could be treated with Gram-negative and Gram-positive bacteria (Coley, 1896). Carswell and co-workers later coined the term "tumor necrosis factor" for the cytokine that caused hemorrhagic necrosis of tumors in animal models injected with LPS (Carswell *et al.*, 1975, Helson *et al.*, 1975). TNFα is produced by cells that have been activated by LPS. LPS (via the lipid A component) is capable of activating a variety of cells including macrophages, monocytes, fibroblast, astrocytes, Kupffer cells, smooth muscle cells, keratinocytes, neutrophils, lymphocytes, platelets, microglial cells, intestinal paneth cells, endothelial, mast cells, natural killer cells, and tumor cells (Dixon and Darveau, 2005). TNFα was purified and characterized by Aggarwal *et al.* (1985). Its crystal structure has been reported (Jones *et al.*, 1992). At present, there are 13 different members in the TNF ligand family, including another factor called TNFβ (also known as lymphotoxin, LTα, or cytotoxin), which has similar cytotoxic activities to TNFα, since it binds to the same receptors, but it is produced only by activated T-lymphocytes.

TNFα is first expressed as a 26 kDa (233 amino acids) transmembrane peptide that is cleaved by the metalloprotease TNF-alpha-converting-enzyme (TACE) in activated cells to the soluble 17 kDa (157 amino acids) mature protein (Black *et al.*, 1997a). The 26 kDa pro-peptide also is functional through cell-to-cell contact. The cleaved

TNFα generates its diverse biological effects through binding to two distinct cell surface receptors of 55-60 kDa (known as p55TNFR, TNFR1, or CD120a) and of 75-80 kDa (p75TNFR, TNFR2, or CD1120b). It is believed that p55TNFR is the principal receptor, whereas p75TNFR is responsible for increasing the local concentration of TNFα and passing it on to p55TNFR once the ligand has been bound. p75TNFR is able to play its supportive role due to its higher affinity for TNFα compared with p55TNFR ($K_d = 0.1 \times 10^{-9}$ M *vs.* 0.5×10^{-9} M, respectively). TNFR1 is found on all nucleated cells, while TNFR2 is only expressed on hematopoietic cells. The binding of TNFα to TNFR1 initiates a kinase cascade that leads to the activation of members of the mitogen-activated protein kinase (MAPK) family: extracellular signal-regulated kinases (ERKs), *jun* terminal kinase-1 (JNK-1 or stress-activated protein kinase-1), and p38/stress-activated protein kinase-2 (p38/SAPK-2). These kinases activate the transcription factors Elk-1, ATF2, c-jun, and NF-κB, which results in the secretion of other cytokines, expression of adhesion molecules, and production of proteases. Both TNF receptor types, after the initial response to cytokines, are shed from host cells and the soluble p55 and p75 receptors become natural inhibitors of TNFα and help clear the cytokine from the host.

The soluble TNFα, which exists as a compact bell-shaped noncovalently linked homotrimer, recruits several other phagocytic cells after release in an attempt to clear the pathogen, and through a series of events involving several cytokines, also induces inflammation, fever, and hypotension. An X-ray defraction analysis of the human TNFβ trimer attached to TNFR1 has revealed that the ligand-receptor interaction occurs mainly at the grooves between the TNFβ monomers, which explains why TNFα and TNFβ monomers are not bioactive (Smith *et al.*, 1994).

6.1.3.2 *Potential for anticancer therapy*

Ever since the initial discovery of TNFα's antitumor activity, numerous groups have reported its use in several Phase I and II clinical trials against a variety of tumor types including sarcoma, colorectal carcinoma, renal cell carcinoma, and melanoma (Kemeny *et al.*, 1990, Rinehart *et al.*, 1990). As mentioned above, TNFα recruits other cytokines such as IL-2 and IFN-γ, and through a complex immune mediated pathway, tumors are disrupted and eliminated. The results of much research support the conclusion that TNFα causes hemorrhagic tumor necrosis through angiogenic endothelial cell apoptosis. Current thinking favors the point of view that TNFα deactivates an important adhesion receptor in tumor angiogenesis, integrin $\alpha_v\beta_3$, which results in a lethal disturbance of the neovasculature. Integrin $\alpha_v\beta_3$ is a continuously active integrin that anchors endothelium cells to the extracellular matrix proteins (ECM). It is essential for neovasculature endothelial cell survival. Altering the $\alpha_v\beta_3$-dependent adhesion of endothelial cells to the ECM proteins causes eventual cell death (Ruegg *et al.*, 1998, Havell *et al.*, 1988). The damaged endothelial cells are then destroyed by T cells. Since integrin $\alpha_v\beta_3$ is only expressed in neovasculature, normal vasculature remains unaffected. TNFα is also known to increase vascular permeability and reduce interstitial fluid pressure within the tumor, allowing easier penetration of anticancer drugs into the tumor site (Folli *et al.*, 1993, Kristensen *et al.*, 1996). This latter effect may partly explain why several chemotherapeutic drugs such as the alkylating agent melphalan (**4**, see Fig. 6.3), marketed as Alkeran, act synergistically with TNFα.

Unfortunately, despite these encouraging findings on the potential for fighting cancer, the adverse effects of systemic TNFα production are too overwhelming to permit its use as a general anticancer therapy. It has been reported that humans can only tolerate about 2 percent of the dose/kilogram needed in mice for tumor regression (about 50 mg/kg) (Spriggs and Yates, 1992, Sidhu and Bollon, 1993). Direct TNFα infusion also has been unsuccessful because the plasma half-life of TNFα is about 15 minutes in humans, insufficient time for complete removal of a tumor (Blick *et al.*, 1987). Under these circumstances, the injected cytokine only causes a transient increase in the host's response before it is cleared by the liver or kidney.

It became clear that localization of the cytokine around the tumor would be the most viable approach to effective and safe treatment. Localized application of TNFα has taken several forms, mostly relying on linking the cytokine to some sort of tumor-selective delivery vehicle.

Currently, a gene therapy product of the human TNFα gene, called TNFerade™, is in phase II trials. TNFerade is an adenoviral vector that expresses the human TNFα cDNA with the ligated radiation-inducible gene promoter *Egr-1*. It is administered through intratumoral injection and is being used in combination with radiation therapy in the treatment of pancreatic cancer (Senzer *et al.*, 2004). In addition, administration of TNFα in combination with IFN-γ during isolated limb perfusion has shown some potential in fighting melanoma (Thom *et al.*, 1995). Targeted delivery of TNFα attached to tumor-homing proteins is under study as well. One coupled system called L19mTNFα contains the peptide scFv L19, which targets the ED-B region of fibronectin, fused to TNFα (Balza *et al.*, 2006). The ED-B domain of fibronectin, an ECM protein, is commonly expressed at angiogenic sites, but is undetectable in normal vasculature. Phase I studies of another adduct called NGR-TNF, a couple between TNFα and a peptide containing a Cys-Asn-Gly-Arg-Cys (CNGRC or NGR) motif, are underway. This particular motif is a ligand of an isoform of CD13 (also known as aminopeptide N), which is highly expressed in angiogenic vessels and in some normal cells, but with a much lower affinity for the NGR ligand (Corti and Ponzoni, 2004, Curnis *et al.*, 2002, Sacchi *et al.*, 2006). NGR-TNF was found to work synergistically with the topoisomerase II blocker doxorubicin (**5**, also known as adriamycin, see Fig. 6.3) by improving substantially its permeability through murine B16 melanoma cell membranes. A third entry, scFvMEL/TNF, is a covalent fusion construct between TNFα and the single-chain antibody scFvMEL that is specific for the surface domain of the gp240 peptide (Liu *et al.*, 2006, Rosenblum *et al.*, 1991, Murray *et al.*, 1987). The gp240 glycoprotein has been found in 80% of melanoma and 67% of lobular breast tumors, but not in normal cells. scFvMEL/TNF was synergistic in its action with the antitumor agents doxorubicin or 5-FU (**6**, Fluorouracil, see Fig. 6.3), a pyrimidine analogue and thymidylate synthase inhibitor, against A375 melanoma cells.

It should be mentioned that recent evidence, which connects chronic synthesis of TNFα to initiation or progression of tumors, has been reported (Szlosarek *et al.*, 2006, Yan *et al.*, 2006). Chronic inflammation leads to over-replication of several cytokine genes in activated cells, increasing the opportunity for replication errors. Furthermore, TNFα recruits inflammatory factors and cells, a process that promotes angiogenesis and provides greater access to other inflammatory factors and cells. This action inadvertently results in feeding and sustaining the tumor and possibly induces eventual malignancy. Several molecular mechanisms by which TNFα may promote cancer growth have been proposed, including increasing cell cycle promoter levels while decreasing levels of cyclin-dependent kinase inhibitors, increasing ras or c-myc (molecular switches of the growth factor receptor signaling pathways), promoting chemo-resistance and androgen insensitivity, activation of NF-κB (suppressing normal cell apoptosis), and inducing DNA damage while also inhibiting DNA repair through up-regulation of nitric oxide production (Szlosarek *et al.*, 2006).

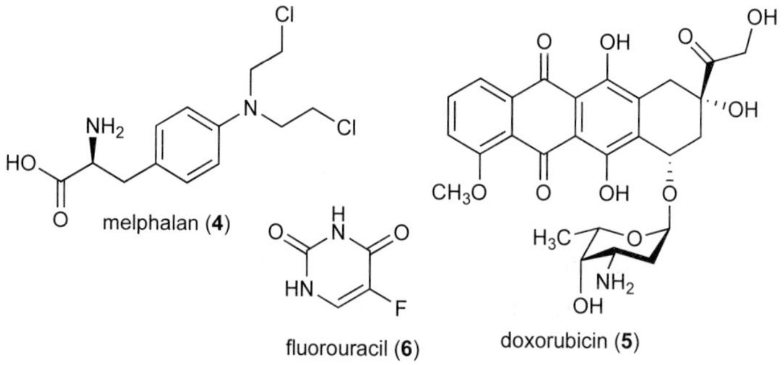

Fig. 6.3 Anticancer agents used in combination with TNFα-based drugs.

6.1.3.3 *Monocyte response pathway*

Bacterial components such as LPS are usually first recognized by monocytes and macrophages through the lipid A receptor system, as will be described in the next section. The lipid A region of LPS is recognized by either a bound receptor cluster found on mononuclear phagocytes (monocytes and macrophages) or by a soluble binding molecule which activates cells that lack the lipid A receptor system. This receptor binding leads to the transcription of more than 120 genes, including those that code for several cytokines (Wang *et al.*, 2000). Among the first expressed cytokines are TNFα, interleukin-1α, interleukin-1β, and interleukin-6. These pleiotropic cytokines are first produced locally but also can be released by cells at sites remote from the infection. Inactivated monocyte cells are known to continuously secrete low levels of TNFα, and LPS stimulation leads to increased levels within minutes. Circulating TNFα levels peak at 1.5 h when the host is challenged with LPS and usually return to normal within 4 h of response (Taveira da Silva *et al.*, 1993). Direct administration of recombinant TNFα to animals has led to symptoms similar to those seen in human septic shock (Maury and Teppo, 1987, Muto *et al.*, 1988). Use of TNFα antibodies prevented the same reaction when mice and primates were exposed to otherwise lethal levels of LPS (Beutler *et al.*, 1985, Tracey *et al.*, 1987). Interestingly, the levels of other cytokines such as IL-1β and IL-6, which are normally associated with sepsis, also were attenuated, an observation that suggests an early role for TNFα in the development of septic shock progression (Fong *et al.*, 1989). In addition to the release of cytokines, activated macrophages also increase their own intracellular stores of antimicrobial agents such as lysozyme, cationic proteins, acid hydrolases, lactoferrin, and oxygen free radical precursors.

Following LPS exposure, several other cytokines have been detected in circulation: interleukin-2, interleukin-8, interleukin-11, interleukin-12, interleukin-15, interleukin-18, interferon-γ (INF-γ), leukemia inhibitory factor, ciliary neurotrophic factor, and colony-stimulating factor (CSF), among others (Cavaillon, 2003). Furthermore, activation of the complement system occurs (release of anaphylatoxins C3a and C5a) and the discharge of other inflammatory mediators, including phospholipase A2, cyclooxygenase (COX), adhesion molecules (selectins E, P, and L), platelet-activating factor (PAF) follows. Release of coagulants such as thrombin, coagulation factors (IXa, Xa, XIa, and XIIa), and tissue factor, also occurs (Sharma and Dellinger, 2003, Demchenko *et al.*, 2003). These inflammatory and adhesion molecules recruit phagocytic polymorphonuclear (PMN) leukocytes and other mononuclear phagocytes. Activated phagocytes then take up and destroy both LPS and whole bacteria via their lysosomal antimicrobial mediators. In addition, active B and T lymphocytes are also attracted to the site of infection and in turn release mediators such as IL-2, INF-γ, and granulocyte-macrophage (GM)-CSF (Akagawa and Tokunaga, 1985, Bone, 1991, Heinzel *et al.*, 1994, Jaeschke, 1996, Yong and Linch, 1993). These factors further activate mononuclear and PMN phagocytes. Unfortunately, several of the produced microcidal agents tend to leak out and cause the severe damage of the surrounding endothelia and vasculature that is often seen in sepsis patients. Blood pressure also decreases with infection as the result of the up-regulation of nitric oxide synthase. These symptoms and others such as fever, coagulation disorders, microthrombi, and myocardial depression may result in multiple organ failure and finally lead to septic shock.

In order to maintain homeostasis, anti-inflammatory cytokines also are released; interleukin-4, interleukin-10, interleukin-13, and transforming growth factor-β, as well as natural cytokine inhibitors that includes soluble TNF receptors, IL-1 receptor antagonist, and soluble IL-1 receptors (Cavaillon, 2003). In addition, liver parenchymal cells produce acute-phase proteins such as C-reactive protein, serum amyloids (A and P), hemopexin, haptoglobin, complement C3 and C9, α_1-acid glycoprotein, α_2-macroglobulin, and proteinase inhibitors to ameliorate the damaged caused by the inflammatory response (Fey *et al.*, 1994, Schumann and Zweigner, 1999, Steel and Whitehead, 1994, Kuipers *et al.*, 1994, Ramadori *et al.*, 1990).

6.1.3.4 *Role of LPBP, CD14, Tlr4, and other proteins*

In humans, there is a highly developed pathogen recognition immune system. In particular, several bacterial products are recognized as xenobiotic including foreign DNA and RNA, lipoproteins, flagellin, zymosan, peptidoglycan, and, of course, LPS. The pathogen-associated molecular patterns are recognized by a group of at least ten membrane receptors called toll-like receptors (TLRs). Among these receptors is TLR4, a 92 kDa protein (841 amino acids), that is partly responsible for recognizing LPS via the lipid A binding site (Fig. 6.4). TLR4 also binds several other foreign ligands including pneumolysin from *Streptococcus pneumoniae*, respiratory syncytial virus (RSV) coat protein, teichuronic acids, bacterial heat shock protein (Hsp) 60, as well as host-born ligands such as lung surfactant protein A, hyaluronin oligosaccharides, heparan sulfate, and fibrinogen fragments (Rossignol and Lynn, 2005). TLR4-deficient mice (C3H/HeJ) are hypo-responsive to the effects of LPS, an observation that will be used to advantage in probing the role of ellagitannins in immunostimulation (see Section 6.4.2) (Hoshino *et al.*, 1999). Interestingly, TLR4 knockout mice showed an immunological response to whole *E. coli* similar to that observed with normal mice, suggesting that there are other receptors involved (Evans *et al.*, 1993). It is now known that there are other bacterial components including lipoprotein, peptidoglycan, bacterial DNA with a CpG sequence, flagellin, and fimbriae, which can act as antigens through other TLR members such as TLR2, TLR5, or TLR9 (Takeuchi *et al.*, 1999, Frendeus *et al.*, 2001, Hayashi *et al.*, 2001, Hemmi *et al.*, 2000). Furthermore, some bacteria such as the Gram-negative *Porphyromonas gingivalis* with highly heterogeneous LPS use both TLR2 and TLR4 for recognition (Shimazu *et al.*, 1999).

LPS circulating in an infected host is first recognized by the serum protein lipopolysaccharide binding protein (LBP), a 60kDa carrier peptide with a very high affinity for the lipid A portion of LPS (K_d varying between 1 to 58 nM) (Schumann *et al.*, 1990, Tobias *et al.*, 1989, Gazzano-Santoro *et al.*, 1994). LBP, an acute-phase protein, functions as an opsonin (binding enhancer for receptors) that aids in the recognition of LPS by macrophages through binding to the surface of bacteria or to the lipid A segment of LPS that is presented by erythrocytes (Darveau *et al.*, 2004). Without infectious microbes in the serum, LBP levels remain low, but after infection, they are rapidly elevated by the liver, lungs, kidneys, and heart.

Once bound, LPS is transported by LBP to either the soluble (sCD14) or the membrane-anchored CD14 receptor (mCD14), which is attached to the cell membrane by a glycosylphosphatidylinositol (GPI) group (Solomon *et al.*, 1998, Wright, 1995, Arditi *et al.*, 1993). The 55kDa glycoprotein CD14 was the first identified component of the LPS receptor system (Wright *et al.*, 1990a). The soluble CD14 glycoprotein mediates recognition by cells such as endothelial cells that lack the membrane bound form of the receptor. The level of sCD14 is substantially increased in septic shock patients. Membrane bound CD14 is expressed mainly on myeloid cells, including macrophages, monocytes, and granulocytes, and some other cells such as liver parenchymal cells, gingival fibroblasts, and B cells. Mice lacking the

CD14 receptor molecule showed little-to-no response when injected with otherwise lethal doses of either LPS or whole *E. coli* (Haziot *et al.*, 1996). In humans injected with LPS, CD14 antibodies partially reduced the clinical symptoms and cytokine response of sepsis (Verbon *et al.*, 2001). These data, and the fact that the CD14 protein lacks an intracellular domain necessary for signal transduction, suggested that there was another transmembrane protein that was part of the LPS receptor system. This elusive signal transmitting protein was, in fact, identified as TLR4 and through an as yet unidentified interaction, CD14 helps transfer LPS to this receptor/signaling molecule. In addition, TLR4 appears to require an accessory protein called MD-2 in order to engage LPS (Shimazu *et al.*, 1999). MD-2, a secreted glycoprotein, forms a complex with the membrane bound TLR4 for effective recognition of this ligand. MD-2 knockout mice did not induce up-regulation of TNFα in response to LPS, demonstrating that MD-2 is a necessary adaptor component for signaling (Nagai *et al.*, 2002).

Malhotra also has suggested that cell-surface adhesion molecules known as selectins may act as low-affinity receptors for LPS, also through the lipid A recognition site, when it is present in high concentrations. CD14 knockout mice have shown a CD14-dependence to macrophage activation at concentrations lower than 10 ng/mL of LPS, but not at greater than 100 ng/mL of LPS (Loppnow *et al.*, 1995, Malhotra and Bird, 1997). At the higher concentrations, the CD14-lacking mice exhibited a response similar to that of the wild-type mice. P-selectin and L-selectin have been reported to bind LPS (Malhotra *et al.*, 1996, 1998). In addition, the use of antibodies against both receptors significantly reduced their ability to bind fluorescently labeled LPS and decreases the levels of both TNFα mRNA and oxygen radicals formed in neutrophils. These observations have lead Malhotra's group to suggest that the selectins act as LPS receptors at high LPS concentrations.

Moesin, a 78 kDa cell-surface protein with both a diverse biological activity profile and signal transduction capability, may also be involved in the recognition of LPS. An antimoesin monoclonal antibody completely blocked LPS induced TNFα release without impeding LPS-to-CD14 binding (Tohme *et al.*, 1999, Iontcheva *et al.*, 2004). Van Dyke and co-workers have shown that moesin is able to bind LPS, CD14, and

TLR4. At this time, it is not known whether moesin acts as an independent LPS receptor or in unison with CD14/TLR4, although the latter appears more likely.

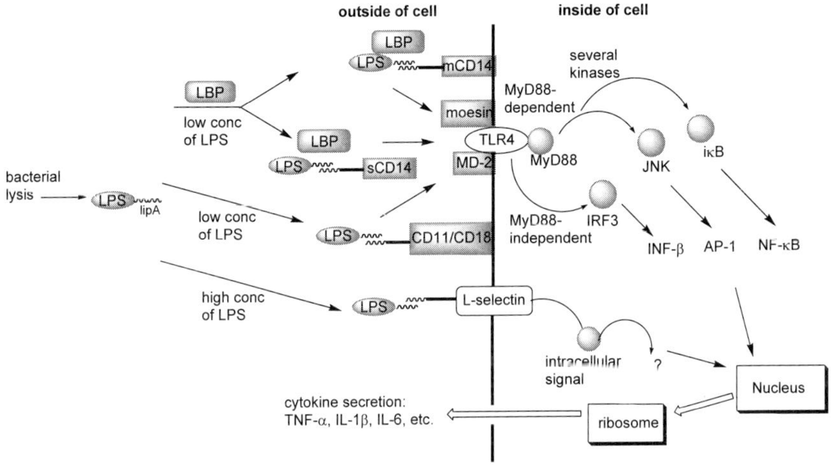

Fig. 6.4 Lipid A recognition leading to cytokine secretion.

Another receptor pathway that has been implicated in the recognition of LPS utilizes CD18 antigens, which are also called CD11/CD18 or β_2-integrins. This group of cell membrane glycoproteins was actually discovered before the CD14 receptor molecule, but due to their lower affinity for LPS, they have been posited to play only a minor role in its recognition (Wright and Jong, 1986). LPS stimulation of mononuclear cells lacking the CD18 receptor indicated that there was another activation pathway, which later led to the discovery of CD14 (Wright *et al.*, 1990b, 1989). The three known CD18 members, which are all expressed on leukocytes, are CD11a/CD18 (also known as $\alpha_1\beta_2$-integrin and LFA-1), CD11b/CD18 (or $\alpha_2\beta_2$-integrin, CR3, and MAC-1), and CD11c/CD18 (or $\alpha_3\beta_3$-integrin, CR4, p150,95). There is some evidence that CD11/CD18, much like CD14, signals through TLR4 and is necessary for optimal production of COX-2, IL-2, and IL-12, but it is not needed for the LPS-induced expression of cytokines such as TNFα (Perera *et al.*, 2001, Ingalls and Golenbock, 1995). Other unidentified molecules may be involved in the recognition of LPS.

Triantifilou and colleagues have reported that LPS was able to bind several proteins like the heat shock proteins Hsp70 and Hsp90, chemokine receptor 4, and growth differentiation factor 5 (Triantafilou *et al.*, 2001, Triantafilou and Triantafilou, 2002). Recent reviews that have focused on LPS/lipid A receptor molecules can be consulted for further details (Dauphinee and Karsan, 2006, Amersfoort *et al.*, 2003).

After LPS interacts with the TLR4/MD-2 complex, several intracellular signaling pathways are activated. The cytoplasmic portion of TLR4 belongs to a signaling domain family called the Toll-interleukin-1 receptor (TIR) domain, which is shared by all TLRs and the IL-1 receptor family. Activation of TLR4 by LPS leads to dimerization of the TIR and recruitment of TIR adaptor proteins (TIRAP) such as myeloid differentiation factor 88 (MyD88), MyD88 adaptor-like protein (Mal or TIRAP), TIR-containing adaptor inducing IFNβ (TRIF, TIRAP-1, or TICAM-1), and TRIF-related adaptor molecule (TRAM, TIRAP-2, or TICAM-2) to enhance signaling. The TIR serves as a dock for these adaptor proteins. Then, one or both of two pathways are activated: the MyD88-dependent pathway and the MyD88-independent pathway. The MyD88-dependent pathway acts through a series of two different cascade signaling channels involving several mediator proteins, starting with MyD88, which leads to the independent activation of c-Jun N-terminal kinase (JNK) and degradation of the inhibitor of NF-κB (iκB). JNK is a mitogen-activated protein kinase (MAPK). Activation of JNK and degradation of iκB results in the release of several inflammatory cytokines including TNFα, IL-1, and IL-6 through the activation of two transcription factors, activator protein-1 (AP-1) and NF-κB, respectively. Also produced are monocyte chemoattractant protein 1 (MCP-1), macrophage inflammatory protein 3α (MIP-3α), IL-8, and cyclooxygenase 2 (COX-2). The MyD88-independent pathway, which was discovered with MyD88 knockout mice, after a series of events, ends in the activation of interferon regulatory factor 3 (IRF3, a transcription factor) and eventually the production of INF-β. Released INF-β can then influence the production of antimicrobial agents such as type 1 INF-α/β, INFα inducible protein-10 (IP-10), monocyte chemoattractant protein-5 (MCP-5), RANTES, and nitric oxide (Kawai *et al.*, 2001, Zughaier *et al.*, 2005). In the MyD88-independent pathway, the production of pro-inflammatory cytokines is drastically reduced, since JNK and NF-κB activation is delayed. Although both the MyD88-dependent and -independent pathways are active in response to LPS, the lipid A structure can apparently influence which signaling pathway is activated (Zughaier *et al.*, 2005). A more thorough discussion of the TLR4 intracellular signaling pathways can be found in some current reviews (Akira and Takeda, 2004, Takeda and Akira, 2004).

Internalization of LPS is a critical step in removing the pathogenic antigen. It has been found that TLR4 does not appear to mediate this step, although much work remains in order to sort out the specific receptor(s) involved. Inhibition of TLR4 with a human antibody or using monocytes from TLR4 knockout mice (C57BL/6) leads to decreased cytokine output, but does not diminish the accumulation of fluorescently-labeled LPS inside of the cells (Zhou *et al.*, 2004, Dunzendorfer *et al.*, 2004). Using CD14-deficient murine monocytes confirmed that mCD14 is necessary for LPS uptake. Recent evidence seems to show that scavenger receptors, known to bind both macromolecules having a negative charge and modified low-density lipoproteins, are involved in the uptake of LPS/LBP complexes (Hampton *et al.*, 1991, Shnyra and Lindberg, 1995, Vishnyakova *et al.*, 2003, Seternes *et al.*, 2001). Monocytes, macrophages, and PMN are among the most active phagocytic cells. Liver cells, especially Kupffer cells (liver macrophages) are also very active in the clearance of both bacterial components and whole bacteria.

Similarly to lipid A, the peptidoglycan (PGN) and lipoteichoic acid (LTA) of Gram-positive bacteria are recognized by CD14, but unlike lipid A, this interaction does not play a prominent role in stimulating TNFα secretion (Haziot *et al.*, 1999). Furthermore, LBP does not mediate the transfer of LTA or PGN to the CD14 receptor. In contrast to Gram-negative bacteria, Gram-positive bacterial components such as LTA and PGN are mainly recognized by TLR2 (Schwandner *et al.*, 1999). However, LTA recognition through TLR4 also has been observed (Takeuchi *et al.*, 2000). Since TLR2 and TLR4 share a substantial part of the TLR signaling pathway, binding to either member results in nearly identical immunological responses involving mostly the same mediators that are generated by LPS stimulation (TNFα, IL-1β, IL-6, and INF-γ).

Other highly active immunogenic components of Gram-positive bacteria are the exotoxins, which act mainly as antigens that activate T cells, leading to the generation of high levels of IL-8 and low levels of TNFα, IL-1, and IL-6. A more thorough comparison between the two responses has been described (Amersfoort *et al.*, 2003).

6.1.4 Small molecule mediators of LPS/TNFα-induced septic shock response

Several small molecules belonging to diverse structural classes either enhance or inhibit the native immunostimulatory response to LPS. Although the exact molecular target has not been discovered for many of these compounds, for some species this point is settled. These small molecule mediators act through several distinct targets within the recognition, signaling pathway, transcription, translation, and post-translational protein modification systems.

Besides small molecule modulators, several biological mediators such as growth hormone, catecholamines, and IL-10 also have been studied. Some peptide-based drug candidates are currently in clinical studies as cytokine inhibitors for the treatment of various inflammatory diseases. These agents include the soluble p75 TNFα receptor Etanercept (Enbrel), the human TNFα antibodies Infliximab (Remicade), Adalimumab (Humira), and Golimumab (CNTO 148), and the human IL-1 receptor antagonist Anakinra. Today, all of the approved anti-cytokine drugs are peptide-based, although some small molecules are currently in clinical trials.

6.1.4.1 *LPS agonists*

There are very few reports of LPS (lipid A) agonists in the literature. One example is the microtubule stabilizer taxol (**7**, Fig. 6.5), which mimics LPS in decreasing TNFR cell surface expression and promoting TNFα release (Ding *et al.*, 1990, Byrd-Leifer *et al.*, 2001). It has been reported to act by binding to TLR4. In addition, some 4-*O*-phosphono-D-glucosamine derivatives such as GLA-60 (**8**) and its analogues have shown agonistic properties. Synthetic acyclic lipid A-like analogues such as ER-112022 (**9**) also have shown agonistic activity through the TLR4 receptor without the need for CD14 (Lien *et al.*, 2001).

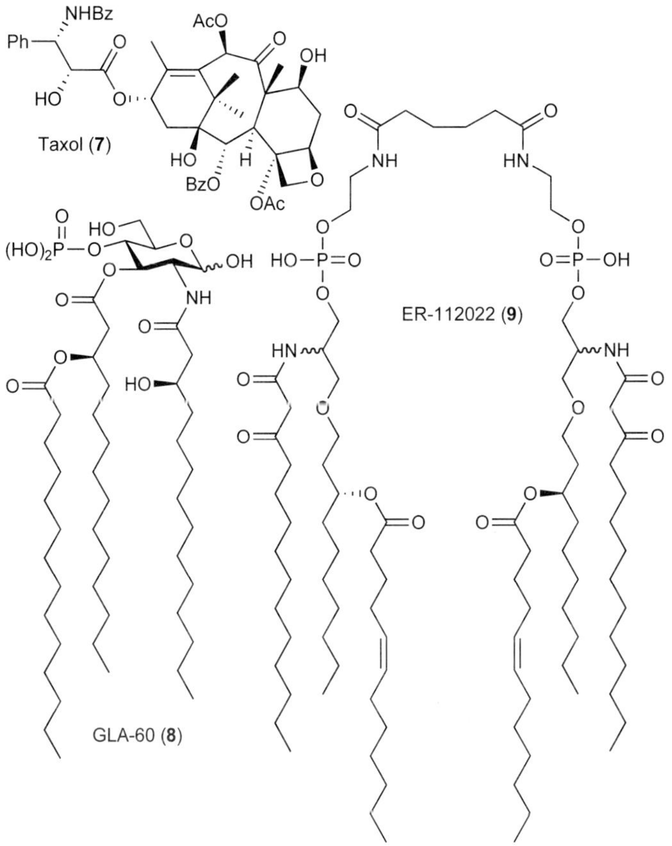

Fig. 6.5 Some LPS agonists.

6.1.4.2 *LPS antagonists*

There are considerably more reports of LPS antagonists than LPS agonists in the literature (Black *et al.*, 1997b, Newton and Decicco, 1999, Paul *et al.*, 2006). One of the earliest approaches to developing LPS antagonists relied on analogues of active lipid A itself. These analogues competitively inhibited LPS/lipid A binding to the TLR4/CD14 receptor complex. For example, the lipid A's of the non-pathogenic Gram-negative bacteria *Rhodobacter capsulatus* (**10**) and

Rhodobacter sphaeroides (**11**, RsDPLA) were used as lead compounds (Fig. 6.6).

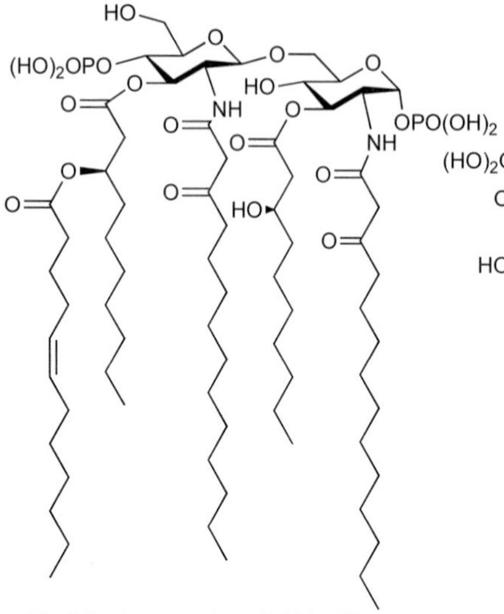

Rhodobacter capsulatus lipid A (**10**)

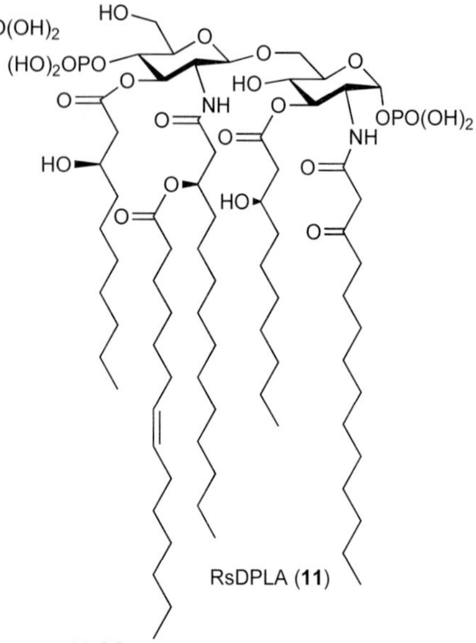

RsDPLA (**11**)

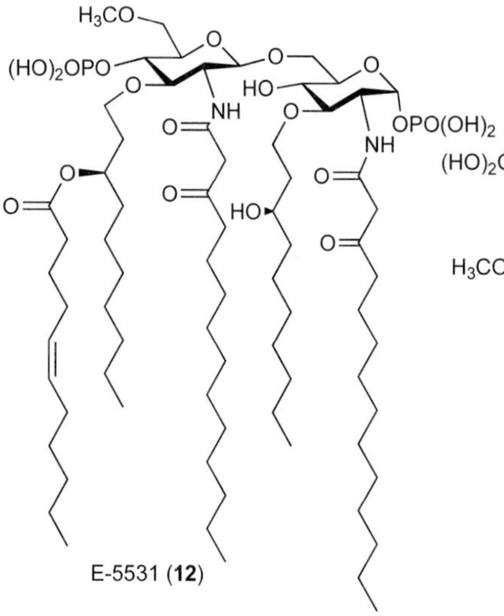

E-5531 (**12**)

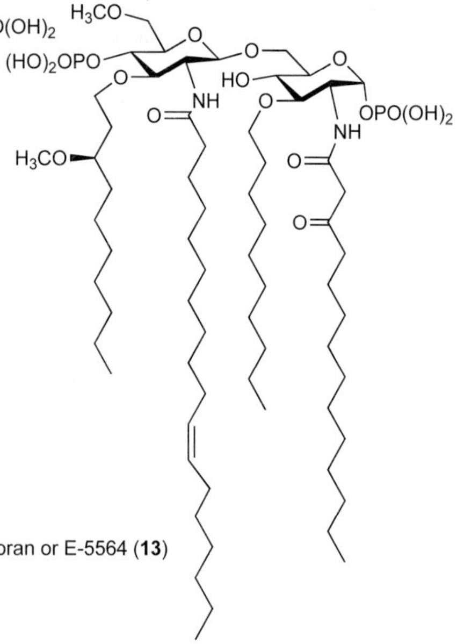

eritoran or E-5564 (**13**)

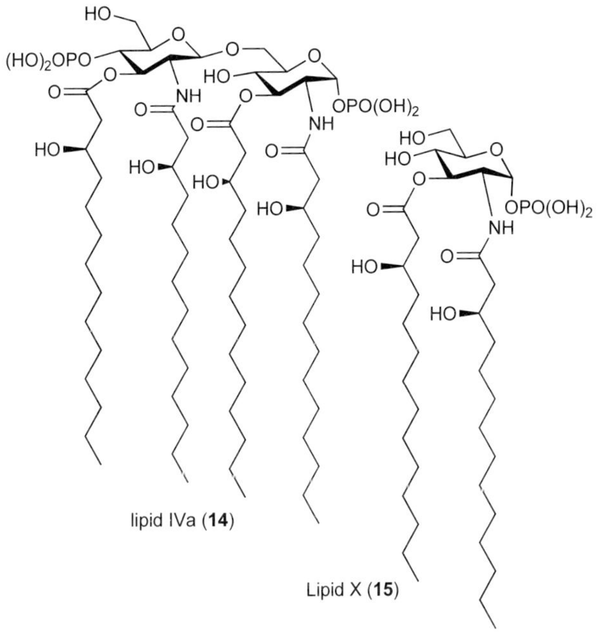

Fig. 6.6 Some lipid A antagonists that act through the TLR4/CD14 receptor.

An early derivative, E-5531 (**12**), has been replaced by the easier-to-synthesize, more robust, and more potent E-5564 (**13**) (also known as Eritoran) (Qureshi *et al.*, 1991, Chilman-Blair *et al.*, 2003). Biosynthetic lipid A precursor lipid Ia (**2**, see Fig. 6.1), and to a lesser extent Lipid X (**14**), also are antagonists. Several other small molecules have also been reported to inhibit TNFα production and to antagonize LPS activity by targeting biological targets downstream from the LPS receptor(s).

Numerous small molecules target the natural inhibitor of NF-κB, iκB. Abolishing the action of NF-κB effectively shuts down the lipid A/TNFα cascade by interrupting intracellular signal transduction. Some of these chemical agents, such as herbimycin A (**15**) and okadaic acid (**16**), act by inhibiting iκB's phosphorylation (Fig. 6.7). The statins, a family of natural and unnatural compounds used to lower cholesterol and low-density lipoprotein (LDL) levels, inhibit the induction of pro-inflammatory cytokines, also through blocking iκB's phosphorylation. In particular, lovastatin (**17**) inhibits the production of TNFα, IL-1, and IL-

6 from LPS-treated macrophages (Pahan *et al.*, 1997). Other lead compounds, like *N*-tosyl-L-phenylalanine chloromethyl ketone (**18**), capsaicin (**19**), and resiniferatoxin (**20**), inhibit the protease(s) that degrade iκB. Antioxidants such as tetrahydropapaveroline (**21**) and 10,11-dihydroxyaporphine (**22**) inhibit the activation of both transcription factors AP-1 and NF-κB.

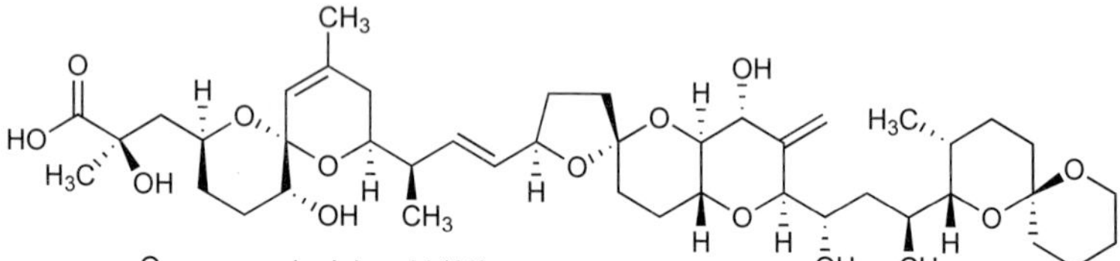

okadaic acid (**16**)

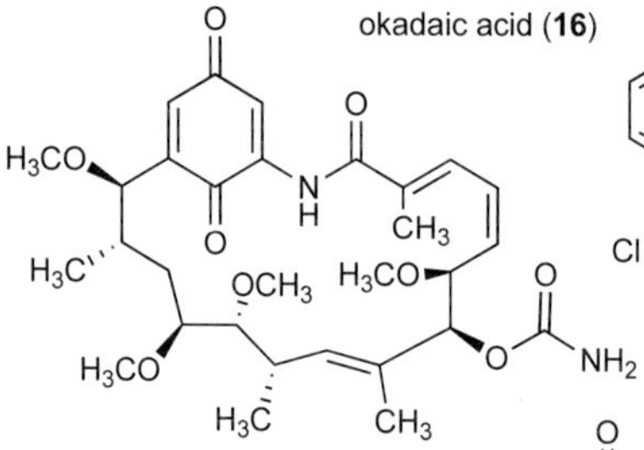

herbimycin A (**15**)

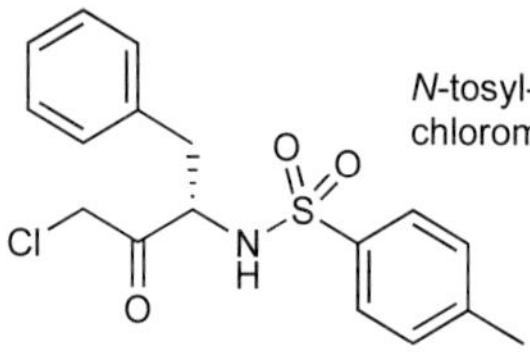

N-tosyl-L-phenylalanine chloromethyl ketone (**18**)

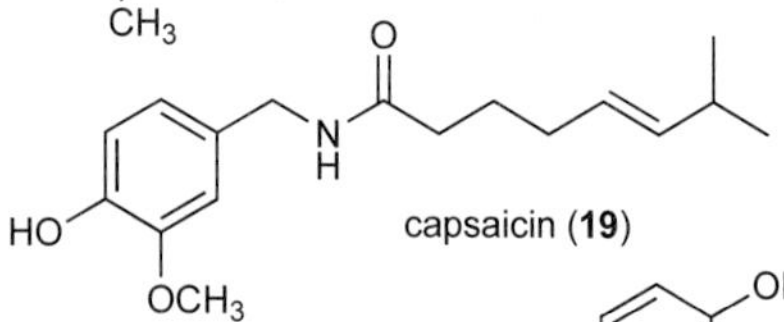

capsaicin (**19**)

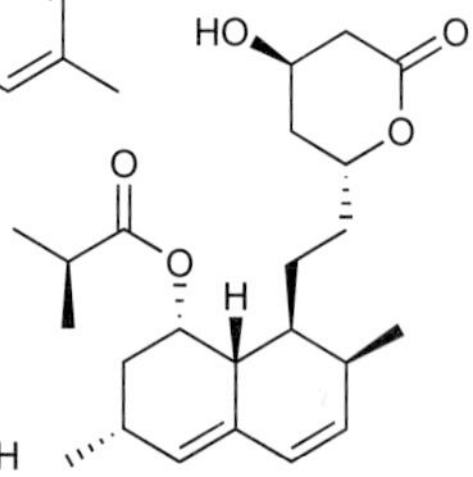

lovastatin (**17**)

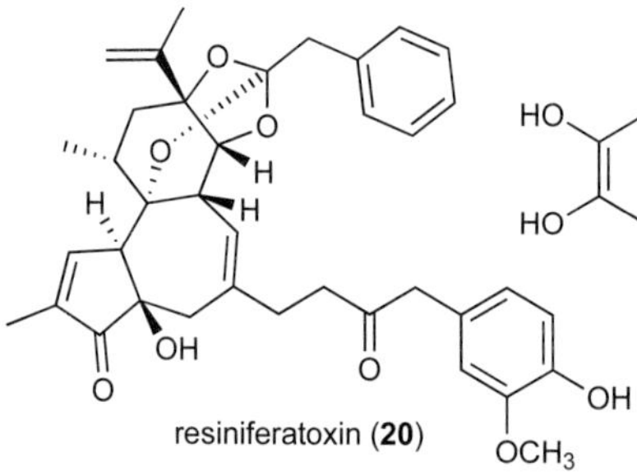

resiniferatoxin (**20**)

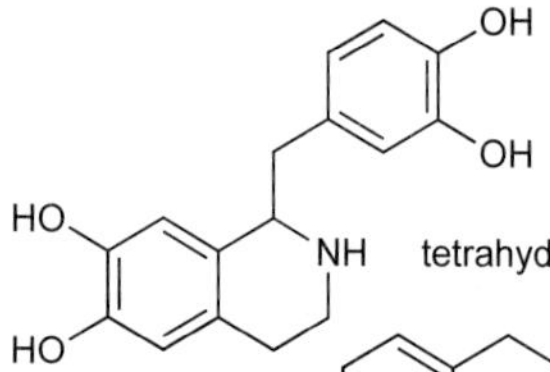

tetrahydropapaveroline (**21**)

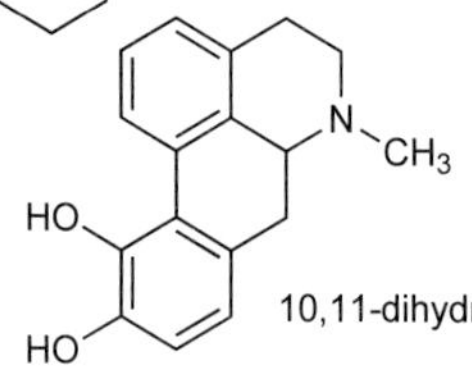

10,11-dihydroxyaporphine (**22**)

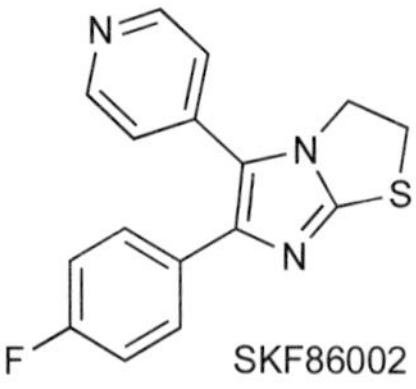

SKF86002 (**24**)

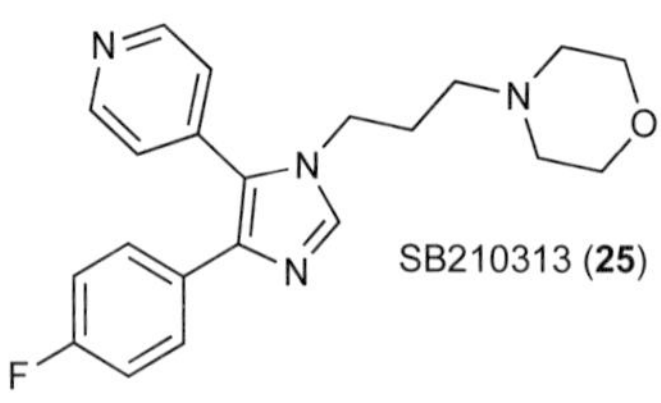

SB210313 (**25**)

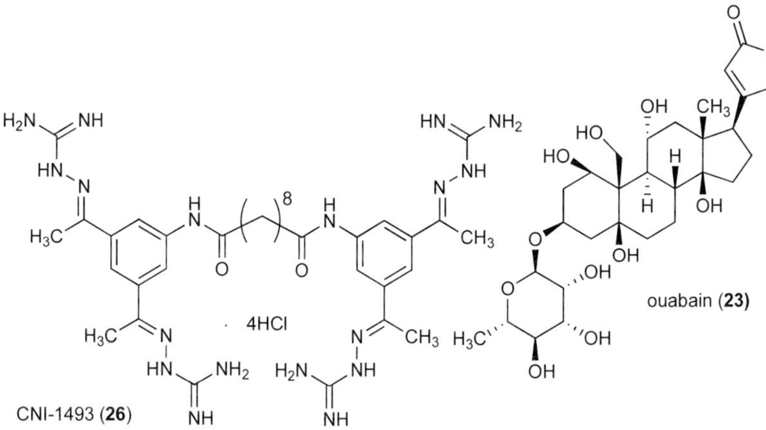

Fig. 6.7 Some LPS antagonists that act through NF-κB or p38 MAPK inhibition.

Ouabain (**23**), a cardiotonic steroidal derivative, was reported to suppress TNFα levels in LPS stimulated mice by targeting the Na,K-ATPase enzyme, which is responsible for controlling intracellular calcium and other cation levels, through an unidentified mechanism involving NF-κB (Matsumori *et al.*, 1997, Bereta *et al.*, 1995). Curiously, in the absence of LPS, ouabain (**23**) induced production of TNFα. Selective inhibitors of NF-κB itself have not been identified, as other kinases and proteases are also affected.

The stress-activated protein kinase p38/SAPK-2, a mitogen-activated protein kinase (MAPK) that is responsible for releasing a translation repressor protein complex from TNFα mRNA to allow protein synthesis, also has been explored as a target for inhibition. The cytokine-suppressing anti-inflammatory drugs (CSAIDS), which include imidazoles SKF86002 (**24**) and SB210313 (**25**), target p38 MAP kinase. Tetravalent guanylhydrazone CNI-1493 (**26**) also suppresses the production of TNFα, perhaps through inhibition of the p38 MAP kinase pathway (see Fig. 6.7).

TNFα-converting-enzyme (TACE), the enzyme responsible for converting pro-TNFα into soluble TNFα, has been a hot target for anti-TNFα therapy. TACE is a member of the same protease family (metzincin), as the matrix metalloproteinases (MMPs). Thus, several known MMP inhibitors have been modified to make selective TACE inhibitors without affecting other MMPs. Some of these drug leads include the sultam hydroxamate **27**, β,β-cyclic β-aminohydroxyxamic acid **28**, macrocycle **29**, and clinical candidate BMS-562392 (**30**) (Cherney *et al.*, 2006) (see Fig. 6.8). The hydroxamate derivate **31** of the natural product gelastatin (**32**), as a mixture of *E*/*Z* isomers, showed potent TACE, as well as MMP inhibition (Park *et al.*, 2006).

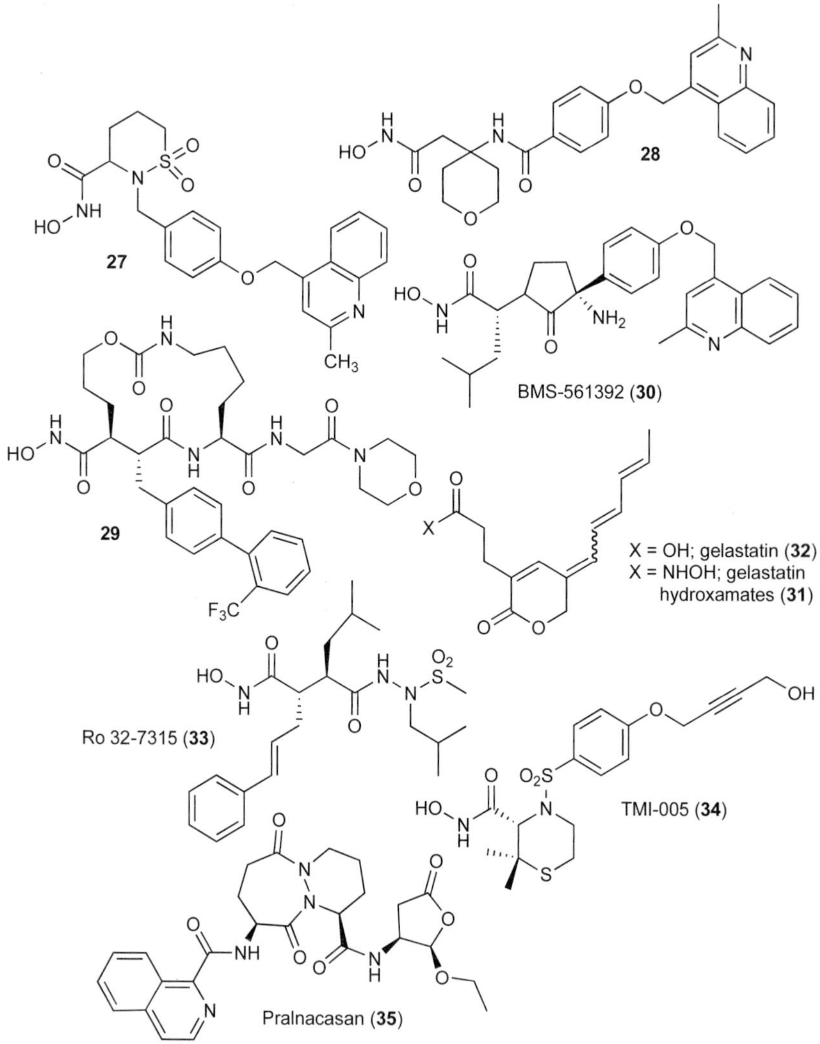

Fig. 6.8 Some LPS antagonists that act through TACE or ICE inhibition.

Other potent but nonselective TACE inhibitors include the clinical candidates Ro 32-7315 (**33**), and TMI-005 (**34**). Another target that has received less attention is interleukin-1-converting-enzyme (ICE). Currently, pralnacasan (**35**), an inhibitor of ICE, is in phase II clinical trials for osteoarthritis (Rudolphi *et al.*, 2003).

Thalidomide (**36**, see Fig. 6.9), a drug originally used for the treatment of insomnia and morning sickness, was removed from the market in the 1960s, because it was discovered to be a teratogen. However, it has recently been resurrected, because it selectively interferes with TNFα synthesis by enhancing degradation of TNFα mRNA, while not interfering with other pro-inflammatory cytokines (Davey and Ashrafian, 2000, Sampaio *et al.*, 1991).

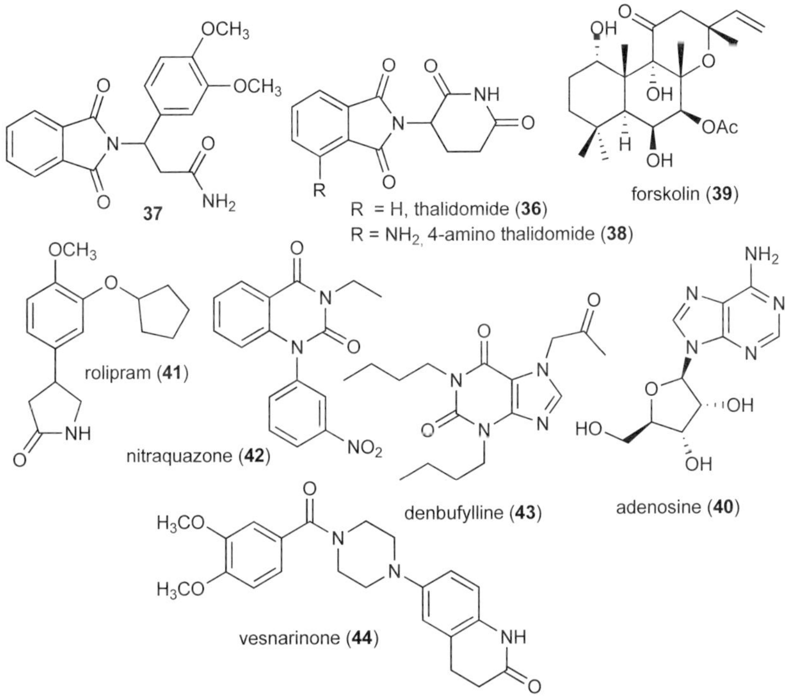

Fig. 6.9 Some natural and synthetic LPS antagonists that degrade TNFα mRNA or increase cAMP levels.

The phenyl substituted thalidomide analogue **37** and the amino substituted thalidomide **38** are more potent inhibitors of LPS-stimulated

TNFα release from human peripheral blood mononuclear cells (hPBMC's) than is the parent **36** (Muller *et al.*, 1999, Moreira *et al.*, 1993) (Fig. 6.9). Agents that bind to the histamine and prostaglandin receptors (among others) increase intracellular levels of cyclic adenosine monophosphate (cAMP), which in turn suppresses TNFα production. The diterpene forskolin (**39**) reversibly activates adenylate cyclase to increase cAMP levels. Adenosine (**40**) also inhibits TNFα expression by increasing cAMP levels after binding to the A_2 receptor (Wagner *et al.*, 1998). Phosphodiesterases (PDEs) are responsible for metabolizing cAMP. Selective inhibition of PDE4 has shown some therapeutic potential, since it is the major isozyme present in inflammatory cells. Thus, the PDE4 inhibitors rolipram (**41**), nitraquazone (**42**), and denbufylline (**43**) also inhibit TNFα generation. Unfortunately, PDE4 inhibitors such as **41** can cross the blood-brain barrier and cause nausea in patients, since PDE4 is also present in the central nervous system. The synthetic quinolinone vesnarinone (**44**, or OPC-8212) has been shown to suppress TNFα production by inhibiting PDE type III in cardiac muscle (Matsumori *et al.*, 1994).

Several classes of natural products have been reported as LPS antagonists besides those mentioned above (Paul *et al.*, 2006). For example, some polyphenolic flavonoid plant secondary metabolites including some flavones, flavonols, and chalcones, are documented TNFα secretion inhibitors (Herath *et al.*, 2003, Xagorari *et al.*, 2001). Luteolin (**45**) and quercetin (**46**) were the most active at inhibiting TNFα release from LPS stimulated cells (Fig. 6.10). Another plant metabolite, scopoletin (**47**) from *Artemisia feddei*, inhibited TNFα, IL-1β, IL-6, and prostaglandin E_2 release (Kim *et al.*, 2004). Other phenolic natural products that have been reported to inhibit TNFα output include the marine metabolite petrosaspongiolide M (**48**), and the phytochemicals curcumin (**49**), 5,6-dehydrokawain (**50**), aucubin (**51**), nardostachin (**52**), magnosalin (**53**), and honokiol (**54**) (Chan, 1995, Hashimoto *et al.*, 2003, Jeong *et al.*, 2002, Ju *et al.*, 2003, Ryu *et al.*, 2002, Tse *et al.*, 2005, Posadas *et al.*, 2003). The plant alkaloid lycorine (**55**) and the diterpenes 2β,5-epoxy-5,10-dihydroxy-6α-angeloyloxy-9β-isobutyloxy-germacran-8α,12-olide (**56**) and tanshinone II A (**57**) were all active against TNFα induction (Yui *et al.*, 2001, Kim *et al.*, 2001, Jang *et al.*, 2003).

Similarly, the fatty acid derivative 13-hydroxy-10-oxo-*trans*-11-octadecenoic acid (**58**, or 13-HOA), the sterol guggulsterol (**59**), and the retinoid retinoic acid (**60**) all attenuated cytokine production (Murakami *et al.*, 2005, Manjula *et al.*, 2006, Dheen *et al.*, 2005).

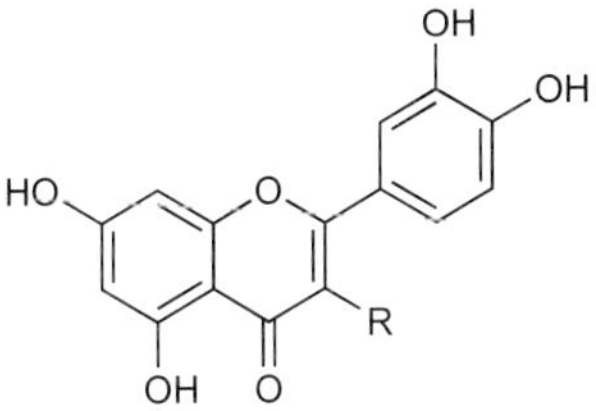

R = H, luteolin (**45**)
R = OH, quercetin (**46**)

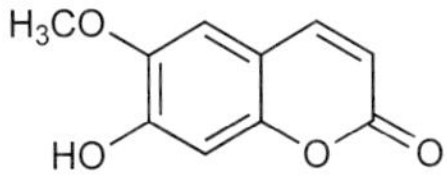

scopoletin (**47**)

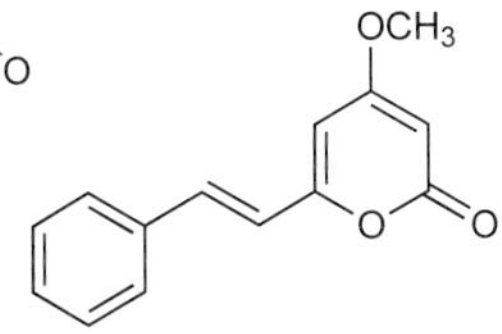

5,6-dehydrokawain (**50**)

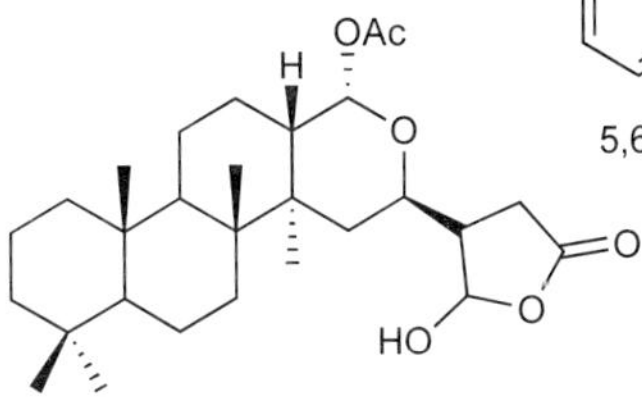

petrosaspongiolide M (**48**)

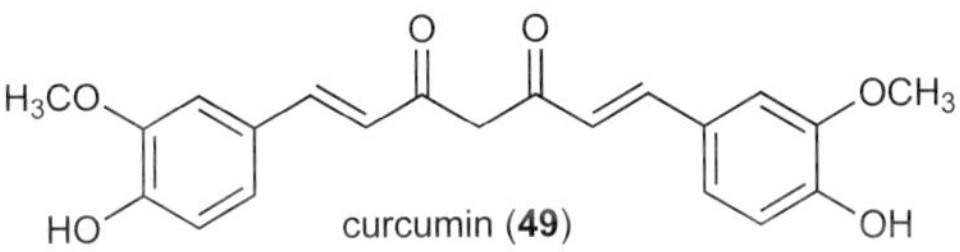

curcumin (**49**)

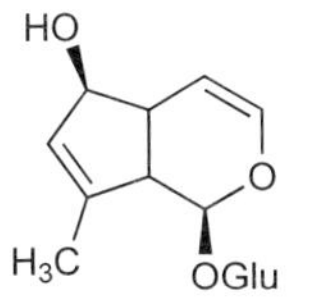

aucubin (**51**)

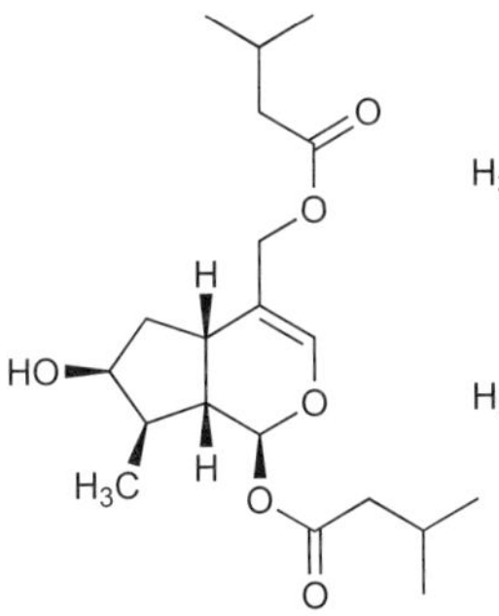

nardostachin (**52**)

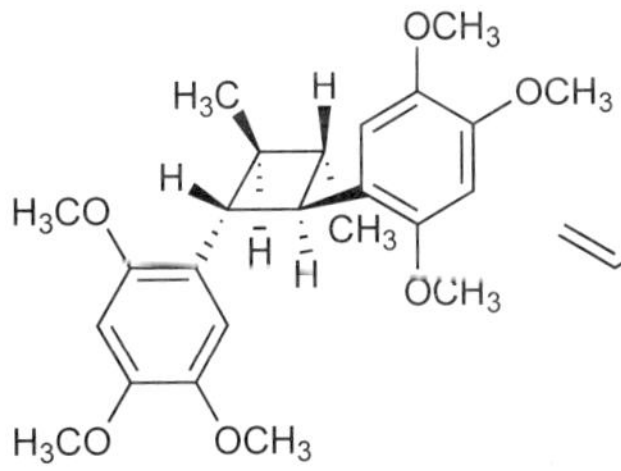

magnosalin (**53**)

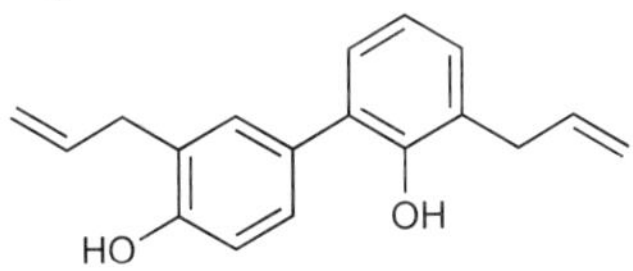

honokiol (**54**)

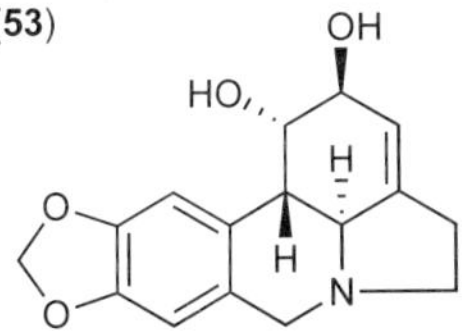

lycorine (**55**)

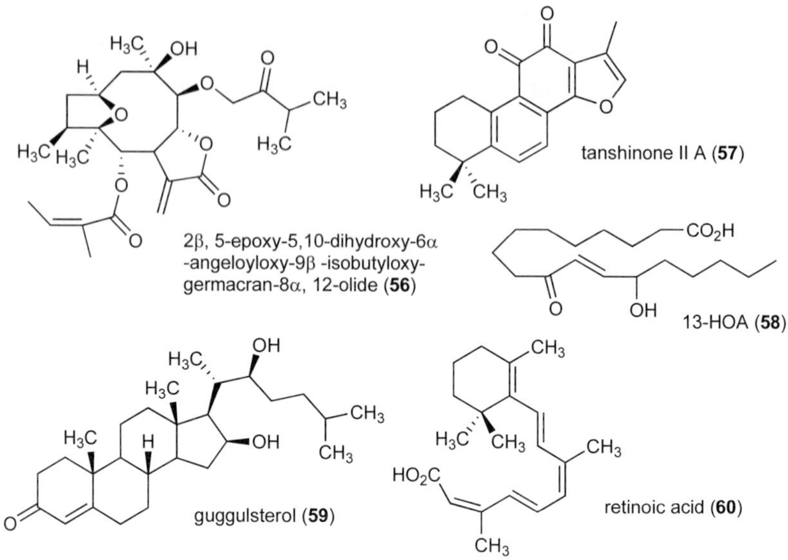

Fig. 6.10 Some naturally occurring LPS antagonists.

6.2 Initial Discovery of Ellagitannins as Immunostimulants

Polyphenol-rich folk medicines from China and Japan, including tannin-containing decoctions, had been used for some time to treat a variety of ailments. In 1987, Okuda, Miyamoto, and colleagues reported that several tannins displayed tumoricidal activity against a model cancer cell line, sarcoma-180 (Miyamoto *et al.*, 1987b). They found that the tannins' anticancer activity was due to their ability to act as immunostimulants in a manner reminiscent of LPS.

6.2.1 Anticancer activity

Some tannin-containing plant extracts became the focus of isolation studies as a consequence of this promising antitumor activity. For example, Miyamoto and co-workers initially reported that the methanol extract from the roots of the plant *Agrimonia pilosa* Ledeb was active in mice against several transplanted tumors (Koshiura *et al.*, 1985). The causative agent was identified tentatively as the dimeric ellagitannin agrimoniin (Miyamoto *et al.*, 1985). Thereafter, a collaboration with Okuda's group has led to the identification of a large number of related tumoricidal ellagitannins.

6.2.1.1 *Agrimoniin*

Although agrimoniin (**61**, Fig. 6.11) had been known for some time, since it had been isolated before from the *Rosaceae* plants *Agrimonia pilosa* Ledeb, *Agrimonia japonica* (MIQ) KOIDZ, and *Potentilla kleiniana* WIGHT *et* ARNOTT (Okuda *et al.*, 1984, 1982c), its antitumor properties were not known before Miyamoto's work. Agrimoniin, the first dimeric ellagitannin to be isolated and characterized, significantly extended the life span of mice inoculated with mammary carcinoma MM2 cells, and in some cases even led to tumor regression. Agrimoniin was found to be effective upon either pre- or posttreatment and showed an IC_{50} of 2.6 μg/mL. In some preliminary experiments, agrimoniin appeared to cause an enhancement in the growth rate of macrophages, lymphocytes, and adherent peritoneal exudate cells, as well as an enlargement of spleen size, observations which supported the idea that tannins act as immunostimulators (Miyamoto *et al.*, 1987a). Agrimoniin (at a 5 mg/kg dose) also was found to be active in mice inoculated with sarcoma-180 by effecting a 75% increase in the life span (%ILS) of the mice, but no tumor regressors out of 6 mice were detected. An increase in dose to 10 mg/kg led to 3/6 cured mice and a 136 %ILS (Miyamoto *et al.*, 1993a). Agrimoniin has been isolated from several other sources such as the rosaceous Chinese medicinal plants *Rosa laevigata* MICHX, *Rosa davurica* Pall., *Potentilla erecta* L., and *Potentilla discolor* Bunge, the North America *Rosa woodsii* Lindle, strawberry plants (cv. Jonsok), the fruit of *Rosa taiwanensis* Nakai, and the Bolivian medicinal plant *Buddleia coriacea* Remy (Yoshida *et al.*, 1989a-c, Hukkanen *et al.*, 2007, Kubo *et al.*, 1995, Feng *et al.*, 1996, Lin and Wang, 1998, Ito *et al.*, 2001, Lund and Rimpler, 1985, Yoshida *et al.*, 1991). This compound also has diverse biological activities including human neutriphil elastase (a serine protease) inhibition (IC_{50} 0.9 μM), antileishmanial properties (EC_{50} 1.1 μg/mL), HIV-1 reverse transcriptase inhibition (IC_{50} <0.88 μM), aryl hydrocarbon receptor (a nuclear transcription factor) inhibition (EC_{70} 6.4 μM), anti KO_2-induced histamine release (IC_{50} 0.68 μM), and tyrosinase (target for insect control) inhibition (ID_{50} 61 μM) (Hrenn *et al.*, 2006, Kolodziej *et al.*, 2001, Amakura *et al.*, 2003, Kanoh *et al.*, 2000, Kubo *et al.*, 1995, Ito *et al.*, 2001).

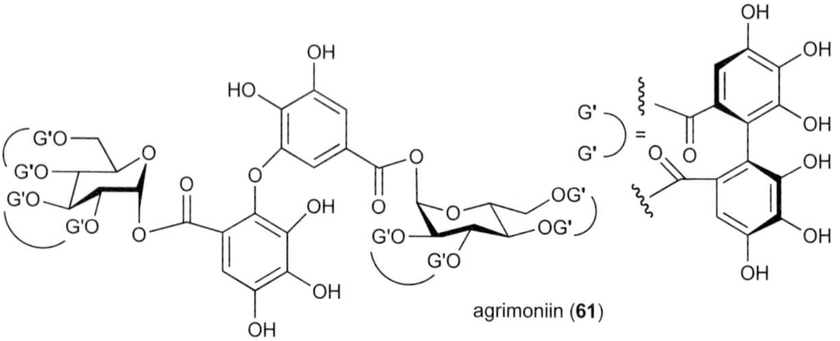

agrimoniin (**61**)

Fig. 6.11 Structure of agrimoniin (**61**).

6.2.1.2 *Coriariin A*

Among the 63 tannins initially surveyed by Okuda's and Miyamoto's groups, coriariin A (**62**, Fig. 6.12) showed the strongest activity against sarcoma-180 tumor cells with a 238 %ILS and tumor regression in 3 out 6 mice at a 5 mg/kg dose (Miyamoto *et al.*, 1987b). Coriariin A, a dimer of tellimagrandin II (**65**), was isolated from the Japanese poisonous plant *Coriaria japonica* A. Gray (Hatano *et al.*, 1986). Coriariin A, along with several oligomeric ellagitannins, was found to be a potent inhibitor of both poly(ADP-ribose) glycohydrolase (IC_{50} 8.5 μM) and of histamine release (IC_{50} 2.97 μM) (Aoki *et al.*, 1993, Maruta *et al.*, 2007, Kanoh *et al.*, 2000). Coriariin A also has anti-herpes simplex virus activity (ED_{50} 0.038 μg/mL) (Fukuchi *et al.*, 1989). The chemical synthesis of coriariin A has been accomplished (Feldman *et al.*, 2000, Feldman and Lawlor, 2000, see Section 5.2.2.6 in Chapter 5). Further examination of other related tannins (now totaling 108) at different doses and through different methods of administration has led to the identification of several other tumoricidal compounds. The best results were found when the tannins were administered to mice through intraperitoneal injection 4 days before treatment with cancer cells (Miyamoto *et al.*, 1987a).

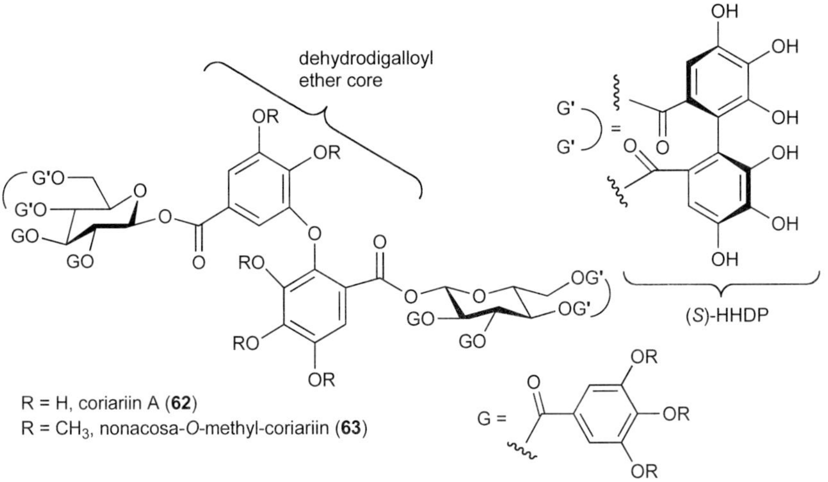

Fig. 6.12 Structure of coriariin A (**62**) and its methylated derivative **63**.

6.2.1.3 *Other ellagitannins*

Besides agrimoniin and coriariin A, the work of Miyamoto and colleagues showed that several other tannins that met some particular structural requirements were active against sarcoma-180 tumor cells (Miyamoto *et al.*, 1987b, 1993a/b). From their structural-activity relationship studies, some general trends emerged. Condensed tannins, caffeic acid derivatives, bergenin derivatives, dehydroellagitannins, and gallotannins exhibited little-to-no anticancer activity. In general, monomeric ellagitannins were found to be less potent than dimeric ones. Exceptions to this generalization are the two particularly active monomers tellimagrandin II (**65**) with 3 out of 6 tumor regressors, although with only a 18 %ILS at a 10 mg/kg dose, and rugosin A (**68**) with one cured mouse out of six and a 110 %ILS at a 5 mg/kg dose (Fig. 6.13). The gallotannin penta-*O*-galloyl-β-D-glucose (β-PGG, **66**) was relatively inactive with no regressors and an 82 %ILS at 10 mg/kg dose. Compounds containing an open-chain glucose unit also were ineffective. Several macrocyclic ellagitannins such as the trimer oenothein A (**69**) (1 regressor/6 mice and 103 %ILS at 10 mg/kg) and the dimer oenothein B (**67**) (4 regressors/6 mice and 196 %ILS at 10 mg/kg) were very active. Exposed phenolic hydroxyls were necessary for activity, as nocacosa-*O*-

methylcoriariin A (**63**), the completely methylated derivative of **62** (Fig. 6.12), was ineffective, whereas coriariin A proved to be the most potent tannin tested at 5 mg/kg. Increases in the degree of oligomerization didn't necessarily correlate to an increase in activity. The dimer [of tellimagrandin I (**64**)] oenothein B was more potent than both the trimer oenothein A and the tetramer woodfordin F, which had no regressors at 10 mg/kg. In some cases, adding more galloyl groups led to an increase in activity, but that trend was not general. Galloyl groups appear to be more potent that the HHDP moiety as seen by rugosin F (**72**, no regressors and 35 %ILS) < rugosin D (**71**, one regressor/6 mice and 172 %ILS), and agrimoniin < coriariin A. However, gemin A (**70**, one regressor/6 mice and 176 %ILS at 10 mg/kg dose) seemed to be less potent than agrimoniin at 10 mg/kg dose. As a comparison to the immunostimulatory effects of tannins, OK-432 (Picibanil), a whole bacterial streptococcal preparation that is a known immunopotentiator, cured 4 out of 12 mice and showed a 79 %ILS at a 10 mg/kg dose.

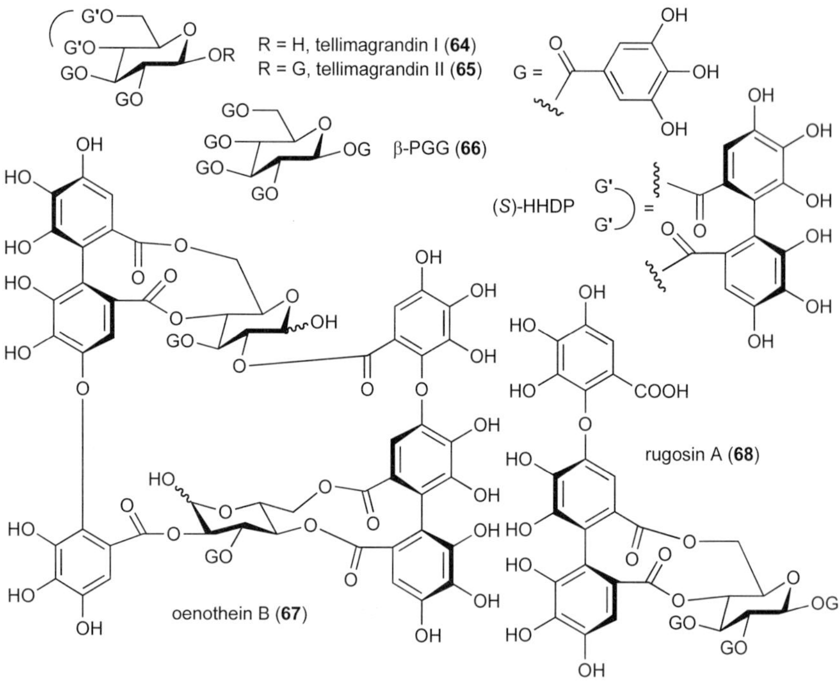

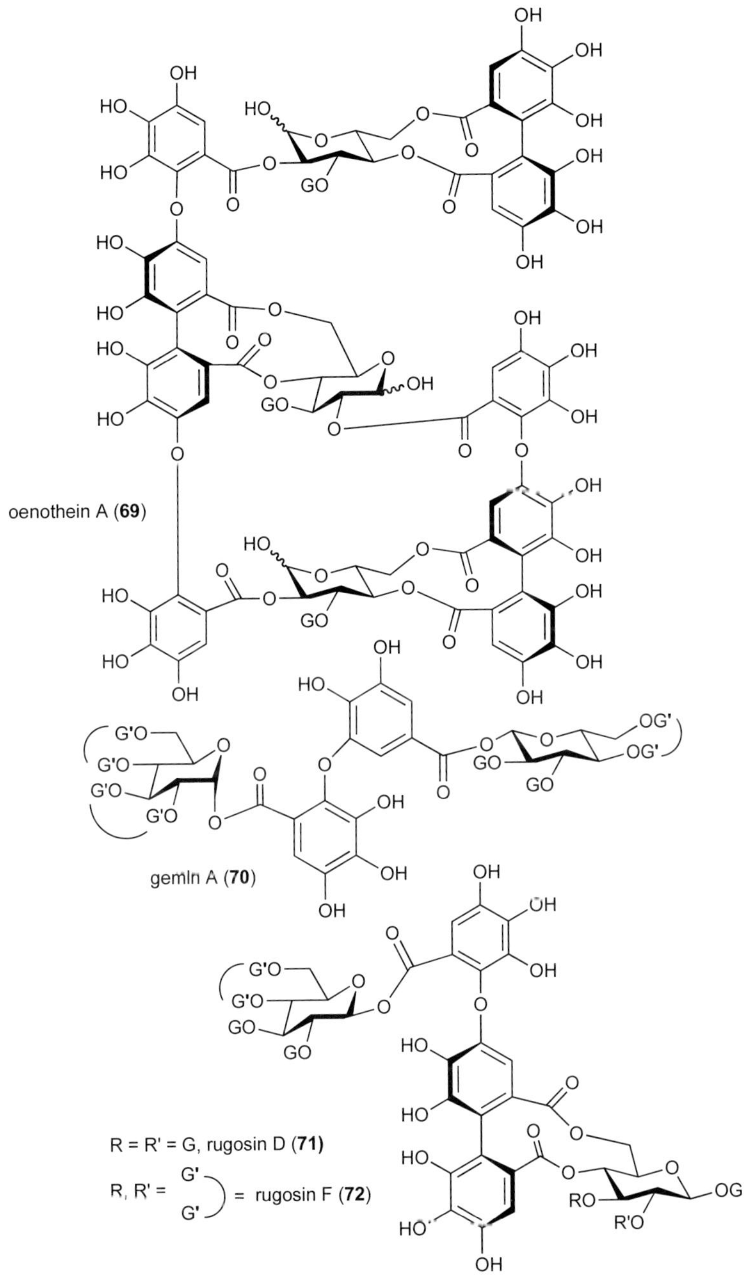

Fig. 6.13 Structures of some anticancer ellagitannins and gallotannins.

6.2.2 Agrimoniin-mouse and hPBMC studies – IL-1β release

Miyamoto and Okuda examined some immune system components in their search for biological mediators of tannin-induced anti-tumor activity. Specifically, treatment of mice with agrimoniin (**61**) stimulated growth of peritoneal exudate cells (PECs) with a peak of about 14×10^6 cells/mouse produced after 4 days, while control was approximately 4×10^6 cells/mouse (Miyamoto *et al.*, 1987a). Of these cells, 58% were monocytes, 23% were polymorphonuclear (PMN) leukocytes, and 20% were lymphocytes. Administration of agrimoniin to isolated murine PECs resulted in increased natural killer (NK) cell activity that peaked after 2 days (Miyamoto *et al.*, 1988). The NK cells were not, however, cytotoxic to MM2 and MH134 tumor cells. In contrast, the adherent PECs (presumably macrophages) did appear to inhibit tumor cell growth (MM2 and MH134). Removal of serum from mice treated with both MM2 tumor cells and agrimoniin, and addition of that serum to a culture of only adherent PEC and MM2 cells, led to complete tumor lysis. The serum was assumed to contain immunostimulatory factors such as cytokines. The cell lytic activity of the serum peaked after 6 days following intraperitoneal injection.

Analysis with an enzyme-linked immunosorbent assay (ELISA) of the supernatant above hPBMC's treated with agrimoniin led to the identification of significant quantities (average of 1203 pg/mL) of the cytokine IL-1β compared to the untreated cells (300 pg/mL) (Murayama *et al.*, 1992). Following the same experimental procedure, treatment of hPBMC's with *E. coli* LPS at 10 μg/mL led to the release of an average of 1350 pg/mL of IL-1β. The supernatant also was examined for the cytokine IL-2, but this species could not be detected by a proliferation assay of IL-2-dependent CTLL2 cells.

6.2.2.1 *Immunostimulation – IL-1β secretion*

Interleukin-1 (IL-1, also known by hemopoietin-1, catabolin, osteoclast-activating factor, and B-cell stimulatory factor), is another pluripotent cytokine that has been identified as a mediator of septic shock. IL-1 was first described in the literature in 1972 and was named in 1979 at the 2nd

International Congress on Lymphokines (Géry and Waksman, 1972). There are two known members in this family of proteins, IL-1α and IL-1β, and both bind to the same receptors and produce similar biological responses. In several studies, IL-1β has been detected consistently in the plasma of sepsis patients, while IL-1α has rarely been seen. IL-1β is first produced as a soluble 31 kDa (269 amino acids) inactive protein, which is converted to the active 18 kDa form by IL-1β–converting-enzyme (ICE, also known as caspase-1). The crystal structure of IL-1β has been obtained and shows that this cytokine consists of a β-barrel with four triangular faces forming a tetrahedron (Priestle *et al.*, 1988, 1989). Similarly to TNFα, IL-1β is produced mostly by monocytes and macrophages, but it can be released by most cell types. IL-1 genes are not expressed in the absence of stimulation. Several agents are able to induce the transcription of these genes including LPS, complement components, and cytokines (TNFα, INF-γ, GM-CSF, and IL-1 itself) (Colotta *et al.*, 1998). As part of homeostasis, Interleukin-1 receptor antagonist (IL-1ra), a naturally occurring cytokine produced by many cell types, binds to the IL-1 receptor without triggering any cellular responses. IL-1ra circulates in the plasma soon after infection along with soluble IL-1 receptors and together both act as anti-inflammatory mediators. There are two known IL-1 receptors: type 1 (or IL-1RI, expressed mostly by fibroblasts and T cells) and type 2 (or IL-1RII, expressed predominantly in monocytes, PMN, and B cells). Although both receptors recognize both IL-1 forms, IL-1RI has a higher affinity for IL-1α ($K_d = 10^{-10}$ M versus 10^{-9} M), whereas IL-1RII binds IL-1β more effectively ($K_d = 10^{-9}$ to 10^{-10} M versus 10^{-8} M) (McMahan *et al.*, 1991). Administration of IL-1 to rodents and primates induced the symptoms of sepsis in the animals (Okusawa *et al.*, 1988, Fischer *et al.*, 1991). Alternatively, blockade of IL-1 receptors with antibodies protected rats and primates from suffering the full symptoms of sepsis such as hypotension and tachycardia (rapid heart rate) during exposure of these animals to *E. coli*, suggesting a critical role for IL-1 in sepsis. IL-1 has been known to activate macrophages and lymphocytes and stimulate secretion of other cytokines such as IL-2, IL-3, IL-6, and interferon (IFN) (Dinarello, 1989). Furthermore, IL-1 had been reported to promote the activation of other cells including helper-T cells, NK cells, and tumor cells (Herman *et al.*, 1985, Durum *et al.*, 1985). Thus, the authors hypothesized that the generated IL-1β was the agent responsible for agrimoniin's tumoricidal activities through stimulation of monocytes, macrophages, and NK cells.

6.2.2.2 *Other ellagitannins*

Oenothein B (**67**, Fig. 6.13) has been isolated from several plants including *Lythrum anceps*, *Oenothera erythrosepala* Bordas, the Central European medicinal plant *Epilobium parviflorum*, the Mexican *Cuphea hyssopifolia* Humb. Bompl. Et Kunth, the American *Quercus rubra* (red oak) and *Epilobium angustifolium* L. (willow herb), and the Asian *Melaleuca leucadendron* L., *Eugenia uniflora*, *Eucalyptus alba* Reinw, and *Woodfordia fruticosa* Kurz (Hatano *et al.*, 1989, 1990, Yoshida *et al.*, 1989d, 1990, 1991, 1992, 1996, Okuda *et al.*, 1993, Lesuisse *et al.*, 1996, Chen *et al.*, 1999, Lee *et al.*, 2000, Barbehenn *et al.*, 2006a/b, Kiss *et al.*, 2004). Similarly to agrimoniin, oenothein B induced an increase in the number of PEC cells and caused the adherent PECs (macrophages) to inhibit the growth of MM2 and Meth-A tumor cells (Miyamoto *et al.*, 1993b/c). Furthermore, macrophages that were incubated with oenothein B released IL-1β. This macrocyclic ellagitannin (10 μg/mL) induced 1010 pg/mL of IL-1β secretion whereas the control (no tannin) induced only 295 pg/mL. As a positive control, *E. coli* LPS at 10 μg/mL led to the release of 1230 pg/mL of IL-1β. A reevaluation of the antitumor effect of oenothein B found that it was significantly more potent that previously thought when given by intraperitoneal injection 4 days before administration of sarcoma-180 (4 regressors out of 6 mice and 196 %ILS) or MM2 (5 regressors out of 6 mice and 9.3 %ILS) tumor cells at a 10 mg/kg dose (Miyamoto *et al.*, 1993a-c).

In addition to its antitumor properties, oenothein B has been reported to inhibit 5-α-reductase (IC_{50} 0.22 μM), neutral endopeptidase (IC_{50} 20 μM), herpes simplex virus absorption into cells (ED_{50} 0.036 μg/mL), Epstein-Barr virus DNA polymerase (IC_{50} 62.3 μM), and poly(ADP-ribose) glycohydrolase (IC_{50} 4.8 μM) (Lesuisse *et al.*, 1996, Kiss *et al.*, 2004, Lee *et al.*, 2000, Fukuchi *et al.*, 1989, Aoki *et al.*, 1993, Maruta *et al.*, 2007). Furthermore, it has exhibited antileishmanial properties

(EC_{50} 0.5 μg/mL), cytotoxicity against human oral squamous cell carcinoma and salivary gland tumor cells, antihistamine, and oxidative activities (Kolodziej *et al.*, 2001, Kanoh *et al.*, 2000, Barbehenn *et al.*, 2006 a/b, Sakagami *et al.*, 2000).

Several other tannins (Figs. 6.13 and 6.14) have been examined for their IL-1β inducing ability with macrophages. Simple tannins such as (-)-epicatechin gallate (**77**, ECG) and (-)-epigallocatechin gallate (**74**) generated only a slight amount of the cytokine (35 and 44% increases over control, respectively). Monomeric ellagitannins such as tellimagrandin I (**64**) (162%), tellimagrandin II (**65**) (198%), and rugosin A (**68**) (209%) were found to stimulate much more IL-1β compared to the control (305 pg/mL). Oligomeric tannins including agrimoniin (**61**) (359%), oenothein A (**69**) (254%), and oenothein B (**67**) (293%) induced 2 to 4 times more of the cytokine over basal levels. Similarly to the direct antitumor activity experiments, higher molecular weight tannins were not necessarily the most potent inducers of IL-1β. For example, the large ellagitannins euphorbin C-Hy (**73**) (108%), laevigatin B (**75**) (125%), and laevigatin C (**76**) (130%) were less potent than the monomers tellimagrandins I and II. In addition, a correlation between anticancer activity and IL-1β generating abilities of the tannins could not be established. It was suggested that perhaps this lack of correlation was due to either tannin instability in the host or differences between human macrophages and whole mice.

In addition, Okuda and co-workers also have reported the release of an IL-1-like cytokine (a more specific characterization was not forthcoming) from human monocytes incubated with 50 μg/mL of one of four condensed tannins: ECG **77** (produced 39 times more IL-1 than the blank control), ECG-dimer **78** (141 times more), ECG-trimer **79** (201 times more), and ECG-tetramer **80** (162 times more) (Sakagami *et al.*, 1992). The amounts of IL-1 generated were indirectly measured by observing the proliferation response ([^{3}H]-thymidine incorporation into cell) of mouse thymocytes incubated with the monocyte supernatant.

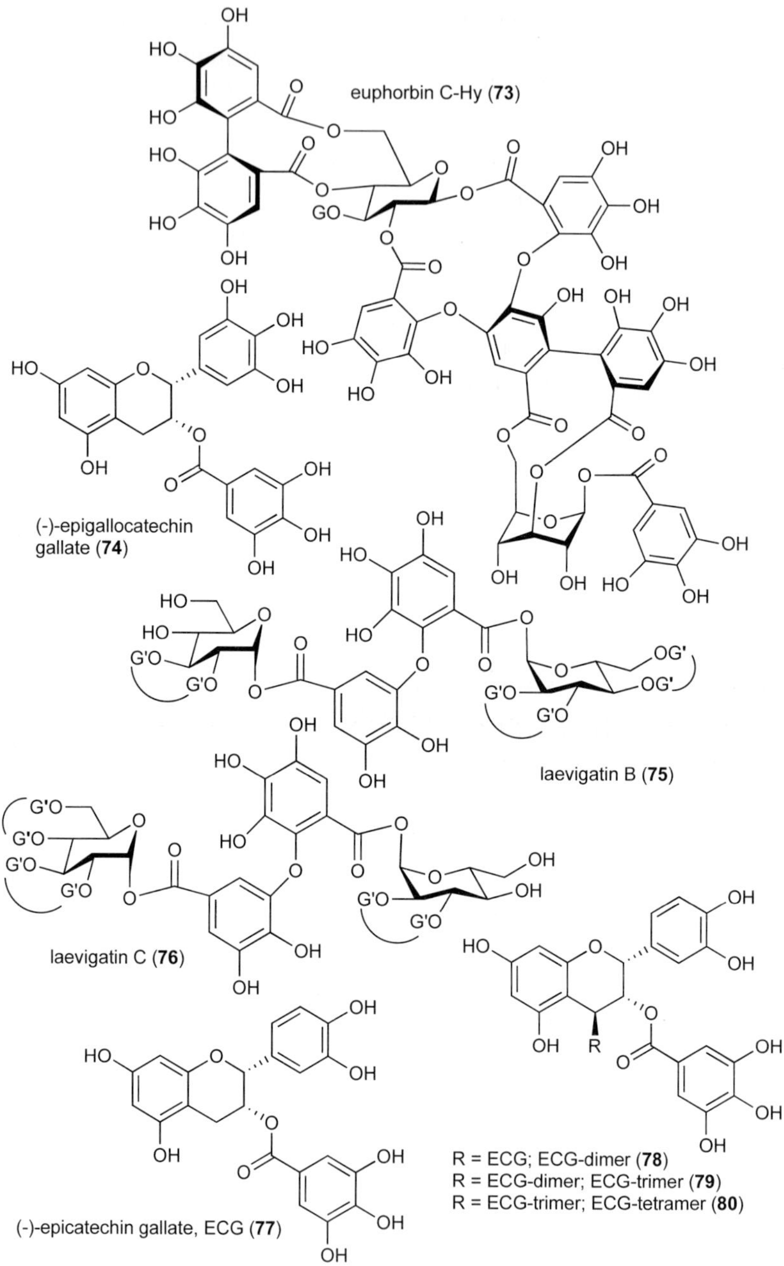

Fig. 6.14 Structures of some tannins that induce IL-1β output in macrophages.

6.3 Ellagitannins as TNFα Secretion Inducers

The ellagitannins' potent antitumor activity through an apparent up-regulation of the immune system is well founded. Miyamoto's group suggested that the secretion of IL-1β seen in murine PEC's and hPBMC's could explain the observed antitumor actions of hydrolyzable tannins. However, at the time that Miyamoto's group was pursuing the immunostimulatory experiments, little was known about the "newly discovered" related cytokine TNFα, and no simple assays were available. Also given IL-1β's late entry into the cytokine cascade, it was conceivable that a mediator different than IL-1β was responsible for the primary activity of the antitumor ellagitannins. Since much post-Miyamoto research established that TNFα was, indeed, a tumor-lethal substance, it seemed like a logical target for further tannin-related study.

Additionally, TNFα is among the first cytokines released in response to immunostimulation and it is known to induce the later production and release of IL-1β (Dinarello *et al.*, 1986). Finally, a readily available commercial ELISA-based TNFα assay had become available. Therefore, some known tumoricidal ellagitannins, an analogue of coriariin A, and the gallotannin 1,2,3,4,6-penta-*O*-galloyl-beta-D-glucose (β-PGG, **66**) were tested for their ability to induce TNFα output from hPBMC's (Feldman *et al.*, 1999).

6.3.1 β-PGG, agrimoniin, and a model dimer analog of coriariin A induce TNFα secretion from hPBMC's

The initial experiments involved monitoring TNFα generation using a qualitative test with the L929 murine fibroblast lysis assay (Feldman *et al.*, 1999). This indirect analysis revealed that significant amounts of TNFα were present in the supernatant of hPBMC's after 4 hours of incubation with agrimoniin at 28 μM. Thus, this crude assay provided the first evidence that the cytokine TNFα was generated by cells treated with an ellagitannin. Using the same experimental procedure, the monomeric gallotannin β-PGG (**66**) and the dimer **81**, an analogue of coriariin A (Fig. 6.15), were assessed quantitatively for their TNFα inducing capabilities utilizing an ELISA kit. A time-course study revealed similarities between both β-PGG and **81**, with low levels of TNFα generated at 4 hours and a maximum release at 24 hours (Fig. 6.16). However, the two tannins were observed to have considerably different TNFα eliciting proficiency, as the dimeric species **81** was found to be significantly more potent than β-PGG at every time point examined.

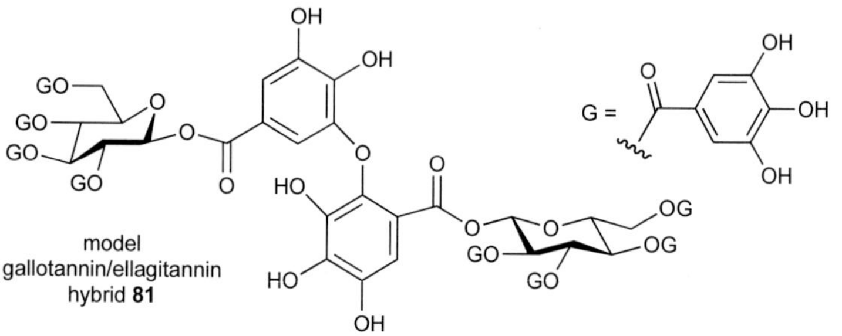

Fig. 6.15 Structures of an analogue of coriariin A.

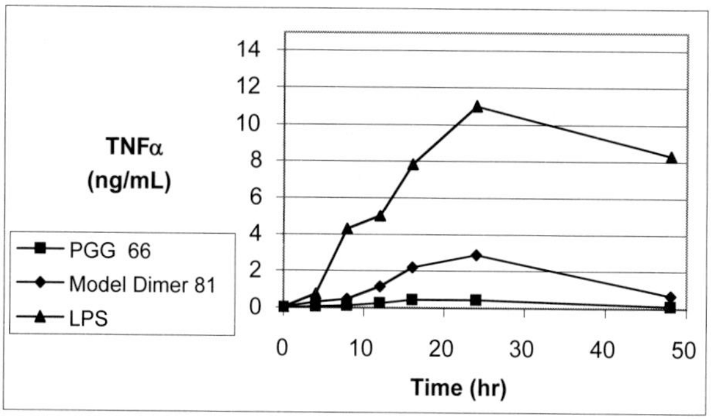

Fig. 6.16 Time course for TNFα release from hPBMC's that were stimulated by β-PGG, model dimer **81**, and LPS.

6.3.2 Structure/activity profile of ellagitannins and gallotannins

A comparison between the dose-response curves of the monomeric gallotannin β-PGG (**66**) and the tannin dimers coriariin A (**62**) and **81** for TNFα output after 24 hours of incubation with hPBMC's showed very different responses. Therefore, the secretion of TNFα appears to be responsive to the size of the tannin. The monomer β-PGG was less potent at generating TNFα from hPBMC's than either of the dimers **62** or **81** at similar concentrations (Fig. 6.17). This observation correlates well with the very different anticancer profiles (Miyamoto *et al.*, 1987b) between coriariin A (3 out of 6 regressors and a 238 %ILS) and β-PGG (zero regressors and 82 %ILS) reported by Miyamoto. Since a measure of IL-1β secretion amounts did not reveal a similar correlation, it remains unlikely that IL-1β is the primary mediator responsible for the tumoricidal activities of tannins. It is also noteworthy that the TNFα-inducing abilities of coriariin A and its model analogue **81** are similar (Fig. 6.17). These data show that the HHDP group is not necessary for activity and that two unconnected galloyl groups can take its place, which is consistent with Miyamoto's *in vivo* SAR studies of ellagitannins against sarcoma-180 (Miyamoto *et al.*, 1987b).

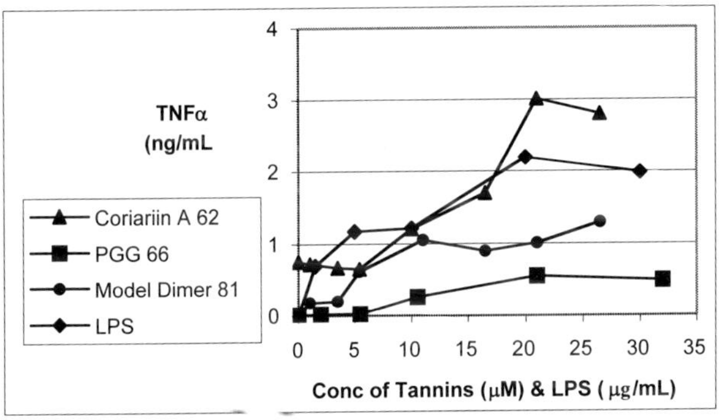

Fig. 6.17 Dose dependence profile of TNFα release (24 hours) upon hPBMC stimulation by several tannins and LPS.

Recently, other hydrolyzable tannins have been reported to effectively modulate the release of TNFα from *Leishmania donovani* infected macrophages (RAW 264.7) *in vitro* (Kolodziej *et al.*, 2001). In total, 27 tannins were tested including agrimoniin, tellimagrandin II, gallic acid, and β-PGG. Contrary to the above results, it was reported that oligomeric ellagitannins (EC_{50} >25 μg/mL) were less potent as TNF inducers (both TNFα and β) than monomeric ellagitannins and gallotannins (EC_{50} 8.5 to >25 μg/mL), and *C*-glycosidic ellagitannins and dehydroellagitannins (EC_{50} 0.6-2.8 μg/mL). In these experiments, TNFα levels were measured by an indirect L929 fibroblast cytotoxicity assay. Since this protocol does not provide either a quantitative or a direct measure of TNFα concentrations, a straightforward comparison with the above results may not be appropriate.

6.4 Ellagitannins as Lipid A Agonists

The similarity in TNFα dose-response data between the immunostimulatory tannins and LPS (with binding via the lipid A component) raises the real possibility that the tannin constructs may be utilizing the previously discussed lipid A receptor system (LBP, CD14, TLR4, etc.) to transduce the ellagitannin binding event into eventual TNFα secretion. In order to test this premise, two types of experiments were pursued. Initially, the role that the LPS-sensitive cell surface receptor CD14 might play in tannin-mediated TNFα secretion from hPBMC's was assayed using monoclonal antibodies for the CD14 receptor as a competitive blocker. Followup studies examined whether the tannin stimulus utilized TLR4 as the signal transduction molecule linking the cell surface binding event to gene transcription.

6.4.1 Use of CD-14 antibodies as a probe for the participation of CD14 in tannin-stimulated TNFα release

Commercially available monoclonal antibodies to the human CD14 receptor have been used previously to test for the participation of this receptor in various biological processes (Verbon *et al.*, 2001, Goyert *et al.*, 1988). This approach has proven effective when there is a competition for the CD14 binding site between a CD14 ligand of interest and the antibody. For the tannin series, use of a positive control (LPS) with hPBMC's and antibody MY4 led to the expected result; nearly complete suppression of TNFα production compared to LPS without antibody (Fig. 6.18, columns 1 and 2). Similar experimental trials with the naturally occurring ellagitannin coriariin A (**62**) and its non-coupled analogue **81** provided the first glimpse that CD14 was indeed implicated in the tannin mediated upregulation of TNFα production. The antibody

MY4 at 10 ug/mL diminished coriariin A's ability to induce TNFα secretion compared to control by about 70%, whereas the gallotannin/ellagitannin analogue **81** was a little better at competing with MY4 (approximately 55% reduction in TNFα formation). These results clearly demand a role for CD14 in the tannin-mediated induction of TNFα secretion from hPBMC's, but the data are not as convincing as with the natural ligand LPS. The intervention of other, non-CD14-utilizing pathways may be indicated for the tannins, perhaps involving the selectin receptors discussed along with Fig. 6.4. Nevertheless, the evidence is unequivocal that at least some of the TNFα-generating pathway stimulated by tannins follows the same receptor system as does the natural ligand LPS/lipid A.

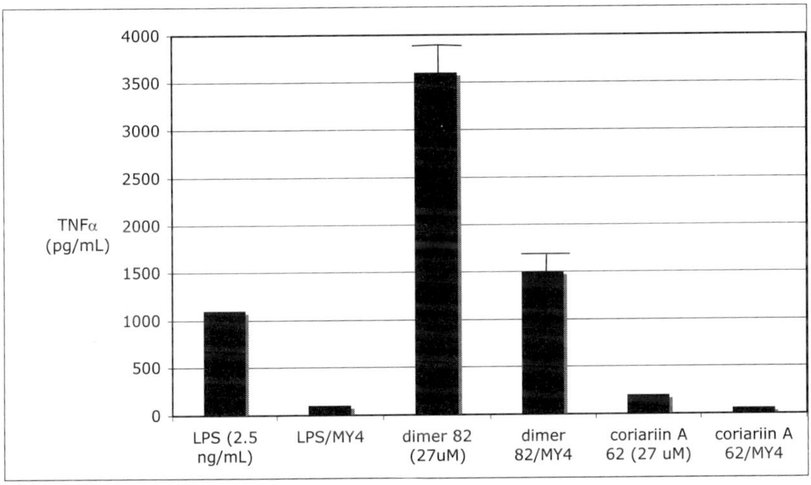

Fig. 6.18 The monoclonal antibody to CD14 suppresses TNFα release from hPBMC's stimulated by (a) LPS, (b) the model dimer **81**, and (c) coriariin A (**62**).

6.4.2 Use of HeJ/OuJ mice as a probe for TLR-4 participation in tannin-stimulated TNFα release

A strain of otherwise healthy mice that was nonresponsive to LPS stimulation were discovered in the 1960s (Nutini *et al.*, 1963, Hoshino *et al.*, 1999). Subsequent studies traced this condition to a mutation in the TLR4 gene, which rendered the receptor inactive. This difference in response to LPS is exemplified in Fig. 6.19, where peritoneal exudate cells (PEC's) from both normal mice (C3H/OuJ) and TLR4-inactive mice (C3H/HeJ) were treated with LPS. The normal mouse PEC's responded as expected, with high levels of TNFα secretion observed at 24 hours within the context of a standard dose-response curve. However, the TLR4-defective HeJ mouse PEC's exhibited no discernable TNFα-generating capability upon LPS stimulation. The identical experiment with the gallotannin/ellagitannin hybrid **81** provided some insight into the role that the TLR4 receptor might play in tannin-mediated secretion of TNFα. There appeared to be a relatively small (compared to the LPS data, see Fig. 6.19) but consistent decrease of TNFα release with the TLR4-inactive HeJ mice compared to the OuJ mice controls (Fig. 6.20).

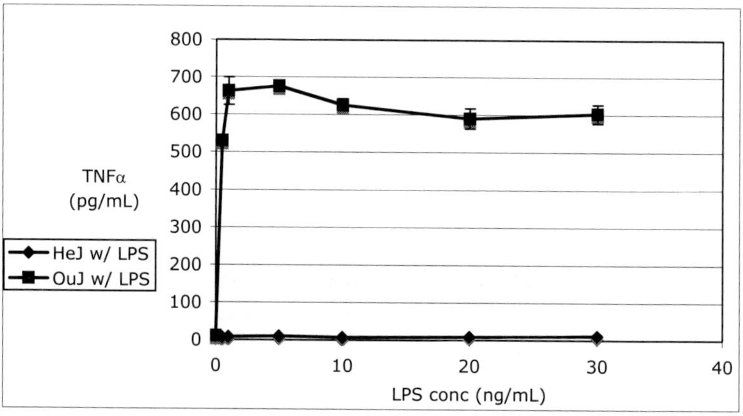

Fig. 6.19 The TNFα secretion response upon treating normal OuJ mice and TLR4-inactive HeJ mice with LPS.

The difference between these two mouse strains was much less dramatic than in the LPS challenge experiment, with TNFα concentrations hovering around 50-60% less in the HeJ series throughout the concentration range. Nevertheless, as with the CD14 competitive binding experiments, there is an unmistakably diminished response with the TLR4-inactive mice, suggesting that at least some of the tannins' effects require a functioning TLR4 receptor. Taken together, these mechanistic experiments provide preliminary data that is entirely consistent with a model whereby the tannin stimulant utilizes, at least in part, the same key receptor components CD14 and TLR4 that the native ligand LPS uses. Therefore, the active tannin species appear to operate, at least in part, as LPS (lipid A) mechanistic agonists.

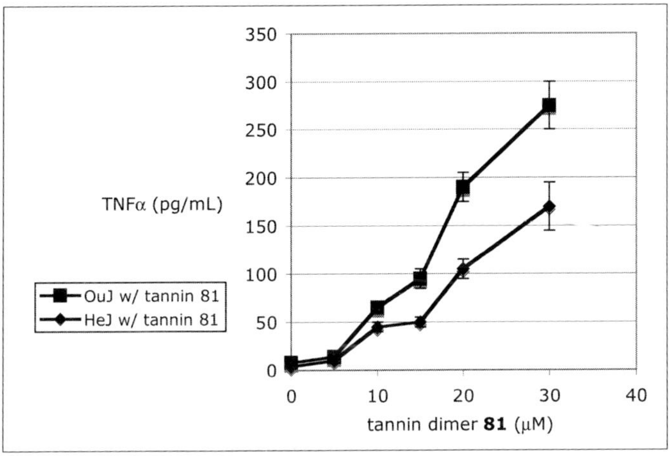

Fig. 6.20 The TNFα secretion response upon treating normal OuJ mice and TLR4-inactive HeJ mice with the dimeric tannin construct **81**.

6.5 Ellagitannins as TNFα Secretion Inhibitors

As discussed earlier, the systemic overproduction of TNFα in response to LPS has been implicated in the symptoms of septic shock. In addition, an assortment of diseases also have been linked to the chronic over-secretion of low levels of TNFα. Since tannins appear to utilize the LPS receptor system, at least in part, to promote TNFα release from monocytes, the possibility of designing tannin constructs that *antagonize* rather than *agonize* LPS activity, was explored. These tannin species would have to meet two stringent criteria for success: (1) bind competitively with LPS (lipid A) at one (or more) of the obligate receptors (CD14, TLR4, others?), and (2) fail to generate a signal to synthesize/secrete TNFα. The role that galloyl groups play in receptor recognition has not been delineated, but the preliminary data clearly speak to their importance. In addition, the dimeric tannin species were invariably better than their smaller monomeric analogues at eliciting

TNFα secretion from hPBMC's. Since the receptor system is composed of several individual and obligate proteins components, perhaps the dimeric nature of the active tannins and the oligomeric nature of the receptor cluster form a match. Merging these observations leads to the testable hypothesis that it might be possible to identify tannin analogues that meet the two criteria listed above by using galloyl groups for binding/recognition, but perhaps spatially misaligning them relative to those of the active tannins in order to prevent the apparently necessary receptor organization between CD14, TLR4 and MD2 that precedes signal transduction. In this way, competitive (to LPS/lipid A) binding without a biological consequence might be realized (= antagonist).

6.5.1 β-PGG as an LPS antagonist

The observation that the monomeric gallotannin β-PGG (**66**) demonstrated almost no TNFα producing ability (see Fig. 6.16) was used as a first pass to test this premise. Apparently, this small tannin molecule cannot provide the necessary binding moieties in the correct orientation for recognizing and organizing the entire receptor complex as is required to initiate the nuclear signaling cascade. However, the hope was that β-PGG, featuring the polygalloyl periphery characteristic of the active tannin constructs, would bind competitively with the lipid A fragment of LPS at one or more of the receptors, thus preventing the endotoxin component from fulfilling this organizational role.

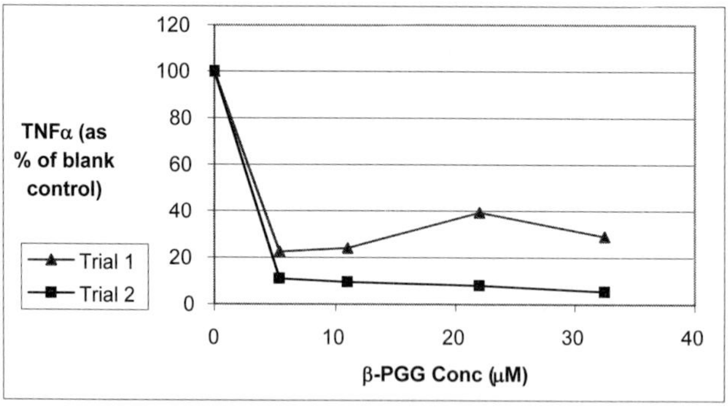

Fig. 6.21 Dose dependent inhibition of LPS-stimulated TNFα release from hPBMC's by β-PGG.

6.5.1.1 *In vitro studies with hPBMC's*

LPS (5 μg/mL) was added to hPBMC's followed by treatment with different concentrations of β-PGG (**66**) to explore this hypothesis (two independent trials, Fig. 6.21). After 4 hours of incubation, the TNFα level in the supernatant, as measured by an ELISA kit, was significantly depleted compared to the control where hPBMC's were treated only with LPS. A concentration of 5 μM for β-PGG appeared to be the point of maximum inhibition (as much as 90% decrease in TNFα levels compared to the control), since higher concentrations did not result in significant further lowering of the TNFα output. Incubation beyond 4 hours (to 24 hours) eliminated any inhibition caused by β-PGG. Thus, there appeared to be a time dependence as well. After these promising *in vitro* studies, β-PGG was tested next with live rats.

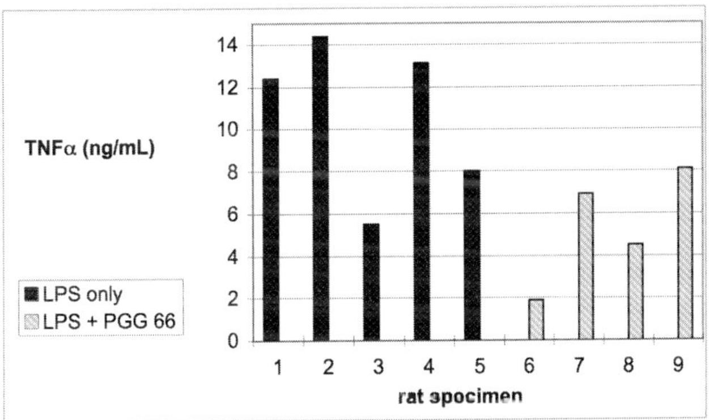

Fig. 6.22 *In vivo* suppression by β-PGG of LPS-stimulated TNFα secretion in rats.

6.5.1.2 *In vivo testing of live rats experiencing septic shock*

Initial *in vivo* experiments determined that a 50-60 mg/rat administration of β-PGG (**66**) led to a lethal level of hypotension. Use of a 30 mg/rat dose led to no changes in blood glucose or blood pressure levels, whereas use of 4 mg/rat caused no inhibition of TNFα production with LPS treated rats. Therefore, a "safe" value of 17 mg/rat was utilized in all subsequent experiments. Injection of the rats with either saline or β-PGG was followed by LPS administration 10 minutes later. The TNFα serum levels 90 minutes after LPS infusion were found to be about 50% lower for the β-PGG-treated rats compared to positive control (just LPS). (Fig. 6. 22).

However, many symptoms of septic shock were still detected in the β-PGG treated rats. The glucose levels for the LPS/β-PGG treated animals (170 mg/dL) were similar to the LPS-only rats (155 mg/dL) and both were higher than the saline-only control rats (55 mg/dL). Also, the animals treated with β-PGG suffered from increased heart rate and hypotension. These results perhaps could be attributed to the IL-1β–inducing properties of β-PGG. Whereas β-PGG managed to lower the LPS-induced TNFα levels, the LPS/β-PGG rats had either similar or higher IL-1β levels compared to the positive controls. In the end, however, a promising new lead compound for the treatment of septic shock may have been identified.

Interestingly, β-PGG also has been described as an efficient inhibitor (IC_{50} 2.1μM) of the binding of vascular endothelial growth factor (VEGF), a critical angiogenic molecule, to its receptor KDR/Flk-1 (Lee *et al.*, 2004). Although β-PGG modified the microvasculature of endothelial cells, it did not directly influence the proliferation of cancer cells (HT1080 human fibrosarcoma and DU-145 human prostate carcinoma).

6.5.2 Dimeric gallotannin analogues as lipid A antagonists

The simple gallotannin β-PGG (**66**) managed to lower TNFα levels *in vitro* and *in vivo*, but did not alleviate the symptoms of septic shock in live rats. It may be possible that these symptoms could be lessened by decreasing the levels of the other proinflammatory cytokine generated, IL-1β, whose concentration was equal to or higher in the β-PGG-administered rats compared with the animals treated only with LPS. To this end, five analogues of the lipid A agonist **81** were studied with the intent of identifying a lead candidate that was ineffective at promoting secretion of both TNFα and IL-1β. Specifically, the linker unit that joins the two tetragalloylglucose end caps (= binding units?) of **81** was targeted for modification. If the tetragalloylglucose components of **81** were responsible for receptor recognition and binding, and if the digalloyl ether linker unit of **81** was complicit in receptor component organization, then altering the latter moiety might meet the requirements of binding without activation. That is, the hypothesis that a change in the spacing and orientation of this linker could bring about a substantial decrease (or increase?) in the ability of the receptor components to interact, was tested.

6.5.2.1 *In vitro testing with hPBMC's*

Five different dimeric gallotannin/ellagitannin analogues were synthesized based on model dimer **81** through coupling of the appropriate diacid chlorides and two molecules of tetragalloylated glucopyranose (Fig. 6.23) (Feldman *et al.*, 2002). Analogue **82**, possessing the most conservative alteration (digalloyl ether hydroxyls removed), was used to test the importance of the phenolic groups in the otherwise identical linker of the parent **81**. Derivative **83** featured a more lipophilic and flexible linker. Analogues **84-86** probed the consequences of modifying both the orientation and spacing of the linker.

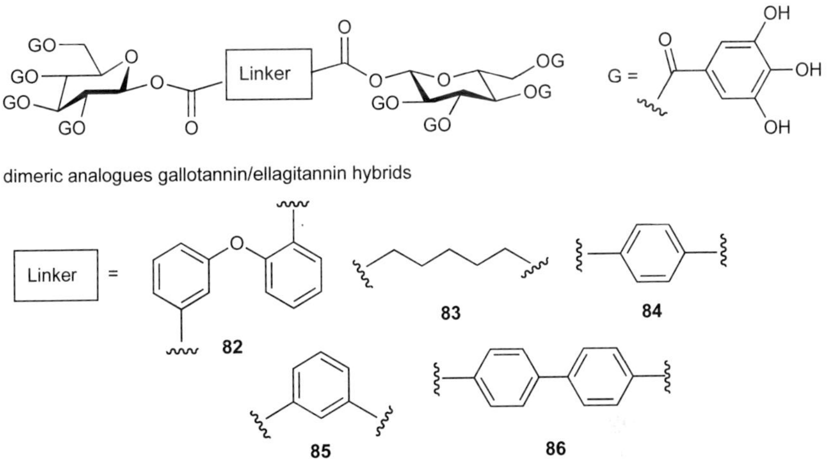

Fig. 6.23 Dimeric gallotannin analogues **82-86** of model dimer **81**.

Derivatives **82-86** were individually and independently added to hPBMC's and assayed for their ability to generate TNFα at the 24-hour mark (Fig. 6.24). Compounds **84**, **85** and **86** produced no detectable TNFα, whereas **83** induced only small amounts of the cytokine. As expected, the most conservatively modified analogue **82** remained active, as it produced approximately 1/3 of the amount of TNFα as the parent **81**. Thus, it appears that the core linker plays a decisive role in receptor recognition, as any misorienting substitute linker greatly suppresses, and even abolishes in some cases, TNFα secretion from hPBMC's compared with the natural ellagitannin linker. Next, analogues **83-86** were tested for their ability to promote secretion of IL-1β. The dimer **85** produced high levels of this cytokine (2.110 ng/mL at 30 μM), while **84** was less potent, but still active (0.690 ng/mL at 30 μM). Gratifyingly, the hPMBC system was almost completely unresponsive to analogues **83** and **86** (< 0.250 ng/mL of IL-1β at 30 μM tannin) in this assay.

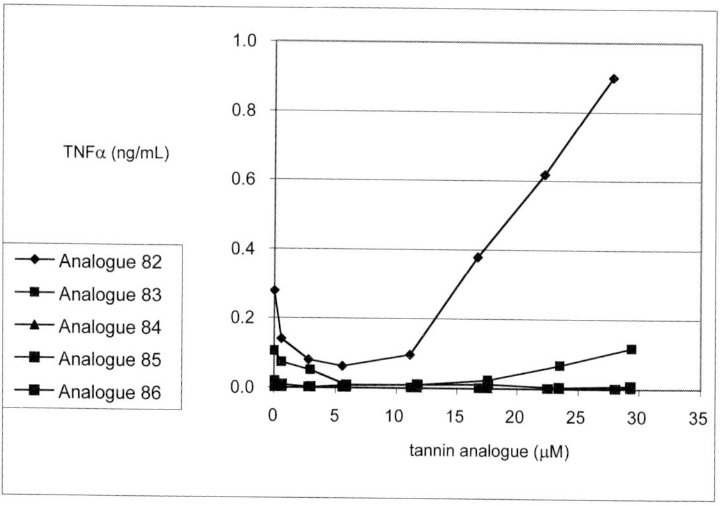

Fig. 6.24 Dose-response profile for TNFα release from hPBMC's treated with the dimeric tannin analogues **82-86**.

Each of these gallotannin analogues **83-86** was added at a range of concentrations to hPBMC's that had been pre-treated for 45 minutes with LPS in order to test whether these species could act as antagonists of lipid A (Fig. 6.25). After 8 hours of incubation (a predetermined optimal time period), compound **86** (at 23.5 μM) was found to be the least inhibitory with a maximum 27% reduction compared with the amount of TNFα produced in LPS-only positive controls. Derivatives **83** (at 23 μM) and **84** (at 3 μM) were found to be somewhat more potent at approximately 44% inhibition. The *meta*-phenyl linked analogue **85** (at 11.8 μM) was the most satisfactory antagonist, with a maximum inhibition of 53% compared to positive control. Since all four compounds were somewhat antagonistic towards LPS (lipid A) in this TNFα secretion assay, analogues **83** and **86** were chosen for further *in vivo* testing because they had proven to be the least capable of promoting undesirable IL-1β release.

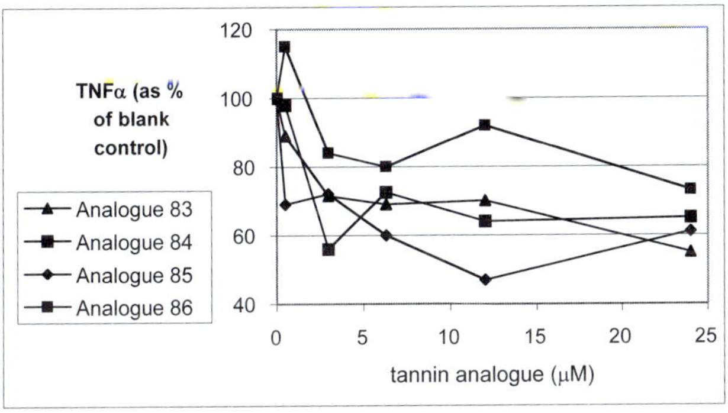

Fig. 6.25 Inhibition of LPS-stimulated secretion of TNFα from hPBMC's by the dimeric tannin analogues **83-86**.

6.5.2.2 *In vivo testing with live rats*

Four restrained rats were dosed with 1 mg/kg of LPS followed by 20 mg/rat of **83** or **86** over a 3-hour period. In addition, saline was administered as a negative control to a 5th rat, and four rats were treated with LPS only. At 90 minutes, the animals showed a significant (≤ 50%) reduction of serum TNFα levels when treated with either one of the tannins **83** or **86** compared to rats dosed only with the positive control LPS (Fig. 6.26). The IL-1β serum levels were also reduced an average of 17% compared to the LPS only animals. Furthermore, rats treated with compound **86** had a lower heart rate (385 beats/min) than the LPS-only rats (483 beats/min). This value is within experimental error of the resting heart rate of the saline-treated control rat (380 beats/min).

Unfortunately, other symptoms of septic shock, such as low blood pressure and high plasma glucose levels, were still detectable. Rats treated with compound **83** showed all of the symptoms of septic shock to some extent. Although these model dimer analogues were less potent LPS antagonists *in vitro* compared to β-PGG, they appeared to exhibit more promising activity with live rats. Perhaps this improvement was simply a consequence of the fact that the dimer species do not induce secretion of IL-1β in and of themselves. Future work may include the search for more powerful antagonists of lipid A using these novel dimeric tannin analogues as lead compounds.

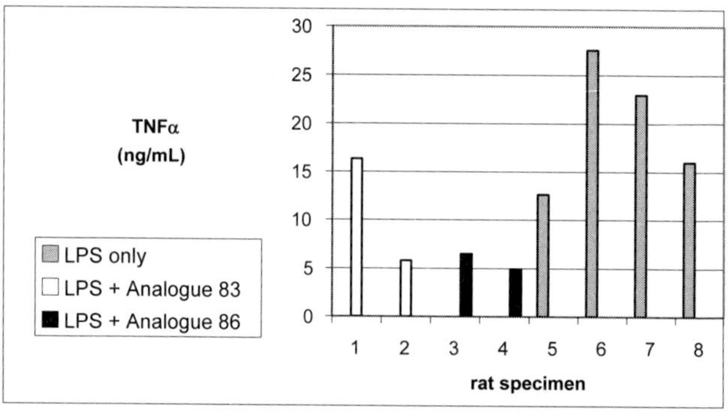

Fig. 6.26 Inhibition of LPS-stimulated TNFα secretion in live rats (at 90 min) by the dimeric tannin analogues **83** and **86**.

6.6 Conclusion

At present, all of the approved anti-TNFα drugs are proteins and given the general disadvantages of peptide-based drugs such as poor biological stability, mandatory subcutaneous or intravenous administration, and high costs to manufacture, small molecules remain very attractive alternatives. Dimeric ellagitannin analogues, with their inherently low toxicity, promising antagonistic action, and novel structures, might represent interesting leads for further development as an anti-sepsis therapy. In a different vein, ellagitannins such as agrimoniin, oenothein B, and coriariin A have documented tumoricidal activities against numerous cancer xenograft models. These tannin species appear to operate through up-regulation of the immune system. In both seemingly opposite therapeutic regimes, the key molecular-level interactions are likely to be the same; the tannin binds (competitively with the lipid A segment of LPS) at the lipid A receptor complex, and either induces or suppresses further signal transduction, depending upon the structural details of the tannin construct. Much refinement of this preliminary and necessarily crude mechanistic picture will be a first step toward realizing the goal of a clinically useful tannin-inspired therapeutic agent in either the antitumor or the anti-sepsis fields.

Acknowledgements

We gratefully acknowledge funding from the National Institutes of Health, Institute of General Medical Science (GM72572).

6.7 Bibliography

Aggarwal, B., Kohr, W., Hass, P., Moffat, B., Spencer, S., Henzel, W., Bringmans, T., Nedwins, G., Goeddels, D. and Harkins, R. (1985). Human tumor necrosis factor. production, purification, and characterization, *J. Biol. Chem.*, 260, pp. 2345–2354.

Akagawa, K. S. and Tokunaga, T. (1985). Lack of binding of bacterial lipopolysaccharide to mouse lung macrophages and restoration of binding by gamma interferon, *J. Exp. Med.*, 162, pp. 1444–1459.

Akira, S. and Takeda, K. (2004). Toll-like receptor signaling, *Nat. Rev. Immunol.*, 4, pp. 499–511.

Amakura, Y., Tsutsumi, T., Sasaki, K., Yoshida, T. and Maitani, T. (2003). Screening of the inhibitory effect of vegetable constituents on the aryl hydrocarbon receptor-mediated activity induced by 2,3,7,8-tetrachlorodibenzo-*p*-dioxin, *Biol. Pharm. Bull.*, 26, pp. 1754–1760.

Amersfoort, E., Van Berkel, T. and Kuiper, J. (2003). Receptors, mediators, and mechanisms involved in bacterial sepsis and septic shock, *Clin. Microbiol. Rev.*, 16, pp. 379–414.

Aoki, K., Nishimura, K., Abe, H., Maruta, H., Sakagami, H., Hatano, T., Okuda, T., Yoshida, T., Tsai, Y., Uchiumi, F. and Tanuma, S. (1993). Novel inhibitors of poly(ADP-ribose) glycohydrolase, *Biochimica et Biophysica Acta*, 1158, pp. 251–256.

Arditi, M., Zhou, J., Dorio, R., Rong, G., Goyert, S. and Kim, K. (1993). Endotoxin-mediated endothelial cell injury and activation: role of soluble CD14, *Infect. Immun.*, 61, pp. 3149–3156.

Balza, E., Mortara, L., Sassi, F., Monteghirfo, S., Carnemolla, B., Castellani, P., Neri, D., Accolla, R., Zardi, L. and Borsi, L. (2006). Targeted delivery of tumor necrosis factor-α to tumor vessels induces a therapeutic T cell-mediated immune response that protects the host against syngeneic tumors of different histologic origin, *Clin. Can. Res.*, 12, pp. 2575–2582.

Barbehenn, R., Jones, C., Hagerman, A., Karonen, M. and Salminen, J. (2006a). Ellagitannins have greater oxidative activities than condensed tannins and galloyl glucoses at high pH: potential impact on caterpillars, *J. Chem. Ecol.*, 32, pp. 2253–2267.

Barbehenn, R., Jones, C., Karonen, M. and Salminen, J. (2006b). Tannin composition affects the oxidative activities of tree leaves, *J. Chem. Ecol.*, 32, pp. 2235–2251.

Bereta, J., Cohen, M. and Bereta, M. (1995). Stimulatory effect of ouabain on VCAM-1 and iNOS expression in murine endothelial cells: involvement of NF-kB, *FEBS Letters*, 377, pp. 21–25.

Beutler, A., Milsark, I. and Cerami, A. (1985). Passive immunization against cachectin/tumor necrosis factor protects mice from the lethal effect of endotoxin, *Science*, 229, pp. 869–871.

Bhakdi, S., Klonisch, T., Nuber, P. and Fischer, W. (1991). Stimulation of monokine production by lipoteichoic acids, *Infect. Immun.*, 59, pp. 4614–4620.

Black, R., Rauch, C. T., Kozlosky, C. J., Peschon, J. J., Slack, J. L., Wolfson, M. F., Castner, B. J., Stocking, K. L., Reddy, P., Srinivasan, S., Nelson, N., Boiani, N., Schooley, K. A., Gerhart, M., Davis, R., Fitzner, J. N., Johnson, R. S., Paxton, R. J., March, C. J. and Cerretti, D. P. (1997a). A metalloproteinase disintegrin that releases tumour-necrosis factor-α from cells, *Nature*, 385, pp. 729–733.

Black, R., Bird, T. and Mohler, K. (1997b). Agents that block TNF-α synthesis or activity, *Ann. Rep. Med. Chem.*, 32, pp. 241–250.

Blick, M., Sherwin, S. A., Roseblum, M. and Gutterman, J. (1987). Phase I study of recombinant tumor necrosis factor in cancer patients, *Cancer Res.*, 47, pp. 2986–2989.

Bone, R. C. (1991). The pathogenesis of sepsis, *Ann. Intern. Med.*, 115, pp. 457–469.

Byrd-Leifer, C., Block, E., Takeda, K., Akira, S. and Ding, A. (2001). The role of MyD88 and TLR4 in the LPS-mimetic activity of taxol, *Eur. J. Immunol.*, 31, pp. 2448–2457.

Carswell, E. A., Old, L. J., Kassel, R. L, Green, S., Fiore, N. and Williamson, B. (1975). An endotoxin-induced serum factor that causes necrosis of tumors, *Proc. Natl. Acad. Sci.*, 72, pp. 3666–3670.

Cavaillon, J.-M. (2003). Proinlammatory and Anti-inflammatory Cytokines as Mediators of Gram-Negative Sepsis. In *Cytokines and Chemokines in Infectious Diseases Handbook*, ed. by Kotb, M. and Calandra, T., Humana Press, Inc.: Totowa, NJ, pp. 33–57.

Chan, M. (1995). Inhibition of tumor necrosis factor by curcumin, a phytochemical, *Biochem. Pharmacol.*, 49, pp. 1551–1556.

Chang, C.-M. and Nowotny, A. (1975). Relation of structure to function in bacterial O-antigens. VII. Endotoxicity of lipid A, *Immunochemistry*, 12, pp. 19–28.

Chen, L., Yen, K., Yang, L., Hatano, T., Okuda, T. and Yoshida, T. (1999). Macrocyclic ellagitannins dimers, cuphiins D1 and D2, and accompanying tannins from *Cuphea hyssopifolia, Phytochemistry*, 50, pp. 307–312.

Cherney, R., King, B., Gilmore, J., Liu, R., Covington, M., Duan, J. and Decicco, C. (2006). Conversion of potent MMP inhibitors into selective TACE inhibitors, *Bioorg. Med. Chem. Lett.*, 16, pp. 1028–1031.

Chilman-Blair, K., Leeson, P. and Bayes, M. (2003). Treatment of septic shock TLR4 (LPS) receptor antagonist, *Drugs of the Future*, 28, pp. 633–639.

Coley, W. B. (1896). The therapeutic value of the mixed toxins of the streptococcus of Erysipelas and Bacillus prodigious in the treatment of inoperable malignant tumors, *Am. J. Med. Sci.*, 112, pp. 251–281.

Colotta, F., Ghezzi, P. and Mantovani, A. (1998). Interleukin 1. In *Cytokines*, ed. by Mire-Sluis, A. and Thorpe, R., Academic Press: San Diego, CA, pp. 1–18.

Corti, A. and Ponzoni, M. (2004). Tumor vascular targeting with tumor necrosis factor α and chemotherapeutic drugs, *Ann. NY Acad. Sci.*, 1028, pp. 104–112.

Curnis, F., Sachi, A. and Corti, A. (2002). Improving chemotherapeutic drug penetration in tumors by vascular targeting and barrier alteration, *J. Clin. Invest.*, 110, pp. 475–482.

Darveau, R., Pham, T., Lemley, K., Reife, R., Bainbridge, B., Coats, S., Howald, W., Way, S. and Hajjar, A. (2004). Porphyromonas gingivalis lipopolysaccharide contains multiple lipid A species that functionally interact with both Toll-like receptors 2 and 4, *Infect. Immun.*, 72, pp. 5041–5051.

Dauphinee, S. and Karsan, A. (2006). Lipopolysaccharide signaling in endothelial cells, *Laboratory Investigation*, 86, pp. 9–22.

Davey, P. and Ashrafian, H. (2000). New therapies for heart failure: is thalidomide the answer? *QJM*, 93, pp. 305-311.

Demchenko, A. V., Wolfert, M. A., Santhanam, B., Moore, J. N. and Boons, G.-J. (2003). Synthesis of biological evaluation of *Rhizobium sin-1* lipid A derivatives, *J. Am. Chem. Soc.*, 125, pp. 6103–6112.

Dheen, S., Jun, Y., Yan, Z., Tay, S. and Ling, E. (2005). Retinoic acid inhibits expression of TNF-α and iNOS in activated rat microglia, *Glia*, 50, pp. 21–31.

Dinarello, C., Cannon, J., Wolff, S., Bernheim, H., Beutler, B., Cerami, A., Figari, I., Palladino, M. and O'Connor, J. (1986). Tumor necrosis factor (cachectin) is an endogenous pyrogen and induces production of interleukin 1, *J. Exp. Med.*, 163, pp. 1433–1450.

Dinarello, C. (1989). Interleukin-1 and its biologically related cytokines. In *Advances in Immunology*, vol. 44, ed. by Dixon, F., Academic Press: San Diego, CA, pp. 153–205.

Ding, A., Porteu, F., Sanchez, E. and Nathan, C. (1990). Shared actions of endotoxin and taxol TNF receptors and TNF release, *Science*, 248, pp. 370–372.

Dixon, D. R. and Darveau, R. P. (2005). Lipopolysaccharide heterogeneity: innate host responses to bacterial modification of lipid A structure, *J. Dent. Res.*, 84, pp. 584–595.

Dunzendorfer, S., Lee, H., Soldau, K. and Tobias, P. (2004). TLR4 is the signaling but not the lipopolysaccharide uptake receptor, *J. Immunol.*, 173, pp. 1166–1170.

Durum, S., Schmidt, J. and Oppenheim, J. (1985). Interleukin 1: an immunological perspective, *Annu. Rev. Immunol.*, 3, pp. 263–287.

Evans, T., Strivens, E., Carpenter, A. and Cohen, J. (1993). Differences in cytokine response and induction of nitric oxide synthase in endotoxin-resistant and endotoxin-sensitive mice after intravenous Gram-negative infection, *J. Immunol.*, 150, pp. 5033–5040.

Feldman, K., Sahasrabudhe, K., Smith, R. and Scheuchenzuber, W. (1999) Immunostimulation by plant polyphenols: a relationship between tumor necrosis factor-α production and tannin structure, *Bioorg. & Med. Chem. Lett.*, 9, pp. 985–990.

Feldman, K., Wilson, S., Lawlor, M., Lang, C. and Scheuchenzuber, W. (2002). In vitro and in vivo inhibition of LPS-induced tumor necrosis factor-α production by dimeric gallotannin analogues, *Bioorg. Med. Chem.*, 10, pp. 47–55.

Feldman, K. and Lawlor, M. (2000). Ellagitannin chemistry: the first total synthesis a dimeric ellagitannin, coriariin A, *J. Am. Chem. Soc.*, 122, pp. 7396–7397.

Feldman, K., Lawlor, M. and Sahasrabudhe, K. (2000). Ellagitannin Chemistry. Evolution of a three-component coupling strategy for the synthesis of the dimeric ellagitannin coriariin A and a dimeric gallotannin analogue, *J. Org. Chem.*, 65, pp. 8011–8019.

Feng, W., Zheng, X., Yoshida, T. and Okuda, T. (1996). Five hydrolyzable tannins from *Potentilla discolor* Bunge, *Tianran Chanwu Yanjiu Yu Kaifa*, 8, pp. 26–30.

Fey, G. H., Hocke, G., Wilson, D., Ripperger, J., Juan, T., Cui, M. and Darlington, G. (1994). Cytokines and the acute phase response of the liver. In *The liver: Biology and Pathobiology*, ed. by Arias, I., Boyer, J., Fausto, N., Jakoby, W., Schachter, D. and Shafritz, D., Raven Press: New York, pp. 113–143.

Fischer, E., Marano, M., Barber, A., Hudson, A., Lee, K., Rock, C.S., Hawes, A. S., Thompson, R. C., Hayes, T. J. and Anderson, T. D. (1991). Comparison between effects of interleukin-1 alpha administration and sublethal endotoxemia in primates, *Am. J. Physiol. Regul. Integr. Comp. Physiol.*, 261, pp. R442–R452.

Flad, H., Loppnow, H., Rietschel, E. and Ulmer, A. (1993). Agonists and antagonists for lipopolysaccharide-induced cytokines, *Immunobiol.*, 187, pp. 303–316.

Folli, S., Pelegrin, A., Chalandon, Y., Yao, X., Buchegger, F., Lienard, D., Lejeune, F. and Mach, J. (1993). Tumor-necrosis factor can enhance radio-antibody uptake in human colon carcinoma xenografts by increasing vascular permeability, *Int. J. Cancer*, 53, pp. 829–836.

Fong, Y., Tracey, K., Moldawer, L., Hesse, D., Manogue, K., Kenney, J., Lee, A., Kuo, G., Allison, A., Lowry S. and Cerami, A. (1989). Antibodies to cachectin/tumor necrosis factor reduce interleukin 1 beta and interleukin 6 appearance during lethal bacteremia, *J. Exp. Med.*, 170, pp. 1627–1633.

Frendeus, B., Wachtler, C., Hedlund, M., Fischer, H., Samuelsson, P., Svensson, M. and Svanborg, C. (2001). *Escherichia coli* P fimbriae utilize the Toll-like receptor 4 pathway for cell activation, *Mol. Microbiol.*, 40, pp. 37–51.

Friedman, G., Silva, E. and Vincent, J. L. (1998). Has the mortality of septic shock changed with time? *Crit. Care Med.*, 26, pp. 2078–2086.

Fukuchi, K., Sakagami, H., Okuda, T., Hatano, T., Tanuma, S., Kitajima, K., Inoue, Y., Inoue, S., Ichikawa, S., Nonoyama, M. and Konno, K. (1989). Inhibition of Herpes simplex virus infection by tannins and related compounds, *Antiviral Res.*, 11, pp. 285–298.

Gazzano-Santoro, H.; Meszaros, K.; Birr, C., Carroll, S., Theofan, G., Horwitz, A., Lim, E., Aberle, S., Kasler, H. and Parent, J. (1994). Competition between rBPI23, a recombinant fragment of bactericidal/permeability-increasing protein, and lipopolysaccharide (LPS)-binding protein for binding to LPS and Gram-negative bacteria, *Infect. Immun.*, 62, pp. 1185–1191.

Géry, I. and Waksman, B. (1972). Potentiation of the T-lymphocyte response to mitogens: II. The cellular source of potentiating mediator(s), *J. Exp. Med.*, 136, pp. 143–155.

Goyert, S., Ferrero E., Rettig, W., Yenamandra, A., Obata, F., and Le Beau, M. (1988). The CD14 monocyte differentiation antigen maps to a region encoding growth factors and receptors, *Science*, 239, pp. 497–500.

Hampton, R., Golenbock, D., Penman, M., Krieger, M. and Raetz, C. (1991). Recognition and plasma clearance of endotoxin by scavenger receptors. *Nature*, 352, pp. 342–344.

Hashimoto, T., Suganuma, M., Fujiki, H., Yamada, M., Kohno, T. and Asakawa, X. (2003). Isolation and synthesis of TNF-α release inhibitors from Fijian kawa (Piper methysticum), *Phytomedicine*, 10, pp. 309–317.

Hatano, T., Hattori, S. and Okuda, T. (1986). Tannins of Coriaria japonica A. Gray. I. Coriariins A and B, new dimeric and monomeric hydrolyzable tannins, *Chem. Pharm. Bull.*, 34, pp. 4092–4097.

Hatano, T., Yasuhara, T., Matsuda, M., Yazaki, K., Yoshida, T. and Okuda, T. (1989). Oenothein B, a dimeric hydrolyzable tannin of cyclic structure, *Chem. Pharm. Bull.*, 37, pp. 2269–2271.

Hatano, T., Yasuhara, T., Matsuda, M., Yazaki, K., Yoshuda, T. and Okuda, T. (1990). Oenothein B, a dimeric, hydrolyzable tannin with macrocyclic structure, and accompanying tannins from *Oenothera erythrosepala*, *J. Chem. Soc., Perkin Trans. 1*, pp. 2735–2743.

Havell, E., Fiers, W. and North, R. (1988). The antitumor function of tumor necrosis factor (TNF), *J. Exp. Med.*, 167, pp. 1067–1085.

Hayashi, F., Smith, K., Ozinsky, A., Hawn, T., Yi, E., Goodlett, D., Eng, J., Akira, S., Underhill, D. and Aderem, A. (2001). The innate immune response to bacterial flagellin is mediated by Toll-like receptor 5, *Nature*, 410, pp. 1099–1103.

Haziot, A., Ferrero, E., Kontgen, F., Hijiya, N., Yamamoto, S., Silve, J., Stewart, C. and Goyert, S. (1996). Resistance to endotoxin shock and reduced dissemination of Gram-negative bacteria in CD14-deficient mice, *Immunity*, 4, pp. 407–414.

Haziot, A., Hijiya, N., Schultz, K., Zhang, F., Gangloff, S. and Goyert, S. (1999). CD14 plays no major role in shock induced by *Staphylococcus aureus* but down-regulates TNF-α production, *J. Immunol.*, 162, pp. 4801–4805.

Heine, H., Muller-Loennies, S., Brade, L., Lindner, B. and Brade, H. (2003). Endotoxic activity and chemical structure of lipopolysaccharides from *Chlamydia trachomatis* serotypes E and L_2 and *Chlamydophyla psittacy* 6BC, *Eur. J. Biochem.*, 270, pp. 440–450.

Heinzel, F. P., Rerko, R., Ling, P., Hakimi, J. and Schoenhaut., D. (1994). Interleukin 12 is produced in vivo during endotoxemia and stimulates synthesis of gamma interferon, *Infect. Immun.*, 62, pp. 4244–4249.

Helson, L., Green, S., Carswell, E. and Old, L. J. (1975). Effect of tumour necrosis factor on cultured human melanoma cells, *Nature*, 258, pp. 731–732.

Hemmi, H., Takeuchi, O., Kawai, T., Kaisho, T., Sato, S., Sanjo, H., Matsumoto, M., Hoshino, K., Wagner, H., Takeda, K. and Akira, S. (2000). A Toll-like receptor recognizes bacterial DNA, *Nature*, 408, pp. 740–745.

Herath, H., Takano-Ishikawa, Y. and Yamaki, K. (2003). Inhibitory effect of some flavonoids on tumor necrosis factor-α production in lipopolysaccharide-stimulated mouse macrophage cell line J774.1, *J. Med. Food*, 6, pp. 365–370.

Herman, J., Dinarello, C., Kew, M. and Rabson, A. (1985). The role of interleukin 1 (IL 1) in tumor-NK cell interactions: correction of defective NK cell activity in cancer patients by treating target cells with IL 1, *J. Immunol.*, 135, pp. 2882–2886.

Hoshino, K., Takeuchi, O, Kawai, T., Sanjo, H., Ogawa, T., Takeda, Y., Takeda, K. and Akira, S. (1999). Cutting Edge: Toll-like receptor 4 (TLR4)-deficient mice are hyporesponsive to lipopolysaccharide: evidence for TLR4 as the *Lps* gene product, *J. Immunol.*, 162, pp. 3749–3752.

Hrenn, A., Steinbrecher, T., Labahn, A., Schwager, J., Schempp, C. and Merfort, I. (2006). Plant phenolics inhibit neutrophil elastase, *Planta Med.*, 72, pp. 1127–1131.

Hukkanen, A., Kokko, H., Buchala, A., McDougall, G., Stewart, D., Karenlampi, S. and Karjalainen, R. (2007). Benzothiadiazole induces the accumulation of phenolics and improves resistance to powdery mildew in strawberries, *J. Agric. Food Chem.*, 55, pp. 1862–1870.

Ingalls, R. R. and Golenbock, D. (1995). CD11c/CD18, a transmembrane signaling receptor for lipopolysaccharide, *J. Exp. Med.*, 181, pp. 1473–1479.

Iontcheva, I., Amar, S., Zawawi, K., Kantarci, A. and Van Dyke, T. (2004). Role for moesin in lipopolysaccharide-stimulated signal transduction, *Infect. Immun.*, 72, pp. 2312–2320.

Ito, H., Nishitani, E., Hatano, T., Nakanishi, T., Inada, A., Murata, H., Inatomi, Y., Matsuura, N., Ono, K., Lang, F., Murata, J. and Yoshida, T. (2001). Tannins and related polyphenols, HIV reverse transcriptase inhibitors, from *Rosa woodsii*, *Nat. Med.*, 55, p. 218.

Jaeschke, H. (1996). Chemokines, neutrophils, and inflammatory liver injury, *Shock*, 6, pp. 403–404.

Jang, S., Jeong, S., Kim, K., Kim, H., Yu, H., Park, R., Kim, H. and You, Y. (2003). Tanshinone IIA from Salvia miltiorrhiza inhibits inducible nitric oxide synthase expression and production of TNF-α, IL-1β and IL-6 in activated RAW 264.7 cells, *Planta Med.*, 69, pp. 1057–1059.

Jeong, H., Koo, H., Na, H., Kim, M., Hong, S., Eom, J., Kim, K., Shin, T. and Kim, H. (2002). Inhibition of TNF-α and IL-6 production by aucubin through blockade of NF-κB activation in RBL-2H3 mast cells, *Cytokine*, 18, pp. 252–259.

Jones, E., Staurt, D. and Walker, N. (1992). Crystal structure of TNF, *Immunol. Ser.*, 56, pp. 93–127.

Ju, H., Baek, S., An, R., Bae, K., Son, K., Kim, H., Kang, S., Lee, S., Son, J. and Chang, H. (2003). Inhibitory effects of nardostachin on nitric oxide, prostaglandin E2, and tumor necrosis factor-α production in lipopolysaccharide activated macrophages, *Biol. Pharm. Bull.*, 26, pp. 1375–1378.

Kanoh, R., Hatano, T., Ito, H., Yoshida, T. and Akagi, M. (2000). Effects of tannins and related polyphenols on superoxide-induced histamine release from rat peritoneal mast cells, *Phytomedicine*, 7, pp. 297–302.

Kawai, T., Takeuchi, O., Fujita, T., Inoue, J., Mühlradt, P., Sato, S., Hoshino, K. and Akira, S. (2001). Lipopolysaccharide stimulates the MyD88-independent pathway and results in activation of IFN-regulatory factor 3 and the expression of a subset of lipopolysaccharide-inducible genes, *J. Immunol.*, 167, pp. 5887–5894.

Keller, R., Fischer, W., Keist, R. and Bassetti, S. (1992). Macrophage response to bacteria: induction of marked secretory and cellular activities by lipoteichoic acids, *Infect. Immun.*, 60, pp. 3664–3672.

Kemeny N., Childs, B., Larchian, W., Rosado, K. and Kelsen, D. (1990). A phase II trial of recombinant tumor necrosis factor in patients with advanced colorectal carcinoma, *Cancer*, 66, pp. 659–663.

Kengatharan, M., De Kimpe, S. and Thiemermann, C. (1996). Analysis of the signal transduction in the induction of nitric oxide synthase by lipoteichoic acid in macrophages, *Br. J. Pharmacol*, 117, pp. 1163–1170.

Kim, E., Jin, H., Kim, Y., Lee, H., Lee, S., Lee, K., Zee, O., Han, J. and Lee, H. (2001). Suppression by a sesquiterpene lactone from Carpesium divaricatum of inducible nitric oxide synthase by inhibiting nuclear factor-kappaB activation, *Biochem. Pharmacol.*, 61, pp. 903–910.

Kim, H.-J., Jang, S., Kim, Y.-J., Chung, H.-T., Yun, Y.-G., Kang, T.-H., Jeong, O.-S. and Kim, Y.-C. (2004). Scopoletin suppresses pro-inflammatory cytokines and PGE_2 from LPS-stimulated cell line, RAW 264.7 cells, *Fitoterapia*, 75, pp. 261–266.

Kiss, A., Kowalski, J. and Melzig, M. (2004). Compounds from Epilobium angustifolium inhibit the specific metallopeptidases ACE, NEP and APN, *Planta Med.*, 70, pp. 919–923.

Kolodziej, H., Kayser, O., Kiderlen, A., Ito, H., Hatano, T., Yoshida, T. and Foo, L. (2001). Antileishmanial activity of hydrolyzable tannins and their modulatory effects on nitric oxide and tumour necrosis factor-α release in macrophages *in vitro*, *Planta Med.*, 67, pp. 825–832.

Koshiura, K., Miyamoto, K., Ikeya, Y. and Taguchi, H. (1985). Antitumor activity of methanol extract from roots of *Agrimonia pilosa* Ledeb, *Jpn. J. Pharmacol.*, 38, pp. 9–16.

Kristensen, C., Nozue, M., Boucher, Y. and Jain, R. (1996). Reduction of interstitial fluid pressure after TNF-α treatment of three human melanoma xenografts, *Br. J. Cancer*, 74, pp. 533–536.

Kubo, I., Yokokawa, Y. and Kinst-Hori, I. (1995). Tyrosinase inhibitors from bolivian medicinal plants, *J. Nat. Prod.*, 58, pp. 739–743.

Kuipers, B., van der Poll, T., Levi, M., van Deventer, S., Ten Cate, H., Imai, Y., Hack, E. and Ten Cate, J. (1994). Platelet-activating factor antagonist TCV-309 attenuates the induction of the cytokine network in experimental endotoxemia in chimpanzees, *J. Immunol.*, 152, pp. 2438–2446.

Lee, M., Chiou, J., Yen, K. and Yang, L. (2000). EBV DNA polymerase inhibition of tannins from *Eugenia uniflora*, *Cancer Letters*, 154, pp. 131–136.

Lee, S., Lee, H., Ji, S., Lee, S., Mar, W. and Gho, Y. (2004). 1,2,3,4,6-Penta-*O*-galloyl-β-D-glucose blocks endothelial cell growth and tube formation through inhibition of VEGF binding to VEGF receptor, *Cancer Letters*, 208, pp. 89–94.

Lehmann, V. and Rupprecht, E. (1977). Microheterogeneity in lipid A demonstrated by a new intermediate in the biosynthesis of 3-deozy-D-manno-octulosonic-acid-lipid A. *Eur. J. Biochem.*, 81, pp. 443–452.

Lesuisse, D., Berjonneau, J., Ciot, C., Devaux, P., Doucet, B., Gourvest, J., Khemis, B., Lang, C., Legrand, R., Lowinski, M., Maquin, P., Parent, A., Schoot, B. and Teutsch, G. (1996). Determination of oenothein B as the active 5-α-reductase-inhibiting principle of the folk medicine *Epilobium parviflorum*, *J. Nat. Prod.*, 59, pp. 490–492.

Lien, E., Chow, J., Hawkins, L., McGuinness, P., Miyake, K., Espevik, T., Gusovsky, F. and Golenbock, D. (2001). A novel synthetic acyclic lipid A-like agonist activates cells via the lipopolysaccharide/Toll-like receptor 4 signaling pathway, *J. Biol. Chem.*, 276, pp. 1873–1880.

Lin, J. and Wang, J. (1998). Hydrolyzable tannins possessing a dehydrodigalloyl group from the fruit of *Rosa taiwanensis* Nakai., *Chin. Pharm. J.*, 50, pp. 167–173.

Liu, W. C., Oikawa, M., Fukase, K., Suda, Y., Winarno, H., Mori, S., Hashimoto, M. and Kusumoto, S. (1997). Enzymatic preparation of (*S*)-3-hydroxytetradecanoic acid and synthesis of unnatural analogues of lipid A containing the (*S*)-acid, *Bull. Chem. Soc. Jpn.*, 70, pp. 1441–1450.

Liu, Y., Zhang, W., Cheung, L., Niu, T., Wu, Q., Li, C., Van Pelt, C. and Rosenblum, M. (2006). The antimelanoma immunocytokine scFvMEL/TNF shows reduced toxicity and potent antitumor activity against human tumor xenografts, *Neoplasia*, 8, pp. 384–393.

Loppnow, H., Stelter, F., Schonbeck, U., Schluter, C., Ernst, M., Schutt, C. and Flad, H. (1995). Endotoxin activates human vascular smooth muscle cells despite lack of expression of CD14 mRNA or endogenous membrane CD14, *Infect. Immun.*, 63, pp. 1020–1026.

Lund, K. and Rimpler, H. (1985). Tormentilla root. Isolation of an ellagitannin and pharmacological screening, *Deutsche Apot. Zeit.*, 125, pp. 105–108.

Malhotra, R., Priest, R. and Bird, M. (1996). Role for L-selectin in lipopolysaccharide-induced activation of neutrophils, *Biochem. J.*, 320, pp. 589–593.

Malhotra, R. and Bird, M. (1997). L-selectin – a signaling receptor for lipopolysaccharide, *Chemistry & Biology*, 4, pp. 543–547.

Malhotra, R., Priest, R., Foster, M. and Bird, M. (1998). P-selectin binds to bacterial lipopolysaccharide, *Eur. J. Immunol.*, 28, pp. 983–988.

Manjula, N., Gayathri, B., Vinaykumar, K., Shankernarayanan, N., Vishwakarma, R. and Balakrishnan, A. (2006). Inhibition of MAP kinases by crude extract and pure compound isolated from Commiphora mukul leads to down regulation of TNF-α, IL-1β and IL-2, *Int. Immunopharmacol.*, 6, pp. 122–132.

Martin, G. S., Mannino, D. M., Eaton, S. and Moss, M. (2003). The epidemiology of sepsis in the United States from 1979 through 2000, *N. Engl. J. Med.*, 348, pp. 1546–1554.

Maruta, H., Okita, N., Takasawa, R., Uchiumi, F., Hatano, T. and Tanuma, S. (2007). The involvement of ATP produced via $(\text{ADP-ribose})_n$ in the maintenance of DNA replication apparatus during DNA repair, *Biol. Pharm. Bull.*, 30, pp. 447–450.

Matsumori, A., Shioi, T., Yamada, T., Matsui, S. and Sasayama, S. (1994). Vesnarinone, a new inotropic agent, inhibits cytokine production by stimulated human blood from patients with heart failure, *Circulation*, 89, pp. 955–958.

Matsumori, A., Ono, K., Nishio, R., Igata, H., Shioi, T., Matsui, S., Furukawa, Y., Iwasaki, A., Nose, Y. and Sasayama, S. (1997). Modulation of cytokine production and protection against lethal endotoxemia by the cardiac glycoside ouabain, *Circulation*, 96, pp. 1501–1506.

Maury, C. and Teppo, A. (1987). Raised serum levels of cachectin/tumor necrosis factor alpha in renal allograft rejection, *J. Exp. Med.*, 166, pp. 1132–1137.

McMahan, C., Slack, J. and Mosley, B. (1991). A novel IL-1 receptor, cloned from B cells by mammalian expression, is expressed in many cell types, *EMBO J.*, 10, pp. 2821–2832.

Meredith, T. C., Aggarwal, P., Mamat, U., Lindner, B. and Woodward, R. W. (2006). Redefining the requisite lipopolysaccharide structure in *Escherichia coli*, *ACS Chem. Biol.*, 1, pp. 33–42.

Miyamoto, K., Koshiura, R., Ikeya, Y. and Taguchi, H. (1985). Isolation of agrimoniin, an antitumor constituent, from the roots of *Agrimonia pilosa* Ledeb, *Chem. Pharm. Bull.*, 33, pp. 3977–3981.

Miyamoto, K., Kishi, N. and Koshiura, R. (1987a). Antitumor effect of agrimoniin, a tannin of Agrimonia pilosa Ledeb., on transplantable rodent tumors, *Jpn. J. Pharmacol.*, 43, pp. 187–195.

Miyamoto, K., Kishi, N., Koshiura, R., Yoshida, T., Hatano, T. and Okuda, T. (1987b). Relationship between the structures and the antitumor activities of tannins, *Chem. Pharm. Bull.*, 35, pp. 814–822.

Miyamoto, K., Kishi, N., Murayama, T., Furukawa, T. and Koshiura, R. (1988). Induction of cytotoxicity of peritoneal exudate cells by agrimoniin, a novel immunomodulatory tannin of *Agrimoniin pilosa* Ledeb. *Cancer Immunol. Immunother.*, 27, pp. 59–62.

Miyamoto, K., Murayama, T., Nomura, M., Hatano, T., Yoshida, T., Furukawa, T., Koshiura, R. and Okuda, T. (1993a). Antitumor activity and Interleukin-1 induction by tannins, *Anticancer Res.*, 13, pp. 37–42.

Miyamoto, K., Nomura, M., Murayama, T., Furukawa, T., Hatano, T., Yoshida, T., Koshiura, R. and Okuda, T. (1993b). Antitumor activities of ellagitannins against sarcoma-180 in mice, *Biol. Pharm. Bull.*, 16, pp. 379–387.

Miyamoto, K., Nomura, M., Sasakura, M., Matsui, E., Koshiura, R., Murayama, T., Furukawa, T., Hatano, T., Yoshida, T. and Okuda, T. (1993c). Antitumor activity of oenothein B, a unique macrocyclic Ellagitannin, *Jpn. J. Cancer Res.*, 84, pp. 99–103.

Miyamoto, K., Murayama, T., Hatano, T., Yoshida, T. and Okuda, T. (1999). Host-mediated anticancer activities of tannins. In Plant Polyphenols 2: Chemistry, Biology, Pharmacology, and Ecology, ed. by Gross, G., Hemingway, R. W., Yoshida, T., Kluwer Academic/Plenum Publishers: New York, pp. 643–663.

Moreira, A., Sampaio, E., Zmuidzinas, A., Frindt, P., Smith, K. and Kaplan, G. (1993). Thalidomide exerts its inhibitory action on tumor necrosis factor alpha by enhancing mRNA degradation, *J. Exp. Med.*, 177, pp. 1675–1680.

Muller, G., Chen, R., Huang, S., Corral, L., Wong, L., Patterson, R., Chen, Y., Kaplan, G. and Stirling, D. (1999). Amino-substituted thalidomide analogs: potent inhibitors of TNF-α production, *Bioorg. Med. Chem. Lett.*, 9, pp. 1625–1630.

Murakami, A., Nishizawa, T., Egawa, K., Kawada, T., Nishikawa, Y., Uenakai, K. and Ohigashi, H. (2005). New class of linoleic acid metabolites biosynthesized by corn and rice lipoxygenases: Suppression of proinflammatory mediator expression via attenuation of MAPK- and Akt-, but not PPARγ-, dependent pathways in stimulated macrophages, *Biochem. Pharmacol.*, 70, pp. 1330–1342.

Murayama, T., Kishi, N., Koshiura, R., Takagi, K., Furukawa, T. and Miyamoto, K. (1992). Agrimoniin, an antitumor tannin of Agrimonia pilosa Ledeb., induces interleukin-1, *Anticancer Res.*, 12, pp. 1471–1474.

Murray, J., Rosenblum, M., Lamki, L., Haynie, T., Glenn, H., Plager, C., Unger, M., Carlo, D. and Hersh, E. (1987). Radio immunoimaging in malignant melanoma patients with the use of indium-111-labeled antimelanoma monoclonal antibody (ZME-018) to high-molecular-weight antigen, *NCI Monogr.*, 3, pp. 3–9.

Muto, Y., Nouri-Aria, K. and Meager, A. (1988). Enhanced tumour necrosis factor and interleukin-1 in fulmiminant hepatic failure, *Lancet*, 332, pp. 72–74.

Nagai, Y., Akashi, S., Nagafuku, M., Ogata, M., Iwakura, Y., Akira, S., Kitamura, T., Kosugi, A., Kimoto, M. and Miyake, K. (2002). Essential role of MD-2 in LPS responsiveness and TLR4 distribution, *Nat. Immunol.*, 3, pp. 667–672.

Newton, R. and Decicco, C. (1999). Therapeutic potential and strategies for inhibiting tumor necrosis factor-α, *J. Med. Chem.*, 42, pp. 2295–2314.

Nikaido, H. and Vaara, M. (1985). Molecular basis of bacterial outer membrane permeability, *Microbiol. Rev.*, 49, pp. 1–32.

Nutini, L., Prince, J., Duarte, A., Juhasz, R. and Cook, S. (1963). Influence of tumor fractions on incidence of spontaneous mammary tumors in C3H/HeJ mice, *Proc. Soc. Exp. Biol. Med.*, 112, pp. 315–318.

Okuda, T., Mori, K., Shiota, M. and Ida, K. (1982a). Effect of the interaction of tannins with coexisting substances. II. Reduction of heavy metal ions and solubilization of precipitates, *Yakugaku Zasshi*, 102, pp. 735–742.

Okuda, T., Mori, K. and Shiota, M. (1982b). Effects of the interaction of tannins with coexisting substances. III. Formation and solubilization of precipitates with alkaloids, *Yakugaku Zasshi*, 102, pp. 854–858.

Okuda, T., Yoshida, T., Kuwahara, M., Memon, M. and Shingu, T. (1982c). Agrimoniin and potentillin, an ellagitannin dimer and monomer having an α-glucose core, *J. Chem. Soc., Chem. Commun.*, pp. 163–164.

Okuda, T., Yoshida, T., Kuwahara, M., Memon, M. and Shingu, T. (1984). Tannins of rosaceous medicinal plants. I. Structures of potentillin, agrimonic acids A and B, and agrimoniin, a dimeric ellagitannin, *Chem. Pharm. Bull.*, 32, pp. 2165–2173.

Okuda, T., Yoshida, T. and Hatano, T. (1993). Classification of oligomeric hydrolysable tannins and specificity of their occurrence in plants, *Phytochemistry*, 32, pp. 507–521.

Okusawa, S., Gelfand, J., Ikejima, T., Connally, R. and Dinarello, C. (1988). Interleukin 1 induces a shock-like state in rabbits. Synergism with tumor necrosis factor and the effect of cyclooxygenase inhibition, *J. Clin. Invest.*, 81, pp. 1162–1172.

Park, S., Han, S., Lee, K., Lee, H., Kho, Y., Chun, H., Choi, Y., Yang, J., Yoon, Y., Lee, C., Kim, H., Choi, H., Tae, H., Lee, H., Nam, K. and Han, G. (2006). Gelastatins and their hydroxamates as dual functional inhibitors for TNF-α converting enzyme and matrix metalloproteinases: synthesis, biological evaluation, and mechanism studies, *Biochem. Biophys. Res. Comm.*, 341, pp. 627–634.

Parrillo, J. (1993). Pathogenic mechanisms of septic shock. *N. Eng. J. Med.*, *328*, pp. 1471–1477.

Pahan, K., Sheikh, F., Namboodiri, A. and Singh, I. (1997). Lovastatin and phenylacetate inhibit the induction of nitric oxide synthase and cytokines in rat primary astrocytes, microglia, and macrophages, *J. Clin. Invest.*, 100, pp. 2671–2679.

Paul, A., Gohil, V. and Bhutani, K. (2006). Modulating TNF-α signaling with natural products, *Drug Discovery Today*, 11, pp. 725–732.

Perera, P. Y., Mayadas, T., Takeuchi, O., Akira, S., Zaks-Zilberman, M., Goyert, S. and Vogel, S. (2001). CD11b/CD18 acts in concert with CD14 and Toll-like receptor (TLR) 4 to elicit full lipopolysaccharide and taxol-inducible gene expression, *J. Immunol.*, 166, pp. 574–581.

Pittet, D., Tarara, D. and Wenzel, R. P. (1994). Nosocomial bloodstream infection in critically ill patients. Excess length of stay, extra costs, and attributable mortality, *JAMA, J. Am. Med. Assoc.*, 271, pp. 1598–1601.

Posadas, I., Terencio, M., Randazzo, A., Gomez-Paloma, L., Payá, M. and Alcaraz, M. (2003). Inhibition of the NF-κB signaling pathway mediates the anti-inflammatory effects of petrosaspongiolide M, *Biochem. Pharmacol.*, 65, pp. 887–895.

Priestle, J., Schar, H. and Grutter, M. (1988). Crystal structure of the cytokine interleukin-1β, *EMBO J.*, 7, pp. 339–343.

Priestle, J., Schar, H. and Grutter, M. (1989). Crystallographic refinement of interleukin 1β at 2.0 angstrom resolution, *Proc. Natl. Acad. Sci. USA*, 86, pp. 9667–9671.

Qureshi, N., Takayama, K. and Kurtz, R. (1991). Diphosphoryl lipid A obtained from the nontoxic lipopolysaccharide of *Rhodopseudomonas sphaeroides* is an endotoxin antagonist in mice, *Infect. Immun.*, 59, pp. 441–444.

Raetz, C. (1990). Biochemistry of endotoxins, *Ann. Rev. Biochem.*, 59, pp. 129–170.

Ramadori, G., Meyer zum Buschenfelde, K., Tobias, P., Mathison, J. and Ulevitch. R. (1990). Biosynthesis of lipopolysaccharide-binding protein in rabbit hepatocytes, *Pathobiology*, 58, pp. 89–94.

Reitschel, E. T., Brade, H., Holst, O., Brade, L., Muller-Loennies, S., Mamat, U., Zahringer, U., Beckmann, F., Seydel, U., Brandenburg, K., Ulmer, A. J., Mattern, T., Heine, H., Schletter, J., Loppnow, H., Schonbeck, U., Flad, H.-D., Hauschildt,

S., Schade, U. F., Di Padova, F., Kusumoto, S. and Schumann, R. R. (1996). Bacterial endotoxin: chemical constitution, biological recognition, host response, and immunological detoxification. In *Pathology of Septic Shock*, ed. by Reitschel, E. T. and Wagner, H., Springer: New York, pp 39–81.

Rietschel, E. T. and Westphal, O. (1999). Endotoxin: historical perspectives. In *Endotoxin in Health and Disease*, ed. by Brade, H., Opal, S., Vogel, S. N. and Morrison, D. C., Marcel Dekker, Inc.: New York, pp. 1–30.

Rinehart, J., Balcerzak, S. and Hersh, M., (1990). Phase II trial of tumor necrosis factor in human sarcoma, *Proc. Am. Soc. Clin. Oncol.*, 9, pp. 317 (abstr 1229).

Roberts, F. J., Geere, I. W. and Coldman, A (1991). A three-year study of positive blood cultures, with emphasis on prognosis, *Rev. Infect. Dis.*, 13, pp. 34–46.

Rosenblum, M., Cheung, L., Murray, J. and Bartholomew, R. (1991). Antibody-mediated delivery of tumor necrosis factor (TNF-α): improvement of cytotoxicity and reduction of cellular resistance, *Cancer Commun.*, 3, pp. 21–27.

Rossignol, D. and Lynn, M. (2005). TLR4 antagonists for endotoxemia and beyond, *Curr. Opin. Inv. Drugs*, 6, pp. 496–502.

Rudolphi, K., Gerwin, N., Verzijl, N., van der Kraan, P., and van den Berg, W. (2003). Pralnacasan, an inhibitor of interleukin-1β converting enzyme, reduces joint damage in two murine models of osteoarthritis, *Osteoart. Cartilage*, 11, pp. 738–746.

Ruegg, C., Yilmaz, A., Bieler, G., Bamat, J., Chaubert, P. and Lejeune, F. (1998). Evidence for the involvement of endothelial cell integrin αVβ3 in the disruption of the tumor vasculature induced by TNF and IFN-γ, *Nat. Med.*, 4, pp. 408–414.

Ryu, J., Son, H., Lee, S. and Sohn, D. (2002). Two neolignans from Perilla frutescens and their inhibition of nitric oxide synthase and tumor necrosis factor-α expression in murine macrophage cell line Raw 264.7, *Bioorg. Med. Chem. Lett.*, 12, pp. 649–651.

Sacchi, A., Gasparri, A., Gallo-Stampino, C., Toma, S., Curnis, F. and Corti, A. (2006). Synergistic antitumor activity of cisplatin, paclitaxel, and gemcitabine with tumor vasculature-targeted tumor necrosis factor-α, *Clin. Cancer. Res.*, 12, pp. 175–182.

Sakagami, H., Asano, K., Tanuma, S., Hatano, T., Yoshida, T. and Okuda, T. (1992). Stimulation of monocyte iodination and IL-1 production by tannins and related compounds, *Anticancer Res.*, 12, pp. 377–387.

Sakagami, H., Jiang, Y., Kusama, K., Atsumi, T., Ueha, T., Toguchi, M., Iwakura, I., Satoh, K., Ito, H., Hatano, T. and Yoshida, T. (2000). Cytotoxic activity of hydrolyzable tannins against human oral tumor cell lines – A possible mechanism, *Phytochemistry*, 7, pp. 39–47.

Sampaio, E., Sarno, E., Galilly, R., Cohn, Z. and Kaplan, G. (1991). Thalidomide selectively inhibits tumor necrosis factor alpha production by stimulated human monocytes, *J. Exp. Med.*, 173, pp. 699–703.

Schumann, R., Leong, S., Flaggs, G., Gray, P., Wright, S., Mathison, J., Tobias, P. and Ulevitch, R. (1990). Structure and function of lipopolysaccharide binding protein, *Science*, 249, pp. 1429–1431.

Schumann, R. and Zweigner, J. (1999). A novel acute-phase marker. Lipopolysaccharide binding protein (LBP), *Clin. Chem. Lab. Med.*, 37, pp. 271–274.

Schwandner, R., Dziarski, R., Wesche, H., Rothe, M., and Kirschning, C. (1999). Peptidoglycan- and lipoteichoic acid-induced cell activation is mediated by Toll-like receptor 2, *J. Biol. Chem.*, 274, pp. 17406–17409.

Senzer, N., Mani, S., Rosemurgy, A., Nemunaitis, J., Cunningham, C., Guha, C., Bayol, N., Gillen, M., Chu, K., Rasmussen, C., Rasmussen, H., Kufe, C., Weichselbaum, R. and Hanna, N. (2004). TNFerade biologic, an adenovector with a radiation-inducible promoter, carrying the human tumor necrosis factor alpha gene: a phase I study in patients with solid tumors, *J. Clin. Oncol.*, 22, pp. 592–601.

Seternes, T., Dalmo, R., Hoffman, J., Bøgwald, J., Zykova, S. and Smedsrød, B. (2001). Scavenger-receptor-mediated endocytosis of lipopolysaccharide in Atlantic cod (*Gadus morhua* L.), *J. Exp. Biol.*, 204, pp. 4055–4064.

Sharma, V. K. and Dellinger, R. P. (2003). Recent developments in the treatment of sepsis. *Expert Opin. Investig. Drugs*, 12, pp. 139–152.

Shimazu, R., Akashi, S., Ogata, H., Nagai, Y., Fukudome, K., Miyake, K. and Kimoto, M. (1999). MD-2, a molecule that confers lipopolysaccharide responsiveness on Toll-like receptor 4, *J. Exp. Med.*, 189, pp. 1777–1782.

Shiozaki, M., Doi, H., Tanaka, D., Shimozato, T. and Kurakata, S. (2006). Syntheses of glucose analogues of E5564 as a highly potent anti-sepsis drug candidate, *Bioorg. Med. Chem.*, 14, pp. 3011–3016.

Shnyra, A., Lindberg, A. (1995). Scavenger receptor pathway for lipopolysaccharide binding to Kupffer and endothelial liver cells in vitro, *Infect. Immun.*, 63, pp. 865–873.

Sidhu, R. and Bollon, A. (1993). Tumor necrosis factor activities and cancer therapy – a perspective, *Pharmacol. Ther.*, 57, pp. 79–128.

Smith, C., Farrah, T. and Goodwin, R. (1994). The TNF receptor superfamily of cellular and viral proteins: activation, costimulation, and death, *Cell*, 76, pp. 959–962.

Solomon, K., Kurt-Jones, E., Saladino, R., Stack, A., Dunn, I., Ferretti, M., Golenbock, D., Fleisher, G. and Finberg, R. (1998). Heterotrimeric G proteins physically associated with the lipopolysaccharide receptor CD14 modulate both in vivo and in vitro responses to lipopolysaccharide, *J. Clin. Invest.*, 102, pp. 2019–2027.

Spriggs, D. and Yates, S. (1992). Cancer chemotherapy, experiences with TNF-administration in humans. In *Tumor Necrosis Factor: The Molecules and Their Emerging Roles in Medicine*, ed. by Beutler, B., Raven Press: New York, pp. 383–406.

Steel, D. and Whitehead, A. (1994). The major acute phase reactants: C-reactive protein, serum amyloid P component and serum amyloid A protein, *Immunol. Today*, 15, pp. 81–88.

Szlosarek, P., Charles, K. and Balkwill, F. (2006). Tumour necrosis factor-α as a tumour promoter, *Eur. J Cancer*, 42, pp. 745–750.

Takeda, K. and Akira, S. (2004). TLR signaling pathways, *Sem. Immunol.*, 16, pp. 3–9.

Takeuchi, O., Hoshino, K., Kawai, T., Sanjo, H., Takada, H., Ogawa, T., Takeda, K. and Akira, S. (1999). Differential roles of TLR2 and TLR4 in recognition of Gram-negative and Gram-positive bacterial cell wall components, *Immunity*, 11, pp. 443–451.

Takeuchi, O., Hoshino, K. and Akira, S. (2000). Cutting edge: TLR2-deficient and MyD88-deficient mice are highly susceptible to *Staphylococcus aureus* infection, *J. Immunol.*, 165, pp. 5392–5396.

Taveira da Silva, A. M., Kaulbach, H. C., Chuidian, F. S., Lambert, D. R., Suffredini, A. F. and Danner, R. L. (1993). Brief report: shock and multiple-organ dysfunction after self-administration of salmonella endotoxin, *N. Engl. J. Med.*, 328, pp. 1457–1460.

Thom, A., Alexander, H. and Andrich, M. (1995). Cytokine levels and systemic toxicity in patients undergoing isolated limb perfusion with high-dose tumor necrosis factor, interferon gamma, and melphalan, *J. Clin. Oncol.* 13, pp. 264–273.

Tobias, P., Soldau, K. and Ulevitch, R. (1989). Identification of a lipid A binding site in the acute phase-reactant lipopolysaccharide binding protein, *J. Biol. Chem.*, 264, pp. 10867–10871.

Tohme, Z., Amar, S. and Van Dike, T. (1999). Moesin functions as a lipopolysaccharide receptor on human monocytes, *Infect. Immun.*, 67, pp. 3215–3220.

Tracey, K., Fong, Y., Hesse, D., Manogue, K., Lee, A., Kuo, G., Lowry, S. and Cerami, A. (1987). Anti-cachectin/TNF monoclonal antibodies prevent septic shock during lethal bacteraemia, *Nature*, 330, pp. 662–664.

Triantafilou, M., Triantafilou, K. and Dedrick, R. (2001). A CD14-independent LPS receptor cluster, *Nat. Immunol.*, 2, pp. 338–346.

Triantafilou, M. and Triantafilou, K. (2002) Lipopolysaccharide recognition: CD14, TLRs and the LPS-activation cluster, *Trends Immunol*, 23, pp. 301–304.

Tse, A., Wan, C., Shen, X., Yang, M. and Fong, W. (2005). Honokiel inhibits TNF-α stimulated NF-κB activation and NF-κB-regulated gene expression through suppression of IKK activation, *Biochem. Pharmacol.*, 70, pp. 1443–1457.

Verbon, A., Dekkers, P., ten Hove, T., Hack, C., Pribble, J., Turner, T., Souza, S., Axtelle, T., Hoek, F., van Deventer, S. and van der Poll, T. (2001). IC14, an anti-CD14 antibody, inhibits endotoxin-mediated symptoms and inflammatory responses in humans, *J. Immunol.*, 166, pp. 3599–3605.

Vishnyakova, T., Bocharov, A., Baranova, I., Chen, Z., Remaley, A., Csako, G., Eggerman, T., Patterson, A. (2003). Binding and internalization of lipopolysaccharide by Cla-1, a human orthologue of rodent scavenger receptor B1, *J. Biol. Chem.*, 278, pp. 22771–22780.

Wagner, D., Combes, A., McTiernan, C., Sanders, V., Lemster, B. and Feldman, A. (1998). Adenosine inhibits lipopolysaccharide-induced cardiac expression of tumor necrosis factor-α, *Circ. Res.*, 82, pp. 47–56.

Wang, J., Jorgensen, P., Almlof, M., Thiemermann, C., Foster, S., Aasen, A. and Solberg, R. (2000). Peptidoglycan and lipoteichoic acid from *Staphylococcus aureus* induce tumor necrosis factor alpha, interleukin 6 (IL-6), and IL-10 production in both T cells and monocytes in a human whole blood model, *Infect. Immun.*, 68, pp. 3965–3970.

Wheeler, A. P. and Bernard, G. R. (1999). Treating patients with severe sepsis, *N. Engl. J. Med.*, 340, pp. 207–214.

Wright, S. D. and Jong, M. (1986). Adhesion-promoting receptors on human macrophages recognize Escherichia coli by binding to lipopolysaccharide, *J. Exp. Med.*, 164, pp. 1876–1888.

Wright, S. D., Tobias, P., Ulevitch, R. and Ramos, R. (1989). Lipopolysaccharide (LPS) binding protein opsonizes LPS-bearing particles for recognition by a novel receptor on macrophages, *J. Exp. Med.*, 170, pp. 1231–1241.

Wright, S. D., Ramos, R., Tobias, P., Ulevitch, R. and Mathison, J. (1990a). CD14, a receptor for complexes of lipopolysaccharide (LPS) and LPS binding protein, *Science*, 249, pp. 1431–1433.

Wright, S. D., Detmers, P., Aida, Y., Adamowski, R., Anderson, D., Chad, Z., Kabbash, L. and Pabst, M. (1990b). CD18-deficient cells respond to lipopolysaccharide in vitro, *J. Immunol.*, 144, pp. 2566–2571.

Wright, S. (1995). CD14 and innate recognition of bacteria, *J. Immunol.*, 155, pp. 6–8.

Xagorari, A., Papapetropoulos, A., Mauromatis, A., Economou, M., Fotsis, T. and Roussos, C. (2001). Luteolin inhibits an endotoxin-stimulated phosphorylation cascade and proinflammatory cytokine production in macrophages, *J. Pharmacol. Exp. Ther.*, 296, pp. 181–187.

Yan, L., Anderson, G., DeWitte, M. and Nakada, M. (2006). Therapeutic potential of cytokine and chemokine antagonists in cancer therapy, *Eur. J Cancer*, 42, pp. 793–802.

Yin, N., Marshall, R. L., Matheson, S. and Savage, P. B. (2003). Synthesis of lipid A derivatives and their interactions with polymyxin B and polymyxin B nonapeptide, *J. Am. Chem. Soc.*, 125, pp. 2426–2435.

Yong, K. L. and Linch. D. (1993). Granulocyte-macrophage-colony-stimulating factor differentially regulates neutrophil migration across IL-1-activated and nonactivated human endothelium, *J. Immunol.*, 150, pp. 2449–2456.

Yoshida, T., Tanaka, K., Chen, X.-M. and Okuda, T. (1989a). Tannins of rosaceous medicinal plants. Part 7. Dimeric ellagitannins, laevigatins E, F, and G, from Rosa laevigata, *Phytochemistry*, *28*, pp. 2451–2454.

Yoshida, T., Tanaka, K., Chen, X.-M. and Okuda, T. (1989b). Tannins of Rosaceous medicinal plants. V. Hydrolyzable tannins with dehydrodigalloyl group from *Rosa laevigata* Michx., *Chem. Pharm. Bull.*, 37, pp. 920–924.

Yoshida, T., Jin, Z. and Okuda, T. (1989c). Tannins of rosaceous medicinal plants. Part 6. Taxifolin apioside and davuriciin M_1, a hydrolysable tannin from *Rosa davurica*, *Phytochemistry*, 28, pp. 2177–2181.

Yoshida, T., Chou, T., Nitta, A. and Okuda, T. (1989d). Woodfordins A, B and C, dimeric hydrolyzable tannins from *Woodfordia fruticosa* flowers, *Heterocycles*, *29*, pp. 2267–2271.

Yoshida, T., Chou, T., Nitta, A., Miyamoto, K., Koshiura, R. and Okuda, T. (1990). Woodfordin C, a macro-ring hydrolyzable tannin dimer with antitumor activity, and accompanying dimers from *Woodfordia fruticosa* flowers, *Chem. Pharm. Bull.*, 38, pp. 1211–1217.

Yoshida, T., Chou, T., Matsuda, M., Yasuhara, T., Yazaki, K., Hatano, T. and Okuda, T. (1991). Woodfordin D and oenothein A, trimeric hydrolyzable tannins of macro-ring structure with antitumor activity, *Chem. Pharm. Bull.*, 39, pp. 1157–1162.

Yoshida, T., Jin, Z. and Okuda, T. (1991). Hydrolyzable tannin oligomers from *Rosa Davurica*, *Phytochemistry*, 30, pp. 2747–2752.

Yoshida, T., Maruyama, T., Nitta, A. and Okuda, T. (1992). Eucalbanins A, B, and C, monomeric and dimeric hydrolyzable tannins from *Eucalyptus alba* Reinw., *Chem. Pharm. Bull.*, 40, pp. 1750–1754.

Yoshida, T., Maruyama, T., Nitta, A. and Okuda, T. (1996). An hydrolysable tannin and accompanying polyphenols from Melaleuca leucadendron, *Phytochemistry*, 42, pp. 1171–1173.

Yui, S., Mikami, M., Mimaki, Y., Sashida, Y. and Yamazaki, M. (2001). Inhibition effect of amaryllidaceae alkaloids, lycorine and lycoricidinol on macrophage TNF-α production, *Yakugaku Zasshi*, 121, pp. 167–171.

Zhou, H., Ding, G., Liu, W., Wang, L., Lu, Y., Cao, H. and Zheng, J. (2004). Lipopolysaccharide could be internalized into human peripheral blood mononuclear cells and elicits TNF-α release, but not via the pathway of Toll-like receptor 4 on the cell surface, *Cell. Mol. Immunol.*, 1, pp. 373–377.

Zughaier, S., Zimmer, S., Datta, A., Carlson, R. and Stephens, D. (2005). Differential induction of the Toll-like receptor 4-MyD88-dependent and -independent signaling pathways by endotoxins, *Infect. Immun.*, 73, pp. 2940–2950.

Chapter 7

Bioavailability and Metabolism of Ellagic Acid and Ellagitannins

Francisco A. Tomás-Barberan,* Juan C. Espín and María T. García-Conesa

*Centro de Edafología y Biología Aplicada del Segura (CSIC), Espinardo, 30100 Murcia, Spain; *corresponding author:* fatomas@cebas.csic.es

7.1 Introduction

Ellagitannins (ETs) are polyphenols that show relevant biological activities *in vitro* that have been associated with pharmacological (ET-rich medicinal plants) and nutritional (ET-rich foods) effects *in vivo*. These are mainly related to the fight and prevention of cardiovascular diseases and cancer. *In vivo* biological effects may be partially due to the free-radical scavenging activity that these compounds exert *in vitro*. However, it is necessary to take into consideration the fate of these ETs in the gastrointestinal (GI) tract, their bioaccessibility, bioavailability, metabolism and tissue distribution of the corresponding metabolites in order to understand the actual physiological role of dietary and medicinal ETs. In this chapter, we review the current knowledge regarding bioavailability and metabolism of ETs and point out various unresolved issues within this process in humans that still need further research.

7.2 How Relevant Are ETs in the Diet?

ETs are present in significant amounts in many berries (see also Chapter 8), including strawberries, red and black raspberries (Zafrilla *et al.*, 2001), blackberries, and nuts including walnuts (Fukuda *et al.*, 2003), pistachio, cashew nut, chestnuts, oak acorns (Cantos *et al.*, 2003) and pecans (Villarreal-Lozoya *et al.*, 2007). They are also abundant in pomegranates (Gil *et al.*, 2000), and muscadine grapes (Lee and Talcot, 2002), and are important constituents of wood, particularly oak wood (Glabasnia and Hofmann, 2006). ETs can be incorporated into several food products such as wines and whiskey, through migration from wood to the food matrix during different ageing processes. Ellagic acid (EA), a structural unit of ETs, has also been found in several types of honey and this phytochemical has been proposed as a honey floral marker for heather honey (Ferreres *et al.*, 1996). Free EA and different glycosidic derivatives are also present in these food products, including glucosides, rhamnosides, arabinosides and the corresponding acetyl esters (Zafrilla *et al.*, 2001).

In a previous review, it was documented that there was no available reliable figure on the ETs dietary burden, but that it would probably not exceed 5 mg day^{-1} (Clifford and Scalbert, 2000). Since then a number of studies have shown that the ETs content of several food products can be quite high (Table 7.1). A glass of pomegranate juice can provide as much as 300 mg of ETs, a raspberry serving (*ca.* 100 g of raspberries) around 300 mg, a strawberry serving 70 mg, and four walnuts, about 400 mg of ETs. As a result, the intake of dietary ETs can be much higher than previously estimated (Clifford and Scalbert, 2000), especially if some of these ETs-rich foods are regularly consumed in the diet.

Representative dietary ellagitannins are shown in Fig. 7.1. Punicalagin is typical of pomegranate, vescalagin of oak-aged wines and spirits, sanguiin H-6 of strawberry and raspberry, sanguiin H-5 of muscadine grapes and pedunculagin of walnuts. All of them release ellagic acid upon hydrolysis, although other metabolites can also be produced and are distinctive of individual ellagitannins (*i.e.*, gallagic and tergallagic acids) (Porter, 1989).

7.3 Do Ellagitannins and Ellagic Acid Show Biological Activity?

ET-rich foods exhibit, in general, high *in vitro* free-radical scavenging activity. Of special interest is the antioxidant activity of pomegranate juice (Gil *et al.*, 2000). In this study, it was shown that pomegranate juice has twice the antioxidant activity of red wine and it was attributed to the ETs extracted from the fruit husk during juice manufacturing (Gil *et al.*, 2000). This remarkable activity has been the driving force for research looking at the biological activity of these powerful antioxidants in pomegranate and is used by food manufacturers to market pomegranate juice products as super-antioxidant foods. ETs are also responsible for a significant part of the antioxidant activity observed in strawberries (Hannun, 2004), raspberries (Zafrilla *et al.*, 2001), blackberries, walnuts (Blomhoff *et al.*, 2006) and pecans (Villarreal-Lozoya *et al.*, 2007). This antioxidant activity may be related to the biological activity reported for these food products.

Table 7.1 Ellagitannins content in various food products.

Food	Content	Reference
Pomegranate juice	1500-1900 mg/L	Gil *et al.*, 2000
Raspberries	263-330 mg/100 g f.w.	Koponen *et al.*, 2007
Raspberry jam	76 mg/100 g f.w.	Koponen *et al.*, 2007
Strawberries	77-85 mg/100 g f.w.	Koponen *et al.*, 2007
Strawberry	25 mg/100 g f.w.	Abby *et al.*, 2007
Strawberry jam	24 mg/100 g f.w.	Koponen *et al.*, 2007
Cloudberry	315 mg/ 100 g f.w.	Koponen *et al.*, 2007
Blackberry	1.5-2.0 mg/g d.w.	Clifford and Scalbert, 2000
Muscadine grape juice	8-84 mg/L	Lee *et al.*, 2002
Muscadine grape wine	2-65 mg/L	Lee *et al.*, 2002
Walnut	802 mg / 50g (8 nuts)	Anderson *et al.*, 2001
Whiskey	1-2 mg/L	Glabasnia and Hofmann, 2006
Cognac	31-55 mg/L	Clifford and Scalbert, 2000
Oak aged red wine	9.4 mg/L	Glabasnia and Hofmann, 2006
Oak aged red wine	50 mg/L	Clifford and Scalbert, 2000

In parallel to the studies on the antioxidant activity of ETs-rich food, several clinical trials have reported relevant biological effects, such as those regarding protection against cardiovascular diseases and cancer, in association with the intake of ETs-rich foods and their antioxidant capacity. However, no direct evidence of the biological activity of these polyphenols has been yet demonstrated.

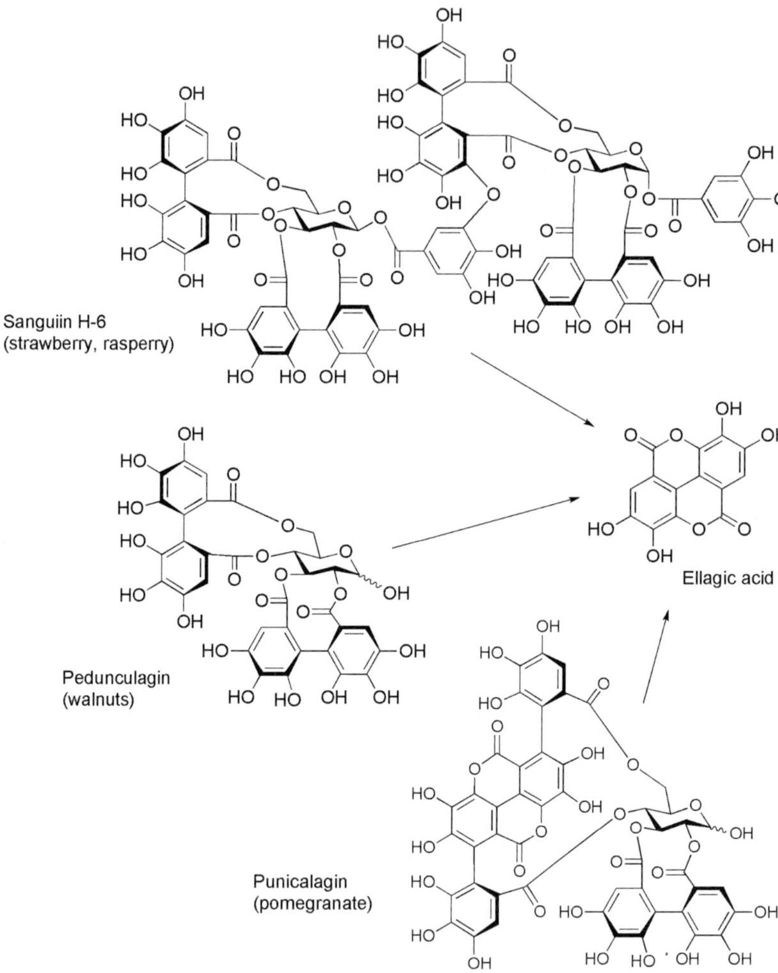

Fig. 7.1 Food ellagitannins release ellagic acid in the GI tract.

Pomegranate juice, one of the richest food products in ETs (Gil *et al.*, 2000), exerted some protection against cardiovascular diseases by diminishing the effect of risk factors for atherosclerosis, such as hypertension, platelet aggregation, blood lipid profiles and oxidative stress (Aviram *et al.*, 2000, Fuhrman and Aviram, 2006). Pomegranate juice supplementation for three years to patients with carotid artery stenosis (CAS) led to decreased atherosclerotic lesion size. The authors attributed these effects to the potent antioxidant characteristics of pomegranate juice (Aviram *et al.*, 2002 and 2004). Pomegranate juice consumption also reduced the systolic blood pressure of CAS patients after one-year intake. In contrast, no significant effect on the patient's diastolic blood pressure was observed. The serum angiotensin converting enzyme (ACE) activity was also significantly decreased after pomegranate juice intake (Aviran and Dornfeld, 2001). The authors suggest that this effect may be secondary to the ability of pomegranate juice antioxidants, such as complex ETs, to inhibit ACE activity. Because ACE inhibitors are metabolized by cytochrome P-450 enzymes, serum ACE activity can be significantly affected by modulation of P-450 enzyme activity. It has also been demonstrated that pomegranate juice decreased the activity of different cytochrome P-450 enzymes by 20-60% (Aviram *et al.*, 1999, Hidaka *et al.*, 2005). In hypertensive patients treated with ACE inhibitors, the inhibitory effect of pomegranate juice consumption on cytochrome P-450 enzymes can possibly decrease P-450-mediated drug (ACE inhibitor) metabolism and hence, serum ACE activity may be further decreased in these treated patients. The authors stated that, as ACE activity is related to enhanced blood pressure and accelerated atherosclerosis, their data may suggest an additional important anti-atherogenic property of pomegranate juice consumption. However, a recent publication has reported the association of rhabdomyolysis with pomegranate juice consumption, due to the cytochrome P-450 inhibition of pomegranate juice (Sorokin *et al.*, 2006). This report suggests that, as pomegranate juice inhibits intestinal cytochrome P-450 enzymes, they can alter statin metabolism increasing its absorption.

Circulating human platelets play an important role in the development of atherosclerosis, and increased platelet aggregation is associated with enhanced atherogenicity. Platelet activation was shown to be associated with oxidative stress. Following 2 weeks of pomegranate juice consumption, a significant reduction (11%) in collagen-induced platelet aggregation was noted (Fuhrman and Aviram, 2006). The authors link the effects to an interaction of pomegranate polyphenols with the platelet surface binding sites for collagen and that their antioxidative properties can attenuate oxidative stress-induced platelet activation.

It has also been published that pomegranate juice polyphenols can protect LDL against cell-mediated oxidation via two pathways, a direct interaction of polyphenols with the lipoprotein, an indirect effect through polyphenols accumulation in arterial macrophages or both (Fuhrman and Aviram, 2006). All these antioxidative and antiatherogenic effects of pomegranate polyphenols have been demonstrated *in vitro*, and *in vivo* in humans and in atherosclerotic apolipoprotein E-deficient mice.

EA and ETs-containing foodstuffs have also been acknowledged with cancer chemopreventive activities. There are a number of *in vitro* studies on various cancer cell lines that show the antiproliferative and pro-apoptotic effects of these compounds. The bioavailability and metabolism of ETs or EA are however very important issues when it comes to select the appropriate cell line for a particular study. From a dietary point of view, if ETs or EA are poorly absorbed from the gut, it is then not appropriate to test these compounds at μM concentrations against cell lines representative of systemic organs or tissues. This would require a high absorption of the compounds into the blood system to reach the target organ (*e.g.*, breast, brain).

The chemopreventive effect of ETs could result from inhibition of nuclear-κβ signalling (Kuo *et al.*, 2007), decrease of intracellular redox status and activation of JNK-1/p38 (Chen *et al.*, 2007), and induction of apoptosis *via* intrinsic pathway (Larrosa *et al.*, 2006a) among other reported tentative pathways. Whether the action is promoted by the ETs themselves or by hydrolysis products such as EA or by valenoic acid dilactone has not been investigated. In this context, in the case of the pomegranate ET, punicalagin, a recent work has reported that EA is the real active component rather than its punicalagin precursor (Larrosa *et al.*, 2006a). Punicalagin and EA provoked the same effects on Caco-2 cells: downregulation of cyclins A and B1 and upregulation of cyclin E, cell cycle arrest in S phase, induction of apoptosis *via* intrinsic pathway (FAS-independent, caspase 8-independent) through bcl-XL downregulation with mitochondrial release of cytochrome *c* into the cytosol, activation of initiator caspase 9 and effector caspase 3. Neither EA nor punicalagin induced apoptosis in normal colon CCD-112CoN cells (no chromatin condensation and no activation of caspases 3 and 9 was detected). In the case of Caco-2 cells, no specific effect can be attributed to punicalagin, since it was hydrolysed in the medium to yield EA, which entered into the cells. This suggests that the anticarcinogenic effect of punicalagin (and possibly of other ETs) could be mostly due to their hydrolysis product EA, which induced apoptosis *via* mitochondrial pathway in colon cancer Caco-2 cells but not in normal colon cells (Larrosa *et al.*, 2006a). The comparison between cancer and normal cell lines is of great interest to discard unspecific (non-desirable) cytotoxic effects. This approach has not often been applied to this type of research due to the difficulty in culturing normal cell lines.

There are also a number of reports regarding the *in vivo* cancer chemopreventive activity of EA and ETs, mostly in animal models. These studies started in the early sixties, when it was shown that EA was able to inhibit induced lung tumours in mice upon oral administration (Lesca, 1983). Other studies also demonstrated EA anticarcinogenic activiy in mice (Pepin *et al.*, 1990, Castonguay *et al.*, 1997). More recent reports have also shown that long term supplementation with dietary doses of pomegranate fruit extracts prevented and inhibited prostate tumours in athymic nude mice (Malik *et al.*, 2005, Malik and Mukhtar, 2006). The potential of ET-containing foodstuffs, such as blackberries, strawberries, black raspberries, as cancer chemopreventive has been reported. There are a number of studies that demonstrate the protection against oral (Casto *et al.*, 2002), esophageal (Carlton *et al.*, 2001, Kresty *et al.*, 2001, Stoner *et al.*, 2006, Chen *et al.*, 2006) and colon (Harris *et al.*, 2001) tumorigenesis in the rat. Although none of these studies clearly identified the compound(s) responsible for these effects, the ETs seems to be the key molecules for the cancer chemopreventive action (Ross *et al.*, 2007). Although the majority of studies show clear effects, a recent study reported the lack of effect of EA and ET-enriched fractions from cloudberries against intestinal tumorigenesis in the Min mouse (Paivarinta *et al.*, 2006). This result raises new questions about the influence of the animal model on the results of a particular study. There may be different responses against a specific treatment from animals with chemically induced tumours and from genetically modified animals that spontaneously develop a tumor such as the *Apc*-mutated Min mice used by Paivarinta *et al.* (2006).

Recently, a Phase II clinical trial administering pomegranate juice to patients with prostate cancer has demonstrated a statistically significant prolongation of the PSA (prostate serum cancer biomarker) doubling time (Pantuck *et al.*, 2006). These results suggest some protective effects of the pomegranate juice in humans. To further understand and reveal the biological protective effects of ETs, it is essential that we clarify the bioavailability and metabolism of these compounds, by performing analysis of ETs and their metabolites *in vivo* in animal and human studies.

7.4 Is the Analytical Methodology Ready for the Analysis of these Metabolites *in vivo*?

The analysis of ETs in food products and in biological fluids presents some limitations. These compounds are present in foods mostly as large molecules with rather unspecific UV spectra, which hampers their identification from the chromatographic profile (UV spectra generally do not show a characteristic maximum in the 250-400 nm range that is common for other phenolic compounds). Acid hydrolysis of ETs releases EA, which can be qualitative and quantitative measured, but this hydrolysis is not always complete and can yield other artefactual compounds (*e.g.*, EA methyl ethers) hindering the quantification of ETs. In addition, ETs can interact with proteins and precipitate, hence adding to the analysis difficulties of ETs and their metabolites in food and in biological fluids.

The improvement of HPLC detection systems with the introduction of MS and MS-MS detectors has allowed the determination of different ETs in food products, and recent publications show that these ETs are more abundant than previously estimated. New HPLC technologies applied to the analysis of ETs and their metabolites in food and biological fluids has led to an increase in the understanding their stability, metabolism (both microbial and mammal cells metabolism) bioavailability and tissue distribution. It has been observed that EA and ETs can disappear when in contact with animal cells or tissues, and this has been attributed to irreversible linkage of these compounds with proteins or other cell structures. For example, intestinal cells become yellow pigmented after incubation with punicalagin, and the colour does not disappear even after extraction with strong organic solvents such as methanol. This agrees with results observed with EA, in which very strong conditions are necessary to remove the EA-protein linkages in animal model intestinal tissues (Whitley *et al.*, 2003). More research is needed to further understand the interactions of ETs and EA with proteins and other cell components in order to understand the fate of these polyphenols *in vivo*.

7.5 Models to Study Bioavailability and Metabolism

In order to understand the metabolism and bioavailability of ETs, different experimental approaches can be used. These include: i) *in vitro* simulation of gastrointestinal (GI) digestion, which may be useful to evaluate the ETs stability and bioaccesibility; ii) *in vitro* studies with human GI tract cell lines to evaluate uptake and metabolism in the GI cells before passing to blood circulation; iii) *in vivo* studies with animal models (from mice and rats to pigs) that are essential for toxicological studies, as well as confirming bioavailability and metabolism in mammals and evaluating tissue distribution; iv) clinical interventions with human volunteers to determine human bioavailability and metabolism using the knowledge generated in animal studies and to confirm the relevant metabolites, as well as their *in vivo* concentrations in humans.

7.5.1 *In vitro* digestion

In general, ETs are quite stable under the physiological conditions of the stomach. The acidic conditions (HCl, pH 1.8-2.0) and the enzymes present did not hydrolyse the native ETs that can release free EA, and no degradation of the ETs was observed. The stomach seems to be a first important location for the absorption of free EA (*vide infra*), but ETs are not absorbed. Under the physiological conditions of the small intestine, however, there is a release of free EA from ETs such as punicalagin. This hydrolysis seems to be due to the pH conditions (neutral to mild alkaline, *i.e.*, pH 7.0-7.3) rather than to the effect of pancreatic enzymes and bile salts (Larrosa *et al.*, 2006a).

7.5.2 *In vitro* cell lines to study uptake and metabolism

Several human GI cell lines can be obtained from culture collections, such as those from the stomach (*e.g.*, KATO-III) or colon (*e.g.*, cancer cells like Caco-2 or normal colon cells) and can be used to evaluate the uptake, metabolism, transport and excretion, *inter alia*, of the ETs and EA. These studies can give evidence of the biochemical changes underwent by the ETs in the cell culture media, and are useful to follow their metabolic fate, once they enter the cells and are transformed into specific metabolites produced by each cell line. These studies may be carried out with pre-digested ETs (*i.e.*, Caco-2 cells may be treated with ETs extracts that have been subjected to stomach and intestinal digestions) to simulate more closely the *in vivo* situation. It has become evident that ETs are not absorbed into these cells, but that they disappear from the culture medium. This may be due to precipitation, degradation or combination with proteins present in the medium. However, EA is absorbed and rapidly transformed into methyl ethers by the action of COMT (catechol *O*-methyl transferases). This enzyme introduces one or two methyl ether groups on the phenolic hydroxyl functions of the *ortho*-dihydroxyphenyl or catecholic moieties of the EA molecule, thus forming EA monomethyl ether and dimethyl ethers. Also, conjugation with glucuronic acid has been observed in these cells. Whitley *et al.* (2003) reported a high accumulation of EA in Caco-2 cells (uptake through Caco-2 monolayer), indicating a facile absorptive transport across the apical membrane. As much as 93% of the cellular EA was irreversibly bound to macromolecules (proteins and DNA). Thus, EA appears to accumulate in the epithelial cells of the aerodigestive tract. In a recent study, the metabolism of punicalagin and ellagic acid by Caco-2 cells was reported (Larrosa *et al.*, 2006a). This study showed that punicalagin is hydrolyzed in the cell medium to yield EA, which enters Caco-2 cells. The first EA-derived metabolite produced by these cells was a dimethylated EA, which involved the active participation of COMT. Second, Caco-2 cells conjugation with glucuronic acid yielded the corresponding dimethylated EA glucuronide, which was the most abundant metabolite detected. Other metabolites detected in smaller amounts included two isomers of dimethylated EA sulphates. All of these metabolites were detected in both the cell culture medium and within the cells (Larrosa *et al.*, 2006a).

EA and ETs are further converted by the intestinal microflora into the dibenzopyranone metabolites, urolithins A and B (Fig. 7.2). These urolithins are largely absorbed in the intestinal cells and glucuronidated. In this case, no methyl ether is produced, for urolithins do not feature any *ortho*-dihydroxylphenyl unit in their structure. In the case of urolithin A (3,8-dihydroxy-6*H*-dibenzo-[b,d]pyran-6-one), a hydroxylation by cytochrome P-450 may also be possible to increase the possibilities of glucuronidation and hence enhance the excretion of the metabolite.

7.5.3 Animal models

Animal studies are usually first recommended to rule out toxicity of a new test compound, for larger doses can often be attained in animals than in humans. The bioavailability of EA and ETs has been directly or indirectly evaluated in animals and results are available.

The bioavailability of orally gavaged ^{3}H-EA was already evaluated in mice by Teel and Martin in 1988. In this study, mice were provided with 0.3 μg of EA/g of body weight (the equivalent of a human dose of 21 mg for a person of 70 kg). Both free EA and some conjugates (sulphate ester, glucuronide and glutathione conjugates) were detected in urine, bile and blood. Absorption of ^{3}H-EA occurred mostly within two hours of oral administration. Levels in blood, bile and tissues were low, and absorbed compounds were excreted in urine. More than half of the administered ^{3}H-EA remained in the gastrointestinal tract after 24 hours. Approximately 19% of ^{3}H-EA was excreted in faeces and 22% in urine at 24 hours.

A rapid absorption and metabolism of EA was reported by Doyle and Griffiths (1980) in rats. These authors detected two derived metabolites in faeces and urine: urolithin A and another unidentified compound, both of microfloral origin, since none of them were found in germ-free animals. Unchanged EA was not detected in urine or faeces of normal rats. In contrast, Smart *et al.* (1986) found low to non-detectable levels of EA in blood, lungs and liver of CD-1 mice after oral administration and this was interpreted as an indication of poor absorption and rapid elimination of the compound in these animals.

The results described in mice by Teel and Martin (1988) and by Smart *et al.* (1986) are in good agreement and suggest that the poor absorption of EA from the gut may lead to very low concentrations in tissues, concentrations that may not be sufficient to exert any *in vivo* anticarcinogenic effects. The poor absorption of EA is supported by a report of the presence of EA calculi in the gastrointestinal tract of monkeys and goats, whose diet naturally contains EA (Van Tassel, 1976). The low bioavailability may be caused by several factors including ionization of EA at physiological pH and formation of poorly soluble complexes with Mg and Ca cations. In addition, extensive binding of EA to the intestinal epithelium could also diminish absorption (Whitley *et al.*, 2003).

A more recent work investigated the bioavailability of pomegranate husk ETs in rats (Cerdá *et al.*, 2003). These ETs are essentially the same as those found in the commercial juice (Gil *et al.*, 2000). The rats were given 6% of their diet as pomegranate ETs from the fruit husk, and the experiment was used to evaluate absorption, tissue distribution and toxicity. Values around 3-6% of the ingested punicalagin were excreted as metabolites in faeces and urine. In faeces, punicalagin was transformed into hydrolysis products and to hydroxy-6*H*-dibenzo-[b,d]-pyran-6-one derivatives (Fig. 7.2) by the rat colonic microflora. In plasma, punicalagin was detected at concentrations of *ca.* 30 μg/mL (0.028 μM). The absorption of intact punicalagin in rats and its detection in plasma is particularly remarkable, for punicalagin (Mw = 1084) is one of the largest polyphenols that has been reported to be absorbed (Cerdá *et al.*, 2003a, Manach *et al.*, 2005). Glucuronides and methyl ether conjugates of EA were also detected in this study. In urine, the main metabolites observed were the mono-, di- and trihydroxy-6*H*-dibenzo-[b,d]-pyran-6-one derivatives, both as aglycones and glucuronide conjugates (Cerdá *et al.*, 2003a). Only 3 to 6% of the ingested punicalagin was detected under its native form or as derived metabolites in urine and faeces, suggesting that the majority of this ET had to be converted to undetectable metabolites or accumulated in tissues. Traces of punicalagin metabolites were further detected in liver and kidney, but punicalagin, EA or other derived metabolites were not found in lung, brain or heart (Cerdá *et al.*, 2003a and 2003b).

In this study, in addition to the very small quantities of punicalagin, the pomegranate ET metabolites detected in plasma included trihydroxy-6*H*-dibenzopyran 6-one diglucuronide and dihydroxy-6*H*-dibenzopyran 6-one glucuronide, gallagic acid, dimethylated ellagic acid glucuronide, dimethylated ellagic acid glucuronide methyl ester and very small amounts of EA. As EA has two *ortho*-dihydroxyphenyl units, it can be expected that, in the liver, *via* COMT (catechol *O*-methyl transferase) activity, one methyl ether group can be introduced per dihydroxyphenyl unit. These metabolites show a UV spectrum nearly identical to that of free EA, and were further conjugated with glucuronic acid to increase water solubility and to facilitate their excretion (Cerdá *et al.*, 2003a).

7.5.4 Humans

In humans, two main types of experiments have been carried out: i) pharmacokinetic studies, in which the absorption of pomegranate polyphenols has been evaluated during several hours following intake; ii) bioavailability and metabolism studies, in which the metabolites present in plasma and urine have been evaluated for long periods of time. In a pharmacokinetic study carried out at the University of California, Los

Angeles, USA, EA was detected at a maximum concentration in the human plasma of one volunteer after 1 h post-ingestion of 180 mL of pomegranate juice ('Wonderful' cultivar; containing 25 mg free EA and 318 mg of ETs). The concentration of EA measured in plasma was 31.9 ng/mL (*ca.* 0.1 μM) and was rapidly eliminated in 4 hours (Seeram *et al.*, 2004). In contrast, in a study carried out in Murcia (Spain), where a healthy volunteer consumed 1 L of pomegranate juice ('Mollar' cultivar), no EA was detected in plasma during the 4 hours following the juice intake (Cerdá *et al.*, 2004). This difference could be simply due to inter-individual variability. Another pharmacokinetic study looking at the absorption of EA from black raspberry has also shown poor but significant absorption of free EA (less than 1% of the ingested EA) during the first 2 hours after the intake of berries (Stoner *et al.*, 2005), in agreement with the results reported (Seeram *et al.*, 2004).

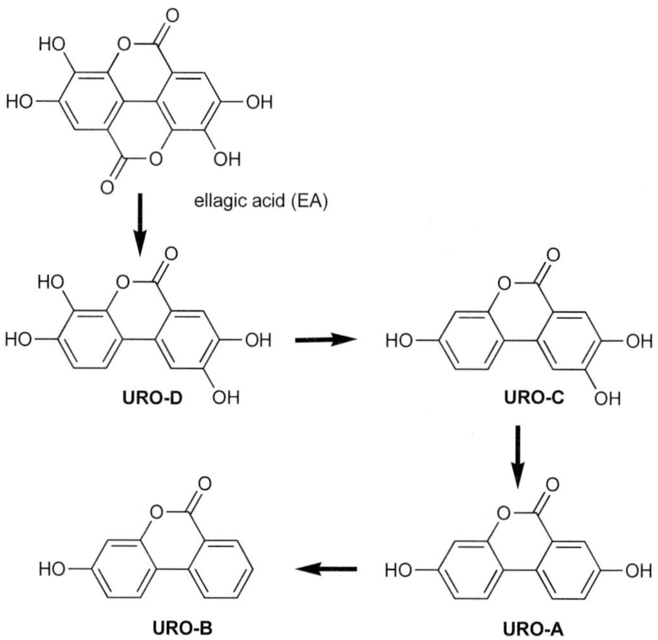

Fig. 7.2 Ellagic acid metabolites produced by intestinal microflora in animals and humans. URO-D = Urolithin D (*i.e.*, tetrahydroxydibenzopyranone); URO-C = Urolithin C (*i.e.*, trihydroxydibenzopyranone); URO-A = Urolithin A; URO-B = Urolithin B (Espín *et al.*, 2007, Ito *et al.*, 2008).

Larger pharmacokinetic studies with a significant number of volunteers are necessary to evaluate the effect of inter-individual variability and of the concentration of free EA in the original juice on the levels of EA detected in plasma. A recently completed pharmacokinetic study at the University of California (*unpublished data*), using a larger number of volunteers (*i.e.*, n = 19), resulted in pharmacokinetic parameters for EA similar to those observed in the study with only one human subject (Seeram *et al.*, 2004).

Regarding long-term evaluation of metabolites in plasma and urine, three main studies should be noted. In one of these studies, healthy volunteers consumed daily 1 L of pomegranate juice (containing 4.37 g/L of punicalagin isomers) for 5 days (Cerdá *et al.*, 2004). Punicalagin or EA that were present in the juice were not detected in the plasma or urine of volunteers. Three microbial ET-derived metabolites (urolithin derivatives) were detected: 3,8-dihydroxy-6*H*-dibenzo[*b,d*]pyran-6-one glucuronide, trihydroxy-6*H*-dibenzo[*b,d*]pyran-6-one and 3-hydroxy-6*H*-dibenzo[*b,d*] pyran-6-one glucuronide. For those volunteers in which the metabolites were detected, the concentration of metabolites found in plasma ranged from 0.5 to 18.6 μM, showing evidence of a large inter-individual variability between participants. In urine, the same metabolites and their corresponding aglycones were detected after 1 day of juice consumption. Total urine excretion of metabolites ranged from 0.7 to 52.7% of the ingested punicalagin. In general, the metabolites found in humans after the pomegranate juice intake coincided with those previously reported to be produced in rats after pomegranate husk intake (Cerdá *et al.*, 2003a). As mentioned above, these metabolites (urolithins) were also reported to be produced by intestinal bacteria in rats after consumption of EA (Doyle and Griffiths, 1980).

A second study looking at ET bioavailability and metabolism was carried out with other ET-rich foodstuffs: strawberry, raspberry, walnut and oak-aged red wine (Cerdá *et al.*, 2005). These foodstuffs differ in the content and type of ETs. However, it is noteworthy that the same metabolite (urolithin A) was detected in all participants (n = 40). This led to the proposal of the microbial metabolite urolithin A (Cerdá *et al.*, 2005a/b) as a biomarker for human exposure to dietary ETs. As already mentioned, a large inter-individual variability was observed amongst volunteers regarding the levels of excretion of this metabolite. The third study reporting on ET bioavailability was carried out in a group (n = 15) of patients with stable chronic obstructive pulmonary disease (COPD). The volunteers were given pomegranate juice supplementation for 5 weeks in a randomized, double blind, placebo-controlled trial (Cerdá *et al.*, 2005c). Urolithins A and B (3,8-dihydroxydibenzopyran-6-one and 3-hydroxydibenzopyran-6-one) were detected in both the plasma and urine of patients and a large inter-individual variability was again observed.

7.6 What Bioavailability and Metabolism Events Take Place in Different Human Body Sites?

ETs are large molecules usually quite polar and therefore, according to the pharmacokinetics general knowledge, they should be poorly absorbed or not absorbed at all. The experimental data available indicates that dietary ETs are generally not detected in plasma or in other biological fluids after the intake of ET-rich foods. Exceptionally, when a very large dose of ETs is provided, low concentrations of these compounds are found in plasma and urine (Cerdá *et al.*, 2003a). In a dietary context in humans or animal models, no ET from ET-rich foods (*e.g.*, strawberries, walnuts, raspberries, acorns) is detected in plasma or urine (Cerdá *et al.*, 2005a).

7.6.1 Absorption of free EA

Free ellagic acid (EA) is already present in most ET-rich foods, and is also produced during food processing and storage. Free EA is quite insoluble in aqueous solvents and precipitates in juices and liquors. Some pharmacokinetic studies show that EA can be absorbed as such within 30 to 60 minutes after the intake of several foods (Seeram *et al.*, 2004). This suggests that absorption of EA already begins in the stomach and can be detected in peripheral blood. EA disappears from plasma after 2 hours of intake. However, other pharmacokinetic studies carried out with different foods containing similar amounts of free EA have shown no absorption of the compound (*i.e.*, EA was not detected in plasma). These results indicate that the food matrix (pH, constituents, processing, etc.) can have a crucial impact on EA absorption in the proximal part of the GI tract. Differences between individuals can also be very important in terms of absorption. The absorption of EA in distal parts of the GI tract seems to be less relevant. Animal studies show that EA is released from ETs in the small intestine, and absorbed in the cells of the small intestine (jejunum, ileon). EA is readily metabolized in the intestinal cells to furnish methyl and dimethyl ethers and glucuronic acid conjugates. These metabolites are detected in bile and also in peripheral plasma and urine, but no free EA was detected.

7.6.2 Intestinal microflora metabolism

One of the main events in ET metabolism and bioavailability is the microbial transformation into a series of hydroxylated dibenzopyranone derivatives. Among them, the best known and characterized derivatives are urolithins A and B, but metabolites with three (urolithin C) and four (urolithin D) phenolic hydroxyl groups are also produced in the small intestine (Fig. 7.2) (Espín *et al.*, 2007; Ito *et al.*, 2008). Urolithins are absorbed in the small intestine, and excreted in the bile after methylation and conjugation with glucuronic acid, suffering entero-hepatic recirculation (Espín *et al.*, 2007). Animal experiments show that these metabolites start to be formed in the small intestine indicating that anaerobic bacteria may be responsible for this. The metabolism continues along the GI tract to end with the production of urolithins A and B. Differences in the production of these metabolites by human volunteers show that they may be produced by the activity of specific microorganisms present in the gut. If these microbial metabolites, which are more bioavailable than native ETs and EA, were the true active principles responsible for the biological activity associated with ETs and EA rich foods consumption, then this would bring up the possibility of developing new functional foods in which specific ET-metabolizing microorganisms could be included together with the ETs.

7.6.3 Phase I and phase II metabolism

In the GI tract and in other tissues (mainly in the liver), EA and ETs microbial metabolites are further metabolized either through Phase I (hydroxylation) or Phase II (methylation, glucuronidation and sulphation) enzymes to render them more soluble for better excretion in urine. Hydroxylation of urolithin A, and probably B, renders them more reactive with additional functions for conjugation, which hence facilitates their excretion. Thus, monohydroxydibenzopyran-6-one (urolithin B) can be hydroxylated to produce a dihydroxylated derivative, and urolithin A (3,8-dihydroxydibenzopyran-6-one) can be further hydroxylated to produce trihydroxylated derivatives. In a recent gene expression study, it was shown that both urolithins A and B, but particularly urolithin B, can induce (15- to 20-fold) the expression of cytochrome P-450 genes in Caco-2 cells, which may explain why the urolithin B-derived dihydroxylated compound is in general more abundant in tissues (liver), plasma and urine. Phase II products are also produced and methyl ethers (products of COMT), as well as different glucuronide conjugates, are detected in different tissues and in urine. Sulphate conjugates of the ETs metabolites are less abundant in animals and humans than the glucuronide conjugates. These conjugates are first produced in the intestinal cells, and further metabolized in the liver before excretion in the urine or the bile.

7.6.4 Tissue distribution

In order to understand the biological activity of ETs and EA, it is essential to determine which metabolites and in which concentrations are present in the different target tissues. In rats, no ET, EA or derived metabolites have been detected in muscle, adipose, heart, lung, or brain tissues, although small amounts of conjugates of the microbial metabolites have been detected in liver and kidney. A study with pigs fed on acorns also revealed a similar distribution of metabolites in systemic tissues. The same study also showed the accumulation of large amounts of different conjugates of the microbial metabolites in the gall bladder, which is indicative of entero-hepatic circulation responsible for the long clearing life of the metabolites (they were detected in urine for as long as 48 to 76 hours after the intake) (Cerdá *et al.*, 2005). The occurrence of these metabolites in sexual tissues (uterus and prostate) has not yet been evaluated and should be considered in further studies.

7.7 The Whole Picture of Ellagitannin Bioavailability

ETs are generally not absorbed. They release EA in the gut and this is poorly absorbed in the stomach and small intestine, and largely metabolized by unidentified bacteria in the intestinal lumen to produce dibenzopyranones (urolithins). This microbial metabolism starts in the small intestine and the first metabolites produced retain four phenolic hydroxyls, and these are further metabolized along the intestinal tract to remove hydroxyls units leading to urolithins A (2 hydroxyls) and B (1 hydroxyl) in the distal parts of the colon (Fig. 7.2). The absorbed metabolites are conjugated with glucuronic acid (one or two units), and/or methylated (when *ortho*-dihydroxyphenyl moieties are present) to give ethers. Urolithin A and B conjugates are the main metabolites detected in plasma and urine, although some trihydroxy dibenzopyranone derivatives (hydroxyurolithin A) or EA-dimethyl ether glucuronide have also been detected in smaller amounts. The tetrahydroxy-dibenzopyranones, trihydroxydibenzopyranones and EA derivatives are not detected in peripheral plasma, but they are absorbed in the small intestine and are transported to the liver, where they are further metabolized and excreted with bile to the small intestine establishing an entero-hepatic circulation that is responsible for the long life of urolithins in plasma and urine. As far as we know, these metabolites do not accumulate in organ tissues, with the exception of gallbladder and urine bladder, where they are accumulated with the excretion fluids.

7.8 Conclusion

The biological activity and physiological effects of ETs and EA have to be consistent with their bioavailability and metabolism. Thus, results obtained from *in vitro* studies using cultured cells that would not be in direct contact *in vivo* with food ETs should be considered with caution. This is for example the case of studies testing activity of ETs on liver or breast cancer cell lines. In cultured cells representing systemic tissues and organs, we should be best evaluating the bioactivity of the actual metabolites circulating in plasma (urolithins and their glucuronic acid conjugates) and at the *in vivo* concentrations reached in plasma (mean values about 1-5 μM) and/or in tissues (detectable but generally below the quantification limit). ETs and EA are more relevant in terms of bioactivity at the gastrointestinal tract, where they are present at significant concentrations. Both *in vitro* and *in vivo* studies show the potential of these polyphenols as chemopreventive agents in several types of cancer.

The metabolic transformation of ETs into urolithins A and B is a widespread phenomenon in nature. These urolithin metabolites have been previously reported to be present in the faeces of the squirrel *Trogopterus xanthippes* and their hyaluronidase inhibitory activity was also demonstrated (Jeong *et al.*, 2000). These metabolites have also been isolated from kidney stones in cattle suffering from the "clover stone" disease, which is most likely associated with a large and chronic intake of a clover species (*Trifolium subterraneum*) (Pope, 1964), probably rich in ETs. Urolithins A and B have also been isolated from beaver excretions and as such are constituents of *Castoreum* (Lederer, 1949). *Castoreum* is a urine-based fluid from castor sacs and/or anal-gland secretion (Rosell *et al.*, 2000).

Apart from the hyaluronidase inhibitory activity (Jeong *et al.*, 2000), no other biological activity has been yet attributed to the microbial metabolites urolithins. However, it is anticipated that these gut microflora metabolites are potential endocrine-disrupting molecules, which could resemble other described "enterophytoestrogens" (microflora-derived metabolites with estrogenic/antiestrogenic activity). They may have potential estrogenic/antiestrogenic effects in the range of that is reported for other well-known estrogenic compounds such as enterolactone, resveratrol, genistein or daidzein (Larrosa *et al.*, 2006b). Further research is warranted to evaluate the possible role of ETs and EA as dietary "pro-phytoestrogens".

Acknowledgements

The authors are grateful to the Spanish MEC and to the EU Commission for financial support of this work (AGL2004-03989 and FOOD-CT-2004-513960 FLAVO).

7.9 Bibliography

Aaby, K., Wrolstad, R. E., Ekeberg, D. and Skrede, G. (2007). Polyphenol composition and antioxidant activity in strawberry purees; Impact of achene level and storage, *J. Agric. Food Chem.*, 55, pp. 5156–5166.

Anderson, K. J., Teuber, S. S., Gobeille, A., Cremin, P., Waterhouse, A. L. and Steinberg, F. M. (2001). Walnut polyphenolics inhibit in vitro human plasma and LDL oxidation, *J. Nutr.*, 131, pp. 2837–2842.

Aviram, M., Kent, U. M. and Hollenbergh, P. F. (1999). Microsomal cytochrome P450 catalyzes oxidation of low density lipoprotein, *Atherosclerosis*, 143, p. 253.

Aviram, M., Dornfeld, L., Rosenblat, M., Volkova, N., Kaplan, M., Coleman, R., Hayek, T., Presser, D. and Fuhrman, B. (2000). Pomegranate juice consumption reduces oxidative stress, atherogenic modifications to LDL, and platelet aggregation: studies in humans and in the atherosclerotic apolipoprotein E-deficient mice, *Am. J. Clin. Nut.*, 71, pp. 1062–1076.

Aviram, M. and Dornfeld, L. (2001). Pomegranate juice consumption inhibits serum angiotensin converting enzyme activity and reduced systolic pressure, *Atherosclerosis*, 158, pp. 195–198.

Aviram, M., Dornfeld, L., Kaplan, M., Coleman, R., Gaitini, D., Nitecki, S., Hofman, A., Rosenblat, M., Volkova, N., Presser, D., Attias, J., Hayek, T. and Fuhrman, B. (2002). Pomegranate juice flavonoids inhibit low density lipoprotein oxidation and cardiovascular diseases: Studies in atherosclerotic mice and humans, *Drugs Exp. Clin. Res.*, 28, pp. 49–62.

Aviram, M., Rosenblat, M., Gaitini, D., Nitecki, S., Hoffman, A., Dornfeld, L., Volkova, N., Presser, D., Attias, J., Liker, H. and Hayek, T. (2004). Pomegranate juice consumption for three years by patients with carotid arterial stenosis reduces common carotid-media thickness, blood pressure and LDL oxidation, *Clin. Nut.*, 23, pp. 423–433.

Blomhoff, R., Carlsen, M. H., Andersen, L. F. and Jacobs, D. R. (2006). Health benefits of nuts: potential role of antioxidants, *Brit. J. Nut.*, 96, pp. S52–60.

Cantos, E., Espín, J. C., López-Bote, C., De la Hoz, L., Ordóñez, J. A. and Tomás-Barberán, F. A. (2003). Phenolic compounds and fatty acids from acrons (*Quercus* sp.), the main dietary constituent of free-ranged Iberian pigs, *J. Agric. Food Chem.*, 51, pp. 6248–6255.

Carlton, P. S., Kresty, L. A., Siglin, J. C., Morse, M. A., Lu, J., Morgan, C. and Stoner, G. D. (2001). Inhibition of N-nitrosomethylbenzylamine-induced tumorigenesis in the rat esophagus by dietary freeze-dried strawberries, *Carcinogenesis*, 22, pp. pp. 441–446.

Casto, B. C., Kresty, L. A., Kraly, C. L., Pearl, D. K., Knobloch, T. J., Schut, H. A., Stoner, G. D., Mallery, S. R. and Weghorst, C. M. (2002). Chemoprevention or oral cancer by black raspberries, *Anticancer Res.*, 22, pp. 4005–4015.

Castonguay, A., Gali, H. U., Perchellet, E. M., GAo, X. M., Boukharta, M., Jalbert, G., Okuda, T., Yoshida, T., Hatano, T. and Perchellet J. P. (1997). Antitumorigenic and antipromoting activities of ellagic acid, ellagitannins and oligomeric anthocyanin and procyanidin, *Int. J. Oncol.*, 10, pp. 367–373.

Cerdá, B., Llorach, R., Cerón, J. J., Espín, J. C. and Tomás-Barberán, F. A. (2003a). Evaluation of the bioavailability and metabolism in the rat of punicalagin, an antioxidant polyphenols from pomegranate juice, *Eur. J. Nutr.*, 42, pp. 18–28.

Cerdá, B., Cerón, J. J., Tomás-Barberán, F. A., Espín, J. C. (2003b). Repeated oral administration of high doses of the pomegranate ellagitannin punicalagin to rats for 37 days is not toxic, *J. Agric. Food Chem.*, 51, pp. 3493–3501.

Cerdá, B., Espín, J. C., Parra, S., Martínez, P. and Tomás-Barberán, F. A. (2004). The potent in vitro antioxidant elagitannins from pomegranate juice are metabolised into bioavailable but poor antioxidant hydroxyl-6*H*-dibenzopyran-6-one derivatives by the colonic microflora of healthy humans, *Eur. J. Nutr.*, 43, pp. 205–220.

Cerdá, B., Tomás-Barberán, F. A. and Espín, J. C. (2005a). Metabolism of antioxidant and chemopreventive ellagitannins from strawberries, raspberries, walnuts, and oak-aged wine in humans: Identification of biomarkers and individual variability, *J. Agric. Food Chem.*, 53, pp. 227–235.

Cerdá, B., Periago, P., Espín, J. C. and Tomás-Barberán, F. A. (2005b). Identification of Urolithin A as a metabolite produced by human colon microflora from ellagic acid and related compounds, *J. Agric. Food Chem.*, 53, pp. 5571–5576.

Cerdá, B.; Soto, M. C., Albaladejo, M. D., Martínez, P., Sánchez-Gascón, F., Tomás-Barberán, F. A. and Espín, J. C. (2006). Pomegranate juice supplementation in COPD: a 5-week randomised, double blind, placebo-controlled trial, *Eur. J. Clin. Nut.*, 60, pp. 245–253.

Chen, T., Hwang, H. J., Rose, M. E., Nines, R. G. and Stoner, G. D. (2006). Chemopreventive properties of black raspberries in *N*-nitrosomethylbenzylamine-induced rat esophageal tumorigenesis: Down-regulation of cyclooxygenase-2, inducible nitric oxide synthase and c-Jun, *Cancer Res.*, 66, pp. 2853–2859.

Chen, P. S. and Li, H. J. (2007). Chemopreventive effect of punicalagin, a novel tannin component isolated from *Terminalia catappa*, on H-ras-transformed NIH3T3 cells, *Toxicology Letters*, 163, pp. 44–53.

Clifford, M. N. and Scalbert, A. (2000). Ellagitannins – nature, occurrence and dietary burden, *J. Sci. Food Agric.*, 80, pp. 1118–125.

Doyle, B. and Griffiths, L. A. (1980). The metabolism of ellagic acid in the rat, *Xenobiotica*, 10, pp. 247–256.

Espín, J. C., González-Barrio, R., Cerdá, B., López-Bote, C., Rey, A.I. and Tomás-Barberán, F.A. (2007). Iberian pig as a model to clarify obscure points in the bioavailability and metabolism of Ellagitannins in humans, *J. Agric. Food Chem.*, 55, pp. 10476–10485.

Ferreres, F., Andrade, P., Gil, M. I. and Tomás-Barberán, F. A. (1996). Floral nectar phenolics as biochemical markers for the botanical origin of heather honey, *Z. Leb. Unt. Forsch.*, 202, pp. 40–44.

Fuhrman, B. and Aviram, M. (2006). Protection against cardiovascular diseases. In *Pomegranates. Ancient Roots to Moderm Medicine*, ed. by Seeram, N. P., Schulman, R. N., Heber, D., CRC Taylor & Francis: Boca Raton, pp. 63–89.

Fukuda, T., Ito, H. and Yoshida, T. (2003). Antioxidative polyphenols from walnuts (*Juglans regia* L.), *Phytochemistry*, 63, pp. 795–801.

Gil, M. I., Tomás-Barberán, F. A., Hess-Pierce, B., Holcroft, D. M. and Kader, A. A. (2000). Antioxidant activity of pomegranate juice and its relationship with phenolic composition and processing, *J. Agric. Food Chem.*, 48, pp. 4581–4589.

Glabasnia, A. and Hofmann, T. (2006). Sensory directed identification of taste-active ellagitannins in American (*Quercus alba* L.) and European oak wood (*Quercus robur* L.) and quantitative analysis in bourbon whiskey and oak-matured red wines, *J. Agric. Food Chem.*, 54, pp. 3380–3390.

Hannum, S. M. (2004). Potential impact of strawberries on human health: a review of the science, *Crit. Rev. Food Sci. Nut.*, 44, pp. 1–17.

Harris, G. K., Gupta, A., Nines, R. G., Kresty, L. A., Habib, S. G., Frankel, W. L., LaPerle, K., Gallaher, D. D., Schwartz, S. J. and Stoner, G. D. (2001). Effects of lyophilized black raspberries on azoxymethane-induced colon cancer and 8-hydroxy-2'-deoxyguanosine levels in the Fisher 344 rat, *Nut. Cancer*, 40, pp. 125–133.

Hidaka, M., Okumura, M., Fujita, K., Ogikubo, T., Yamasaki, K., Iwakiri, T. and Setoguchi, N. (2005). Effects of pomegranate juice on human cytochrome P450 3A (CYP3A) and carbamazepine pharmacokinetics in rats, *Drug Met. Disposition*, 33, pp. 644–648.

Ito, H., Okuda, T., Fukuda, T., Hatano, T. and Yoshida, T. (2007). Two novel dicarboxylic acid derivatives and a new dimeric hydrolyzable tannin from walnuts, *J. Agric. Food. Chem.*, 55, pp. 672–679.

Ito, H., Iguchi, A., and Hatano, T. (2008). Identification of urinary and intestinal bacterial metabolites of ellagitannin geraniin in rats, *J. Agric. Food Chem.*, 56, pp. 393–400.

Jeong, S. J., Kim, N. Y., Kim, D. H., Kang, T. H., Ahn, N. H., Miyamoto, T., Higuchi, R. and Kim, Y. C. (2000). Hyaluronidase inhibitory active 6*H*-dibenzo[b,d]pyran-6-ones from the faeces of *Trogopterus xanthippes*, *Planta Med.*, 66, pp. 76–77.

Kim, H., Yoon, Y. J., Shon, J. H., Cha, I. J., Shin, J. G. and Liu, K. H. (2006). Inhibitory effects of fruit juices on CYP3A activity, *Drug Met. Disp.*, 34, pp. 521–523.

Koponen, J. M., Happonen, A. M., Mattila, P. H. and Torronen, A. R. (2007). Contents of anthocyanins and ellagitannins in selected foods consumed in Finland, *J. Agric. Food Chem.*, 55, pp. 1612–1619.

Kuo, P. L., Hsu, Y. L., Lin, T. C., Tzeng, W. S., Chen, Y. Y. and Lin, C. C. (2007). Ruosin E, an ellagitannin, inhibits MDA-MB-231 human breast cancer cell proliferation and induces apoptosis by inhibiting nuclear factor-κβ signalling, *Cancer Letters*, 248, pp. 280–291.

Larrosa, M., Tomás-Barberán, F. A. and Espín, J. C. (2006a). The hydrolysable tannin punicalagin releases ellagic acid which induces apoptosis in human colon adenocarcinoma Caco-2 cells by using the mitochondrial pathway, *J. Nut. Biochem.*, 17, pp. 611–625.

Larrosa, M., González-Sarrías, A., García-Conesa, M. T., Tomás-Barberán, F. A. and Espín, J. C. (2006b). Urolithins, ellagic acid-derived metabolites produced by human colonic microflora, exhibit estrogenic and antiestrogenic activities, *J. Agric. Food Chem.*, 54, pp. 1611–1620.

Lederer, E. (1949). Chemistry and biochemistry of some mammalian secretions and excretions, *J. Chem. Soc.*, pp. 2115–2125.

Lee, J. H., Johnson, J. V. and Talcott, S. T. (2005). Identification of ellagic acid conjugates and other polyphenolics in muscadine grapes by HPLC-ESI-MS, *J. Agric. Food Chem.*, 53, pp. 6003–6010.

Lesca, P. (1983). Protective effects of ellagic acid and other plant phenols on benzo(a)pyrene-induced neoplasia in mice, *Carcinogenesis*, 4, pp. 1651–1653.

Kresty, L. A., Morse, M. A., Morgan, C., Carlton, P. S., Lu, J., Gupta, A., Blackwood, M. and Stoner, G. D. (2001). Chemoprevention of esophageal tumorigenesis by dietary administration of lyophilized black raspberries, *Cancer Research*, 61, pp. 6112–6119.

Malik, A., Afaq, F., Sarfaraz, S., Adhami V. M., Syed D. N. and Mukhtar H. (2005). Pomegranate fruit juice for chemoprevention and chemotherapy of prostate cancer, *Proc. Nat. Acad. Sci. USA*, 102, pp. 14813–14818.

Malik, A. and Mukhtar, H. (2006). Prostate cancer prevention through pomegranate fruit, *Cell Cycle*, 5, pp. 371–373.

Manach, C., Williamson, G., Morand, C., Scalbert, A. and Remesy, C. (2005). Bioavailability and bioefficacy of phytonutrients in humans. I. Review of 97 bioavailability studies, *Am. J. Clin. Nutr.*, 81, pp. 230S–242S.

Mertens-Talcott S. U., Jilma-Stohlawetz, P., Rios, J., Hingorani, L., and Derendorf, H. (2006). Absorption, metabolism, and antioxidant effects of pomegranate (*Punica granatum* L.) polyphenols after ingestion of a standardized extract in healthy human volunteers, *J. Agric. Food Chem.*, 54, pp. 8956–8961.

Paivarinta, E., Pajari, A. M., Torronen, R. and Mutanen, M. (2006). Ellagic acid and natural sources of ellagitannins as possible chemopreventive agents against intestinal tumorigenesis in the Min mouse, *Nut. Cancer*, 54, pp. 79–83.

Pantuck, A. J., Leppert, J. T., Zomoridian, N., Aronson, W., Hong, J., Barnard, R. J., Seeram, N. P., Liker, H., Wang, H. J., Elashoff, R., Heber, D., Aviram, M., Ignarro, L. and Belldegrun, A. (2006). Phase II study of pomegranate juice for men with rising prostate-specific antigen following surgery or radiation for prostate cancer, *Clin. Cancer Res.*, 12, pp. 4018–4026.

Pepin, P., Rossignol, G. and Castonguay, A. (1990). Inhibition of NNK-induced lung tumorigenesis in A/J mice by ellagic acid and butylated hydroxyanisole, *Cancer Journal*, 3, pp. 266–273.

Pope, G. S. (1964) Isolation of two benzocoumarins from ‘clover stone’, a type of renal calculus found in sheep, *Biochem. J.*, 93, pp. 474–477.

Porter, L. J. (1989). Tannins. In *Methods in Plant Biochemistry, vol. 1, Plant Phenolics*, ed. by Harborne, J. B., Academic Press: London, pp. 389–419.

Rosell, F., Johansen, G. and Parker, H. (2000). Eurasian beavers (Castor fiber) behavioural response to simulated territorial intruders, *Can J. Zool.*, 78, pp. 931–935.

Ross, H.A., McDougall, G. J., Stewart, D. (2007). Antiproliferative activity is predominantly associated with ellagitannins in raspberry extracts, *Phytochemistry*, 68, pp. 218–228.

Seeram, N. P., Lee, R. and Heber, D. (2004). Bioavailability of ellagic acid in human plasma after consumption of ellagitannins from pomegranate (*Punica granatum* L.) juice, *Clin. Chim. Acta*, 348, pp. 63–68.

Seeram, N. P., Henning, S. M., Zhang, Y., Suchard, M., Li, Z. and Heber, D. (2006). Pomegranate juice ellagitannin metabolotes are present in human plasma and some persist in urine up to 48 hours, *J. Nut.*, 136, pp. 2481–2485.

Smart, R. C., Huang, M. T., Chang, R. L., Sayer, J. M., Jerina, D. M. and Conney, A. H. (1986). Disposition of the naturally occurring antimutagenic plant phenol, ellagic acid, and its synthetic derivatives, 3-O-decylellagic acid and 3,3'-di-O-methylellagic acid in mice, *Carcinogenesis*, 7, pp. 1663–1667.

Sorokin, A. V., Duncan, B., Panetta, R. and Thompson, P. D. (2006). Rhabdomyolysis associated with pomegranate juice consumption, *Am. J. Cardiol.*, 98, pp. 705–706.

Stoner, G. D., Sardo, C., Apseloff, G., Mullet, D., Wargo, W., Pound, V., Singh, A., Sanders, J., Aziz, R., Casto, B. and Sun, X. (2005). Pharmacokinetics of anthocyanins and ellagic acid in healthy volunteers fed freeze-dried black raspberries daily for 7 days, *J. Clin. Pharmacol.*, 45, pp. 1153–1164.

Stoner, G. D., Chen, T., Kresty, L. A., Aziz, R. M., Reinemann, T. and Nines, R (2006). Protection against esophageal cancer in rodents with lyophilized berries: potential mechanisms, *Nut. Cancer*, 54, pp. 33–46.

Teel, R. W. and Martin, R. W. (1988). Disposition of the plant phenol ellagic acid in the mouse following oral administration by gavage, *Xenobiotica*, 18, pp. 397–405.

Van Tassel, R. (1973). Bezoars, *Janus*, 60, pp. 241–259.

Whitley, A. C., Stoner, G. D., Darby, M V. and Walle, T. (2003). Intestinal epithelial cell accumulation of the cancer preventive polyphenol ellagic acid – Extensive binding to protein and DNA, *Biochem. Pharmacol.*, 66, pp. 907–915.

Villarreal-Lozoya, J. E., Lombardini, L. and Cisneros-Zevallos, L. (2007). Phytochemical constituents and antioxidant capacity of different pecan (*Carya illinoinensis*) cultivars, *Food Chem.*, 102, pp. 1241–1249.

Zafrilla, P., Ferreres, F. and Tomás-Barberán, F. A. (2001). Effect of processing and storage on the antioxidant ellagic acid derivatives and flavonoids of red raspberry (*Rubus idaeus*) jams, *J. Agric. Food Chem.*, 49, pp. 3651–3655.

Chapter 8

Sources and Health Effects of Dietary Ellagitannins

Riitta Törrönen

Food and Health Research Centre, Department of Clinical Nutrition, University of Kuopio, 70211 Kuopio, Finland; riitta.torronen@uku.fi

8.1 Introduction

Polyphenols are an integral part of the human diet. Although they are not regarded as essential nutrients, they may affect human health by many mechanisms. Antioxidant, antimicrobial, anti-inflammatory and anticarcinogenic and other biological activities demonstrated for various polyphenols suggest that they could have beneficial health effects and provide protection against chronic diseases such as cardiovascular diseases, neurodegenerative diseases and cancers (Scalbert *et al.*, 2005, Erdman *et al.*, 2007). Scientific evidence on the role of polyphenol consumption in disease prevention is promising but not conclusive, and more human clinical trials and epidemiologic studies are needed. For evaluation of the health benefits of polyphenols, comprehensive data on the composition and contents of the most common compounds in foods and on their dietary intake in different populations are required. During the past two decades, research on food phenolics has mainly focused on flavonoids, phenolic acids and proanthocyanidins, but much less on ellagitannins. This is probably due to the fact that – according to the current knowledge – the occurrence of ellagitannins and ellagic acid in commonly consumed foods is limited to a few berry, fruit and nut species and oak-aged wines (Clifford and Scalbert, 2000). This chapter complements the previous one and summarises the research data on ellagitannins and ellagic acid in the human diet, while outlining the current evidence on the potential health benefits of foods and beverages rich in ellagitannins.

8.2 Ellagitannins and Ellagic Acid in Foods

8.2.1 Analytical methods

Ellagitannins are complex derivatives of ellagic acid. They contain one or more hexahydroxydiphenic acid (HHDP) moieties esterified to a polyol, usually glucose (Clifford and Scalbert, 2000). Hydrolysis of ellagitannins with acids or bases yields HHDP, which is spontaneously lactonized to ellagic acid (Fig. 8.1). This reaction has been commonly utilized for the detection and quantification of food ellagitannins as ellagic acid equivalents by HPLC after acid hydrolysis (*e.g.*, Daniel *et al.*, 1989, Häkkinen *et al.*, 2000, Lee and Talcott, 2004, Vrhovsek *et al.*, 2006, Koponen *et al.*, 2007). Since food samples also contain small amounts of free ellagic acid and its glycosides, these should be analyzed without hydrolysis and subtracted from the total ellagic acid value to obtain the true content of ellagitannins.

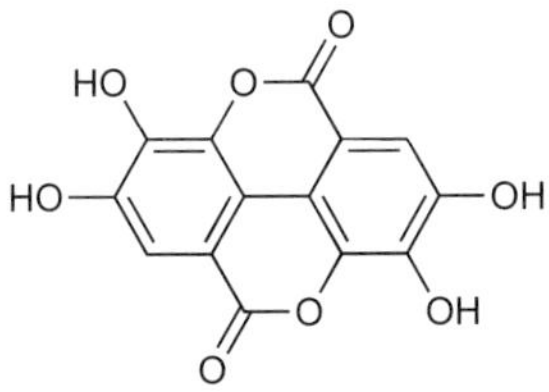

Fig. 8.1 Ellagic acid, the hydrolysis product of ellagitannins. Food ellagitannins can be quantified as ellagic acid equivalents after acid hydrolysis.

The method based on acid hydrolysis is the most practical choice for routine quantification of complex ellagitannins in food samples. It can be used for various food materials and allows simultaneous analysis of all forms of ellagitannins. It uses commercially available, inexpensive ellagic acid as standard, while such standards for ellagitannins are lacking. Comparable levels of ellagitannins in berries were obtained by the analysis of soluble ellagitannins as gallic acid equivalents and by the analysis of ellagic acid equivalents released by acid hydrolysis (Määttä-Riihinen *et al.*, 2004).

The contents of ellagitannins reported in literature are variable (Table 8.1), since different conditions of extraction and acid hydrolysis have been used and these significantly affect the yield of ellagic acid. Attention should be paid to optimization of the hydrolysis conditions and selection of the solvent for solubilization of the released ellagic acid. Poorly soluble ellagic acid may form unrecognized precipitates resulting in underestimation of the content of ellagitannins.

Daniel *et al.* (1989) used methanol extraction and hydrolysis with trifluoroacetic acid for 2 h at 100 °C, but this method was later on found to result in poor recovery of ellagic acid (Häkkinen *et al.*, 2000). For simultaneous screening of flavonols and ellagic acid in berries, the samples were hydrolyzed for 16 h at 35 °C in 50% methanol and 1.2 M HCl (Häkkinen *et al.*, 1999). Later, the method was optimized for ellagic acid analysis (Häkkinen *et al.*, 2000), and this procedure (20 h at 85 °C in 50% methanol and 1.2 M HCl) has also been applied with slight modifications in other studies (Häkkinen and Törrönen, 2000, Mattila and Kumpulainen, 2002, Määttä-Riihinen *et al.*, 2004, Koponen *et al.*, 2007). However, Vrhovsek *et al.* (2006) noted that this treatment resulted in incomplete hydrolysis even after 20 h. Instead, they suggested a 6-hour hydrolysis with 4 M HCl and extraction with aqueous acetone for a method of choice for ellagitannin analyses.

In addition to ellagic acid, its conversion products are detected in acid-hydrolyzed samples. In the earlier studies (Daniel *et al.*, 1989, Häkkinen *et al.*, 2000, Mattila and Kumpulainen, 2002), only ellagic acid was quantified. Määttä-Riihinen *et al.* (2004) observed that berry ellagitannins were degraded by acid hydrolysis into ellagic acid and a less polar derivative (*ca.* 20-26% of the content), and they quantified both conversion products. Vrhovsek and co-workers (2006) showed that besides ellagic acid, two ellagic acid derivatives were detected in raspberry samples after acid hydrolysis of ellagitannins. The second main peak in the chromatogram was identified as methyl sanguisorboate. Also Koponen *et al.* (2007) detected ellagic acid and two derivatives with UV-visible spectra similar to ellagic acid, and these three compounds were taken into account in quantification of ellagitannins in Finnish foods (Fig. 8.2).

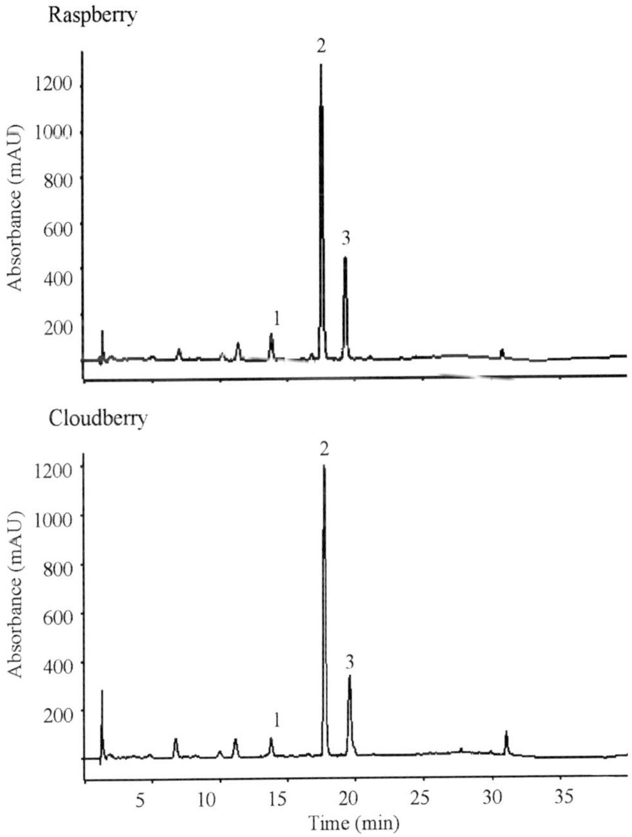

Fig. 8.2 HPLC chromatograms (254 nm) of raspberry and cloudberry after acid hydrolysis (20 h at 85 °C in 70% methanol and 2.0 M HCl). Peak 2 is ellagic acid and peaks 1 and 3 are ellagic acid derivatives. Peak 3 of raspberry was identified as methyl sanguisorboate by Vrhovsek *et al.* (2006). The chromatograms were kindly provided by J. Koponen, University of Kuopio, Finland.

8.2.2 Fresh and processed berries

Daniel *et al.* (1989) determined the ellagic acid content of 21 fruits and 5 nuts after acid hydrolysis of the samples, and found measurable levels only in strawberries, raspberries, blackberries, cranberries, walnuts and pecans. The highest level was detected in berries of the genus *Rubus* (raspberries and blackberries), intermediate levels in strawberries and nuts, and a low level in cranberries (Table 8.1). High levels were also detected in other *Rubus* berries, *i.e.*, in black raspberries, marionberries and boysenberries (Wada and Ou, 2002). When 19 berries consumed in Finland were screened for the presence of ellagic acid after acid hydrolysis, high levels were detected in the *Rubus* berries (cloudberries, arctic brambles, red raspberries) and strawberries, whereas in all other berries only traces were detected (Häkkinen *et al.*, 1999). In a subsequent study, Häkkinen *et al.* (2000) used an optimised method for quantification of ellagic acid in strawberries and *Rubus* berries (Table 8.1).

Free and glycosylated forms of ellagic acid, as well as ellagitannins, have been identified and quantified in berries of *Fragaria* (strawberries) and *Rubus* species (cultivated and wild, red and yellow raspberries, cloudberries and arctic brambles) (Määttä-Riihinen *et al.*, 2004). In this study, ellagitannins were quantified in their native forms as gallic acid equivalents and after acid hydrolysis as ellagic acid equivalents (Table 8.1). Free ellagic acid was detected in all berries (the highest levels in cloudberries and wild red raspberries), whereas ellagic acid glycosides were detected only in raspberries (the highest level in wild red raspberries). Two major ellagitannins were detected in all berries studied at the following levels: arctic bramble > cloudberry > wild red raspberry > cultivated yellow raspberry > cultivated red raspberry > strawberry. The main ellagitannins were similar in all *Fragaria* and *Rubus* species, and have been identified in raspberries and cloudberries as the dimer sanguiin H-6 (Fig. 8.3) and the trimer lambertianin C (Mullen *et al.*, 2003, Beekwilder *et al.*, 2005, Heinonen *et al.*, 2007). Sanguiin H-6 is also present in strawberries (Cerdá *et al.*, 2005). Free ellagic acid and its glycosides are minor constituents as compared to ellagitannins (Zafrilla *et al.*, 2001, Määttä-Riihinen *et al.*, 2004). In strawberries, achenes have much higher concentrations of ellagitannins, free ellagic acid and its glycosides than the flesh (Aaby *et al.*, 2005).

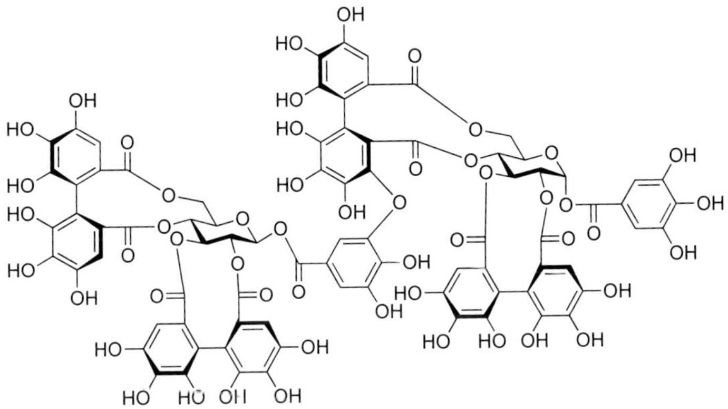

Fig. 8.3 Sanguiin H-6, an ellagitannin in raspberries, cloudberries and strawberries.

Koponen *et al.* (2007) analyzed the content of ellagitannins (as ellagic acid equivalents after acid hydrolysis) in selected foods consumed in Finland, including berries, fruits, vegetables and processed products. Ellagic acid was found in 9 of 33 selected food items, and the total ellagic acid content varied from 10 to 3300 mg/kg (Table 8.1). Ellagic acid was mostly present as ellagitannins, and the relative amount of free ellagic acid and its glycosides (*i.e.*, non-tannin ellagic acid) was less than 6%, and in most cases only 1-2%. Berries of the family *Rosaceae* (cloudberries, raspberries, strawberries and rose hips) contained high levels of ellagic acid equivalents, whereas minor levels were found in sea buckthorn (family *Elaeagnaceae*). The total ellagic acid concentrations in commercial raspberry and strawberry jams were 23-36% of those found in the unprocessed berries. Ellagic acid compounds were detected only in these five berries and processed products, and not in fruits, vegetables and peanuts. These data, together with data on several other groups of polyphenols, will be entered into the Finnish Food Composition Database Fineli (http://www.fineli.fi), maintained by the National Public Health Institute of Finland. To our knowledge, there are no other food composition databases containing ellagitannin data.

Table 8.1 Ellagitannin content in berries
(expressed as ellagic acid equivalents after acid hydrolysis).

Berry	**mg/kg f.w.**	**Reference**
Strawberry,	630 ± 90 (d.w.)	Daniel *et al.*, 1989
Fragaria x *ananassa*	430-4640 (d.w.)	Maas *et al.*, 1991
	320-380	Häkkinen *et al.*, 2000
	340-590	Häkkinen and Törrönen, 2000
	310	Mattila and Kumpulainen, 2002
	250-560	Skupien and Oszmianski, 2004
	650-850	Määttä-Riihinen *et al.*, 2004
	680-800	Koponen *et al.*, 2007
Red raspberry,	1500 ± 100 (d.w.)	Daniel *et al.*, 1989
Rubus idaeus	510-660	Häkkinen *et al.*, 2000
	1600	Mattila and Kumpulainen, 2002
	1900	Määttä-Riihinen *et al.*, 2004
	380-1180	Anttonen and Karjalainen, 2005
	2640-3310	Koponen *et al.*, 2007
Red raspberry, wild	2700	Määttä-Riihinen *et al.*, 2004
Blackberry, *Rubus ursinus*	1500 ± 140 (d.w.)	Daniel *et al.*, 1989
Cloudberry,	560-610	Häkkinen *et al.*, 2000
Rubus chamaemorus	3600	Määttä-Riihinen *et al.*, 2004
	3150	Koponen *et al.*, 2007
Arctic bramble,	690	Häkkinen *et al.*, 2000
Rubus arcticus	3900	Määttä-Riihinen *et al.*, 2004
Rose hip, *Rosa rugosa*	1100	Koponen *et al.*, 2007
Sea buckthorn, *Hippóphaë rhamnoides*	10	Koponen *et al.*, 2007
Cranberry, *Vaccinium*	120±4 (dw)	Daniel *et al.*, 1989

f.w. = fresh weight, d.w. = dry weight

Freezing and long-term (9-12 months) storage of strawberries and red raspberries at –20 °C reduced their total ellagic acid content by 15-40% (de Ancos *et al.*, 2000, Häkkinen *et al.*, 2000). Cooking strawberries with sugar to make jam only slightly reduced their total ellagic acid content, and storage of jam at +5 or –20 °C up to 9 months had no significant effect (Häkkinen *et al.*, 2000). However, the content of free ellagic acid increased during industrial raspberry jam processing, and continued to increase during the first months of storage at room temperature and then decreased slightly over the rest of the 6-month storage period (Zafrilla *et al.*, 2001). These changes may be explained by the release of ellagic acid from ellagitannins during processing and storage, or by an easier extractability of ellagic acid from processed products due to degradation of cell structures. In the commercial berry jams investigated by Koponen *et al.* (2007), the relative amount of free ellagic acid derivatives was as high as 11% in strawberry jam, in contrast to 3% in raspberry jam. This may suggest that during strawberry processing, ellagitannins are decomposed and converted to ellagic acid more easily than during raspberry processing.

Fig. 8.4 Cloudberry (*Rubus chamaemorus*), an arctic delicacy and an excellent source of ellagitannins. This boreal plant grows wild in the circumpolar regions. Photo: Arctic Flavours Association, Finland.

8.2.3 Pomegranate juice

Pomegranate (*Punica granatum*) fruits are ancient medicinal foods, which have been used for centuries in folk medicine. They are consumed fresh and as juice, which is an excellent source of ellagitannins. Commercial juices contain gallagyl-type ellagitannins, including punicalagin isomers (1500-1900 mg/L), undefined hydrolyzable tannins (400-500 mg/L), and ellagic acid and its glycosides (120-260 mg/L) (Gil *et al.*, 2000). Punicalagins, ellagitannins in which gallagic and ellagic acids are linked to a glucose molecule, are abundant in pomegranate peel. Punicalagin isomers and ellagic acid derivatives are not present in the aril juice, but during industrial juice processing, they are extracted from the husk and released in large quantities into the juice.

8.2.4 Grape juice and wine

Muscadine grapes (*Vitis rotundifolia*) are cultivated in southeastern U.S. and are consumed fresh or used for juice and wine production. Unique among *Vitis* species, muscadine grapes contain ellagitannins and ellagic acid derivatives (Lee and Talcott, 2004, Lee *et al.*, 2005). In three red and white cultivars, the following concentrations were detected: free ellagic acid, 14-50 mg/kg, ellagic acid glycosides, 44-87 mg/kg, and total ellagic acid after acid hydrolysis, 360-910 mg/kg (Lee *et al.*, 2005). In ripe grapes, more than 80% of the total ellagic acid compounds is localized in the skin, and ripe grapes have higher levels than unripe grapes (Lee and Talcott, 2004).

In muscadine grape juices prepared from different red and white grape cultivars, total ellagic acid concentrations of 100-300 mg/L were measured (Lee and Talcott, 2004). Concentrations of free ellagic acid and ellagic acid glycosides were 4-23 mg/L and 10-30 mg/L, respectively. Processing technologies (extraction and fermentation techniques) and cultivar influence the concentrations of ellagic acid and ellagitannins in juice and wine. Juices have higher levels compared to wines, indicating loss of ellagitannins during wine production and storage (Lee and Talcott, 2002).

During storage, muscadine juice and wine may produce a water-insoluble precipitate of yellowish to red crystals. Sediments are usually considered as quality defects and influence consumer acceptability of muscadine grape products. Ellagic acid was identified as the predominant component of the sediment (Boyle and Hsu, 1990). Due to hydrolysis of ellagitannins during storage and poor solubility of free ellagic acid, the content of free ellagic acid increases in the sediment (Lee and Talcott, 2002). Only about 12% of the weight of the sediment is ellagic acid, with the rest consisting of unidentified compounds and traces of ellagitannins.

8.2.5 Oak-aged wine

Although the content of ellagitannins in common grapevine (*Vitis vinifera*) is low, ellagitannins are common in wines aged in oak barrels. Ellagitannins of oak heartwood, such as vescalagin and castalagin, are potentially extracted into wine during aging and contribute to the sensory properties of the wine (Puech *et al.*, 1999). The oak ellagitannins also react with grape phenolics and ethanol producing flavano-ellagitannins and other derivatives. In a Bordeaux red wine aged for 18 months in oak barrels, the concentrations of oak ellagitannins were about 10 mg/L and those of flavano-ellagitannins about 2 mg/L (Saucier *et al.*, 2006).

8.2.6 Walnuts

Walnut trees (*Juglans regia*) are cultivated in many European countries, and the seeds are an excellent source of many nutrients. Compared to other nuts, walnuts are unique due to their high content of α-linolenic acid, an essential omega-3 polyunsaturated fatty acid. They also contain non-flavonoid polyphenols, with the highest concentration found in the thin tan-brown pellicle. The total content of phenols (as determined by the Folin-Ciocalteu assay) in 50 g of shelled walnuts was 802 mg as gallic acid equivalents (Anderson *et al.*, 2001). Daniel *et al.* (1989) reported an ellagic acid content of 590 mg/kg of dry weight after acid hydrolysis of *Juglans nigra* walnuts. Ellagic acid, valoneic acid dilactone, pedunculagin and 15 other ellagi- and gallotannin compounds have been identified from walnut kernels (Anderson *et al.*, 2001, Fukuda *et al.*, 2003, Cerdá *et al.*, 2005, Ito *et al.*, 2007).

8.3 Dietary Intake

The intake of phenolic acids, including ellagic acid, by Bavarian adults was studied in Germany (Radtke *et al.*, 1998). A database containing the phenolic acid contents of foods (data from literature) was built and 7-day dietary protocols of 63 women and 56 men of the German National Food Consumption Survey were evaluated. The intake of ellagic acid was 5.2 mg/day. Berries (strawberries, red raspberries) and nuts (walnut) provided 38% and 54%, respectively, of the ellagic acid intake.

In the Finnish diet, strawberries, red raspberries and cloudberries are the main potential sources of ellagitannins. The daily intake of total ellagic acid (including ellagitannins and free ellagic acid) from berries was calculated by using the food consumption data obtained from the Finnish Household Survey 1990 (8258 households) and 1998 (4359 households). This survey records the purchase of foods for the household. Fresh strawberries and cloudberries, but not raspberries, were reported in the survey. In 1990 and 1998, strawberries and cloudberries alone provided 5.8 mg (Häkkinen *et al.*, 2000) and 8.7 mg (Häkkinen *et al.*, unpublished data), respectively, of ellagic acid per day per person. The most recent estimation of intake is based on the compositional data on ellagitannins (as ellagic acid equivalents) in selected foods consumed in Finland (Koponen *et al.*, 2007) and food consumption data from 48-hour dietary recalls of 2007 adults of the FINDIET 2002 Study. The dietary intake of ellagitannins and free ellagic acid derivatives was 12 mg/day (women 15, men 8 mg/day), of which 99% was obtained from berries and berry dishes (Ovaskainen *et al.*, 2008).

8.4 Potential Health Benefits of Foods Rich in Ellagitannins

Several biological activities and health-related properties have been reported for ellagitannin-containing foods, suggesting that they could provide protection against, *e.g.*, oxidative stress, pathogenic bacteria and cancer. Although these properties may be attributed to their high content of ellagitannins, these foods also contain a variety of other polyphenolic compounds as well as common nutrients, which probably contribute to these bioactivities.

8.4.1 Antioxidant activity

Oxidative stress is caused by imbalance between production of free radicals and the body's antioxidation system, and has been implicated in the pathogenesis of many human diseases. Antioxidants can terminate free radical chain reactions by removing radical intermediates, or they can slow or prevent oxidation of other compounds, *e.g.*, lipids, proteins or DNA, by being oxidized themselves. Antioxidant activity is the most well-known bioactive property of polyphenols and polyphenol-rich foods.

Antioxidant activities of food extracts rich in ellagitannins have been determined by using various *in vitro* assays, and the high activities of strawberries (Meyers *et al.*, 2003, Aaby *et al.*, 2005, 2007), raspberries (Liu *et al.*, 2002, Beekwilder *et al.*, 2005), cloudberries (Kähkönen *et al.*, 2001) and other *Rubus* berries (Wada and Ou, 2002), pomegranates (Gil *et al.*, 2000) and walnuts (Anderson *et al.*, 2001) and their ellagitannins have been extensively reported. These foods also rank high when compared to other plant-based foods.

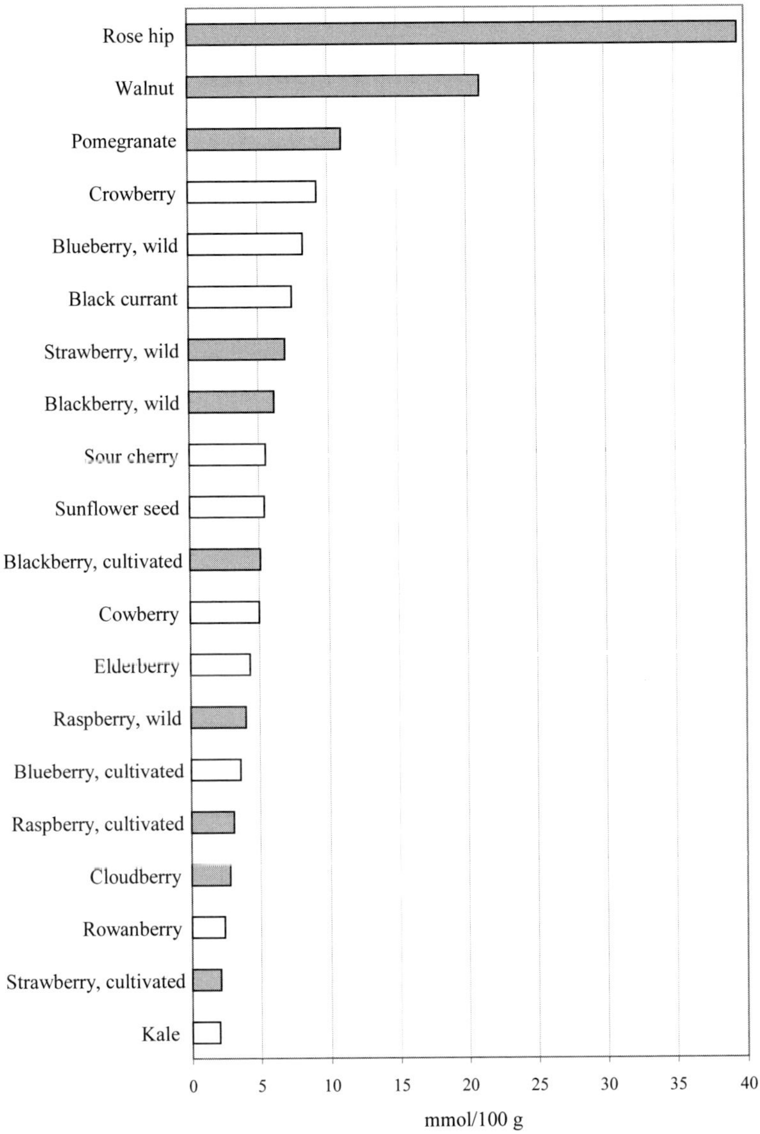

Fig. 8.5 Dietary plants with highest total antioxidant concentrations as determined by the FRAP assay (Halvorsen *et al.*, 2002). The presence of ellagitannins is indicated in grey.

A study conducted in Norway examined a large number of dietary plants, including fruits, berries, vegetables, cereals, nuts and pulses, using the FRAP assay (Halvorsen *et al.*, 2002). Interestingly, dietary plants with high ellagitannin content are well represented among those with the highest level of total antioxidants (see Fig. 8.5 for the 20 best antioxidant sources).

Less is known about the effects of consumption of ellagitannin-rich foods on the antioxidant status *in vivo*. In elderly women, the total antioxidant capacity of serum increased by about 10% during the 4-hour period after consumption of 240 g of strawberries (Cao *et al.*, 1998). A single dose of standardized pomegranate extract (Mertens-Talcott *et al.*, 2006a) and long-term consumption of pomegranate juice (Rosenblat *et al.*, 2006) also improved several antioxidant parameters in human volunteers. However, the daily consumption of walnuts for three weeks had no effect on the antioxidant status of subjects with metabolic syndrome (Davis *et al.*, 2007).

8.4.2 Inhibition of growth of human pathogenic microbes

Phenolic extracts of many berries inhibit the growth of human pathogenic microbes *in vitro.* Among all the berries studied, berries with high contents of ellagitannins, *i.e.*, cloudberries, raspberries and strawberries appear the most promising, showing antimicrobial activity against *Salmonella*, *Escherichia coli*, *Bacillus cereus*, *Campylobacter jejuni*, *Clostridium perfringens*, *Helicobacter pylori*, *Staphylococcus aureus*, *Staphylococcus epidermis* and *Candida albicans*, but not against beneficial probiotic lactic acid bacteria (Puupponen-Pimiä *et al.*, 2001, 2005a/b, Nohynek *et al.*, 2006). *Campylobacter jejuni* and *Candida albicans* are sensitive only to these berries. *Bacillus cereus*, *Clostridium perfringens* and *Staphylococcus aureus* cause food poisonings by producing toxins in food, and *Salmonella*, *Escherichia coli* and *Campylobacter jejuni* cause gastrointestinal infections. *Helicobacter pylori* infection may lead to chronic gastritis, peptic ulcer and stomach cancer. *Staphylococcus aureus* and *Staphylococcus epidermis* are found in normal flora of human skin, but may cause diseases from minor skin infections to severe life-threatening infections in patients with impaired immune system. *Candida albicans* is present in normal human flora, but it is also the most common yeast causing infections in various parts of the body. In addition to berry extracts, also ellagitannin fractions isolated from cloudberries and raspberries are strong inhibitors of *Staphylococcus* bacteria (Puupponen-Pimiä *et al.*, 2005a). Major part of the growth inhibition of *Salmonella* bacteria, however, originates from other compounds such as organic acids. The antimicrobial activity of cloudberries and raspberries is stable during long-term frozen storage of berries (Nohynek *et al.*, 2006).

In addition to ellagitannin-rich berries, pomegranates also contain antimicrobial compounds. Tannin-rich fractions and ellagitannins are effective against *Escherichia coli*, *Pseudomonas aeruginosa*, *Candida albicans*, *Cryptococcus neoformans*, *Aspergillus fumigatus*, *Mycobacterium intracellulare*, and even methicillin-resistant *Staphylococcus aureus* (Machado *et al.*, 2002, Reddy *et al.*, 2007).

Berries and pomegranates possess antimicrobial activity against a wide spectrum of bacteria and yeast, and could possibly be used as therapeutic or disinfective agents against those bacteria. Several applications for food industry and medicine have been suggested, such as functional foods or beverages for gut well-being and balanced microflora, natural food preservatives for foods that are easily contaminated by bacteria, products to prevent gastrointestinal and urinary tract infections, and products to prevent and treat antibiotic resistant bacterial infections (Puupponen-Pimiä *et al.*, 2005b). However, more research is needed to verify the effects and stability of the active components in various food matrices. For substantiation of the therapeutic potential, the efficacy of such foods or other type of products should be demonstrated in clinical trials.

8.4.3 Inhibition of cancer cell growth

Cancer cell growth is dependent on the balance between proliferation and apoptosis. Unregulated cell proliferation and suppression of apoptosis are key steps in initiation and progression of cancer. There is a lot of evidence that extracts of ellagitannin-rich foods reduce the growth of cancer cells *in vitro*, by inhibiting cell proliferation, inducing apoptotic cell death, and modulating cell cycle kinetics and signal transduction pathways.

In vitro studies carried out with cancer cell lines have shown that strawberries (Meyers *et al.*, 2003, Olsson *et al.*, 2004, Ramos *et al.*, 2005, Wang *et al.*, 2005, Wu *et al.*, 2007), raspberries (Liu *et al.*, 2002, Olsson *et al.*, 2004, Wu *et al.*, 2007), cloudberries (Wu *et al.*, 2007) and rose hips (Olsson *et al.*, 2004) inhibit cell proliferation, induce apoptosis and cause cell cycle arrest in human colon, liver, lung, breast or cervical cancer cells. In the studies mentioned above, the contribution of ellagitannins on the activities of berry extracts was not assessed. However, a recent study (Ross *et al.*, 2007) suggests that the antiproliferative activity of raspberries is predominantly associated with ellagitannins.

Also pomegranate juice and its ellagitannins inhibit proliferation, induce apoptosis and suppress inflammatory cell signaling in colon cancer cell lines (Seeram *et al.*, 2005, Adams *et al.*, 2006, Larrosa *et al.*, 2006). The potency of pomegranate juice is higher than its purified ellagitannins. This indicates synergistic action of ellagitannins and other bioactive components, such as anthocyanins and flavonols abundant in pomegranate juice. Likewise, polyphenols in muscadine grape skin inhibit the growth of colon cancer cells and induce apoptosis (Yi *et al.*, 2005). Fractions isolated from red muscadine grapes and rich in ellagic acid, ellagic acid glycosides and ellagitannins induce apoptosis, decrease cell number, and cause alterations in cell cycle kinetics in colon carcinoma cells (Mertens-Talcott *et al.*, 2006b).

Pomegranate fruit juice is effective against prostate cancer cells *in vitro*, but not against normal prostate epithelial cells. Treatment of highly aggressive human prostate cancer cells with pomegranate fruit extract resulted in inhibition of cell growth and viability and induction of apoptosis (Malik *et al.*, 2005, Malik and Mukhtar, 2006).

The effects of ellagitannins observed in these studies are probably mainly due to their hydrolysis product, ellagic acid. Pomegranate and raspberry ellagitannins produce ellagic acid in the cell culture media (Larrosa *et al.*, 2006, Ross *et al.*, 2007), and ellagic acid reduces cell proliferation and induces cell cycle arrest and apoptosis (Narayanan *et al.*, 1999, Losso *et al.*, 2004, Larrosa *et al.*, 2006, Wu *et al.*, 2007). Also in the human gastrointestinal tract, dietary ellagitannins are not absorbed as such, but are hydrolyzed to yield ellagic acid, which is then further metabolized by colonic microflora to more bioavailable derivatives (Cerdá *et al.*, 2004, 2005).

8.4.4 Cancer prevention

A food-based approach utilizing berries concentrated by freeze-drying has been extensively investigated in the prevention of esophageal and colon cancer in rodent model systems (Stoner *et al.*, 2006, 2007). These studies have used berries rich in ellagic acid (black raspberries and strawberries), since pure ellagic acid had been shown to inhibit chemically induced tumorigenesis in rodent esophagus. The freeze-dried berries (10% in the diet) inhibited chemically induced cancer in the rat esophagus by 30-60% and in the colon up to 80%. The berries affect both the initiation and promotion/progression stages of tumor development. So far, the results point to anthocyanins as the active components, and the role of ellagitannins and other berry components are under investigation. Several clinical trials are being conducted in patients at high risk for esophagus and colon cancer. Patients with Barrett's esophagus, esophageal dysplasia or colonic polyps consume freeze-dried black raspberry powder daily, and various biomarkers of development of the diseases are determined.

The effects of cloudberries on intestinal tumorigenesis have been investigated in *Apc*-mutated Min mice. Although these mice develop tumors mainly in the small intestine, they are used as an experimental model for colon carcinogenesis. Freeze-dried cloudberries (10%) in a modified rodent diet, with a composition and content of fat as in a typical Western-type diet, prevented adenoma formation (Misikangas *et al.*, 2007). However, when cloudberry seeds and pulp were studied separately, no effect on adenoma formation was observed (Päivärinta *et al.*, 2006). One explanation for the lack of effect is the lower content of ellagitannins in the seed and pulp diets; they were only half of that in the whole cloudberry diet. Pure ellagic acid in the diet (1.5 g/kg) had no effect on the number or size of adenomas in the distal small intestine, but it increased adenoma size in the duodenum. Animal and human studies have provided promising evidence on the potential of pomegranate juice in the chemoprevention of prostate cancer. Athymic nude mice were implanted with androgen-sensitive cells, which form rapid and reproducible prostate carcinoma tumors in nude mice and secrete PSA (prostate-specific antigen, a marker of prostate cancer progression) in the blood of the host. Oral administration of a human acceptable dose of pomegranate fruit extract to the animals resulted in inhibition of tumor growth and concomitant decrease in the serum PSA levels (Malik *et al.*, 2005, Malik and Mukhtar, 2006).

A clinical trial has been conducted to investigate the effects of pomegranate juice consumption in patients with recurrent prostate cancer (Pantuck *et al.*, 2006). Men with rising PSA levels after surgery or radiotherapy consumed 240 mL of commercial pomegranate juice daily until disease progression. A significant prolongation of PSA doubling time was observed, suggesting delayed progression of the disease. Serum antioxidant status was also improved, and *in vitro* assays of the patients' sera showed decreased proliferation and increased apoptosis of prostate cancer cells.

8.5 Conclusions

According to the present knowledge, berries of *Rubus*, *Fragaria* and *Rosa* species, as well as pomegranate and muscadine grape products and walnuts are the best dietary sources of ellagitannins. Due to limited food sources, the dietary intake of ellagitannins and ellagic acid is probably low compared to many other polyphenols. Experimental studies on ellagitannin-rich foods have demonstrated potential for health benefits, and dietary interventions are being conducted in patients at increased risk for cancer development. In addition to high levels of ellagitannins, berries, pomegranates and walnuts also contain a variety of other phytochemicals, essential fatty acids and other nutrients, and may therefore be regarded as "health foods" or even "superfoods". Research is progressing in many laboratories and will eventually provide new evidence on the health effects of dietary ellagitannins. Meanwhile, increased consumption of the delicious ellagitannin-rich foods and beverages as a part of a balanced diet can be recommended.

8.6 Bibliography

Aaby, K., Skrede, G. and Wrolstad, R. E. (2005). Phenolic composition and antioxidant activities in flesh and achenes of strawberries (*Fragaria ananassa*), *J. Agric. Food Chem.*, 53, pp. 4032–4040.

Aaby, K., Ekeberg, D. and Skrede, G. (2007). Characterization of phenolic compounds in strawberry (*Fragaria × ananassa*) fruits by different HPLC detectors and contribution of individual compounds to total antioxidant capacity, *J. Agric. Food Chem.*, 55, pp. 4395–4406.

Adams, L. S., Seeram, N. P., Aggarwal, B. B., Takada, Y., Sand, D. and Heber, D. (2006). Pomegranate juice, total pomegranate ellagitannins, and punicalagin suppress inflammatory cell signaling in colon cancer cells, *J. Agric. Food Chem.*, 54, pp. 980–985.

Anderson, K. J., Teuber, S. S., Gobeille, A., Cremin, P., Waterhouse, A. L. and Steinberg, F. M. (2001). Walnut polyphenols inhibit in vitro human plasma and LDL oxidation, *J. Nutr.*, 131, pp. 2837–2842.

de Ancos, B., González, E. M. and Cano, M. P. (2000). Ellagic acid, vitamin C, and total phenolic contents and radical scavenging capacity affected by freezing and frozen storage in raspberry fruit, *J. Agric. Food Chem.*, 48, pp. 4565–4570.

Anttonen, M. J. and Karjalainen, R. O. (2005). Environmental and genetic variation of phenolic compounds in red raspberry, *J. Food Compos. Anal.*, 18, pp. 759–769.

Beekwilder, J., Jonker, H., Meesters, P., Hall, R. D., van der Meer, I. M. and de Vos, C. H. R. (2005). Antioxidants in raspberry: on-line analysis links antioxidant activity to a diversity of individual metabolites, *J. Agric. Food Chem.*, 53, pp. 3313–3320.

Boyle J. A. and Hsu, L. (1990). Identification and quantification of ellagic acid in muscadine grape juice, *Am. J. Enol. Vitic.*, 41, pp. 43–47.

Cao, G., Russell, R. M., Lischner, N. and Prior, R. L. (1998). Serum antioxidant capacity is increased by consumption of strawberries, spinach, red wine or vitamin C in elderly women, *J. Nutr.* 128, pp. 2383–2390.

Cerdá, B., Espín, J. C., Parra, A., Martinez, P. and Tomás-Barberán, F. A. (2004). The potent *in vitro* antioxidant ellagitannins from pomegranate juice are metabolised into bioavailable but poor antioxidant hydroxy-6*H*-dibenzopyran-6-one derivatives by the colonic microflora of healthy humans, *Eur. J. Nutr.*, 43, pp. 205–220.

Cerdá, B., Tomás-Barberán, F. A. and Espín, J. C. (2005). Metabolism of antioxidant and chemopreventive ellagitannins from strawberries, raspberries, walnuts, and oak-aged wine in humans: identification of biomarkers and individual variability, *J. Agric. Food Chem.*, 53, pp. 227–235.

Clifford, M. N. and Scalbert, A. (2000). Ellagitannins – nature, occurrence and dietary burden, *J. Sci. Food Agric.*, 80, pp. 1118–1125.

Daniel, E. M., Krupnick, A. S., Heur, Y.-H., Blinzler, J. A., Nims, R. W. and Stoner, G. D. (1989). Extraction, stability, and quantification of ellagic acid in various fruits and nuts, *J. Food Compos. Anal.*, 2, pp. 338–349.

Davis, L., Stonehouse, W., Loots, D. T., Mukuddem-Petersen, J., van der Westhuizen, F. H., Hanekom, S. M. and Jerling, J. C. (2007). The effect of high walnut and cashew nut diets on the antioxidant status of subjects with metabolic syndrome, *Eur J. Nutr.*, 46, pp. 155–164.

Erdman, J. W. Jr., Balentine, D., Arab, L., Beecher, G., Dwyer, J. T., Folts, J., Harnly, J., Hollman, P., Keen, C. L., Mazza, G., Messina, M., Scalbert, A., Vita, J., Williamson, G. and Burrowes, J. (2007). Flavonoids and heart health, *Proceedings of the ILSI North America Flavonoids Workshop*, May 31-June 1, 2005, Washington, DC. *J. Nutr.*, 137 (Suppl), pp. 718S–737S.

Fukuda, T., Ito, H. and Yoshida, T. (2003). Antioxidative polyphenols from walnuts (*Juglans regia* L.), *Phytochemistry*, 63, pp. 795–801.

Gil, M. I., Tomás-Barberán, F. A., Hess-Pierce, B., Holcroft, D. M. and Kader, A. A. (2000). Antioxidant activity of pomegranate juice and its relationship with phenolic composition and processing, *J. Agric. Food Chem.*, 48, pp. 4581–4589.

Halvorsen, B. L., Holte, K., Myhrstad, M. C. W., Barikmo, I., Hvattum, E., *et al.* (2002). A systematic screening of total antioxidants in dietary plants, *J. Nutr.*, 132, pp. 461–471.

Heinonen, M. (2007). Antioxidant activity and antimicrobial effect of berry phenolics – a Finnish perspective, *Mol. Nutr. Food Res.*, 51, pp. 684–691.

Häkkinen, S., Heinonen, M., Kärenlampi, S., Mykkänen, H., Ruuskanen, J. and Törrönen, R. (1999). Screening of selected flavonoids and phenolic acids in 19 berries, *Food Res. Int.*, 32, pp. 345–353.

Häkkinen, S. H., Kärenlampi, S. O., Mykkänen, H. M., Heinonen, I. M., Törrönen, A. R. (2000). Ellagic acid content in berries: Influence of domestic processing and storage, *Eur. Food Res. Technol.*, 212, pp. 75–80.

Häkkinen, S. H. and Törrönen, A. R. (2000). Content of flavonols and selected phenolic acids in strawberries and *Vaccinium* species: Influence of cultivar, cultivation site and technique, *Food Res. Int.*, 33, pp. 517–524.

Ito, H., Okuda, T., Fukuda, T., Hatano, T. and Yoshida, T. (2007). Two novel dicarboxylic acid derivatives and a new dimeric hydrolysable tannin from walnuts, *J. Agric. Food Chem.*, 55, pp. 672–679.

Koponen, J. M., Happonen, A. M., Mattila, P. H. and Törrönen, A. R. (2007). Contents of anthocyanins and ellagitannins in selected foods consumed in Finland, *J. Agric. Food Chem.*, 55, pp. 1612–1619.

Kähkönen, M. P., Hopia, A. I. and Heinonen, M. (2001). Berry phenolics and their antioxidant activity, *J. Agric. Food Chem.*, 49, pp. 4076–4082.

Larrosa, M., Tomás-Barberán, F. A. and Espín, J. C. (2006). The dietary hydrolysable tannin punicalagin releases ellagic acid that induces apoptosis in human colon adenocarcinoma Caco-2 cells by using the mitochondrial pathway, *J. Nutr. Biochem.*, 17, pp. 611–625.

Lee, J.-H. and Talcott, S. T. (2002). Ellagic acid and ellagitannins affect on sedimentation in muscadine juice and wine, *J. Agric. Food Chem.*, 50, pp. 3971–3976.

Lee, J.-H. and Talcott, S. T. (2004). Fruit maturity and juice extraction influences ellagic acid derivatives and other antioxidant polyphenolics in muscadine grapes, *J. Agric. Food Chem.*, 52, pp. 361–366.

Lee, J.-H., Johnson, J. V. and Talcott, S. T. (2005). Identification of ellagic acid conjugates and other polyphenolics in muscadine grapes by HPLC-ESI-MS, *J. Agric. Food Chem.*, 53, pp. 6003–6010.

Liu, M., Li, X. Q., Weber, C., Lee, C. Y., Brown, J. and Liu, R. H. (2002). Antioxidant and antiproliferative activities of raspberries, *J. Agric. Food Chem.*, 50, pp. 2926–2930.

Losso, J.N., Bansode, R. R., Trappey, A., Bawadi, H. A. and Truax, R. (2004). In vitro anti-proliferative activities of ellagic acid, *J. Nutr. Biochem.*, 15, pp. 672–678.

Maas, J. L., Wang, S. Y. and Galletta, G. J. (1991). Evaluation of strawberry cultivars for ellagic acid content, *Hort. Sci.*, 26, pp. 66–68.

Machado, T. B., Leal, I. C. R., Amaral, A. C. F., dos Santos, K. R. N., da Silva, M. G. and Kuster, R. M. (2002). Antimicrobial ellagitannin of *Punica granatum* fruits, *J. Braz. Chem. Soc.*, 13, pp. 606–610.

Malik, A. and Mukhtar, H. (2006). Prostate cancer prevention through pomegranate fruit, *Cell Cycle*, 5, pp. 371–373.

Malik, A., Afaq, F., Sarfaraz, S., Adhami, V. M., Syed, D. N. and Mukhtar, H. (2005). Pomegranate fruit juice for chemoprevention and chemotherapy of prostate cancer, *Proc. Nat. Acad. Sci. USA*, 102, pp. 14813–14818.

Mattila, P. and Kumpulainen, J. (2002). Determination of free and total phenolic acids in plant-derived foods by HPLC with diode-array detection, *J. Agric. Food Chem.*, 50, pp. 3660–3667.

Mertens-Talcott, S. U., Jilma-Stohlawetz, P., Rios, J., Hingorani, L. and Derendorf, H. (2006a). Absorption, metabolism, and antioxidant effects of pomegranate (*Punica granatum* L.) polyphenols after ingestion of a standardized extract in healthy human volunteers, *J. Agric. Food Chem.*, 54, pp. 8956–8961.

Mertens-Talcott, S. U., Lee, J.-H., Percival, S. S. and Talcott, S. T. (2006b). Induction of cell death in Caco-2 human colon carcinoma cells by ellagic acid rich fractions from muscadine grapes (*Vitis rotundifolia*), *J. Agric. Food Chem.*, 54, pp. 5336–5343.

Meyers, K. J., Watkins, C. B., Pritts, M. P. and Liu, R. H. (2003). Antioxidant and antiproliferative activities of strawberries, *J. Agric. Food Chem.*, 51, pp. 6887–6892.

Misikangas, M., Pajari, A.-M., Päivärinta, E., Oikarinen, S. I., Rajakangas, J., Marttinen, M., Tanayama, H., Törrönen, R. and Mutanen, M. (2007). Three Nordic berries inhibit intestinal tumorigenesis in multiple intestinal neoplasia/+ mice by modulating β-catenin signaling in the tumor and transcription in the mucosa, *J. Nutr.*, 137, pp. 2285–2290.

Mullen, W., Yokota, T., Lean, M. E. J. and Crozier, A. (2003). Analysis of ellagitannins and conjugates of ellagic acid and quercetin in raspberry fruits by LC-MS^n, *Phytochemistry*, 64, pp. 617–624.

Määttä-Riihinen, K. R., Kamal-Eldin, A. and Törrönen, A. R. (2004). Identification and quantification of phenolic compounds in berries of *Fragaria* and *Rubus* species (family Rosaceae), *J. Agric. Food Chem.*, 52, pp. 6178–6187.

Narayanan, B. A., Geoffroy, O., Willingham, M. C., Re, G. G. and Nixon, D. W. (1999). p53/p21 (WAF1/CIP1) expression and its possible role in G1 arrest and apoptosis in ellagic acid treated cancer cells, *Cancer Lett.*, 136, pp. 215–221.

Nohynek, L. J., Alakomi, H.-L., Kähkönen, M. P., Heinonen, M., Helander, I. M., Oksman-Caldentey, K.-M. and Puupponen-Pimiä, R. H. (2006). Berry phenolics: antimicrobial properties and mechanisms of action against severe human pathogens, *Nutr. Cancer*, 54, pp. 18–32.

Olsson, M. E., Gustavsson, K.-E., Andersson, S., Nilsson, Å. and Duan, R.-D. (2004). Inhibition of cancer cell proliferation in vitro by fruit and berry extracts and correlations with antioxidant levels, *J. Agric. Food Chem.*, 52, pp. 7264–7271.

Ovaskainen, M.-L., Törrönen, R., Koponen, J. M., Sinkko, H., Hellström, J., Reinivuo, H. and Mattila, P. (2008). Dietary intake and major food sources of polyphenols in Finnish adults, *J. Nutr.*, 138, pp. 562–566.

Pantuck, A. J., Leppert, J. T., Zomorodian, N., Aronson, W., Hong, J. Barnard, R. J., Seeram, N., Liker, H., Wang, H., Elashoff, R., Heber, D., Aviram, M., Ignarro, L. and Belldegrun, A. (2006). Phase II study of pomegranate juice for men with rising prostate-specific antigen following surgery or radiation for prostate cancer, *Clin. Cancer Res.*, 12, pp. 4018–4026.

Puech, J.-L., Feuillat, F. and Mosedale, J. R. (1999). The tannins of oak heartwood: Structure, properties, and their influence on wine flavor, *Am. J. Enol. Vitic.*, 50, pp. 469–478.

Puupponen-Pimiä, R., Nohynek, L., Meier, C., Kähkönen, M., Heinonen, M., Hopia, A. and Oksman-Caldentey, K.-M. (2001). Antimicrobial properties of phenolic compounds from berries, *J. Appl. Microbiol.*, 90, pp. 494–507.

Puupponen-Pimiä, R., Nohynek, L., Hartmann-Schmidlin, S., Kähkönen, M., Heinonen, M., Määttä-Riihinen, K. and Oksman-Caldentey, K.-M. (2005a). Berry phenolics selectively inhibit the growth of intestinal pathogens, *J. Appl. Microbiol.*, 98, pp. 991–1000.

Puupponen-Pimiä, R., Nohynek, L., Alakomi, H.-L. and Oksman-Caldentey, K.-M. (2005b). The action of berry phenolics against human intestinal pathogens, *Biofactors*, 23, pp. 243–251.

Päivärinta, E., Pajari, A.-M., Törrönen, R. and Mutanen, M. (2006). Ellagic acid and natural sources of ellagitannins as possible chemopreventive agents against intestinal tumorigenesis in the Min mouse, *Nutr. Cancer*, 54, pp. 79–83.

Radtke, J., Linseisen, J. and Wolfram, G. (1998). Phenolic acid intake of adults in a Bavarian subgroup of the national food consumption survey, *Z. Ernährungswiss.*, 37, pp. 190–197.

Ramos, S., Alía, M., Bravo, L. and Goya, L. (2005). Comparative effects of food-derived polyphenols on the viability and apoptosis of a human hepatoma cell line (HepG2), *J. Agric. Food Chem.*, 523, pp. 1271–1280.

Reddy, M. K., Gupta, S. K., Jacob, M. R., Khan, S. I. and Ferreira, D. (2007). Antioxidant, antimalarial and antimicrobial activities of tannin-rich fractions, ellagitannins and phenolic acids from *Punica granatum* L., *Planta Med.*, 73, pp. 461–467.

Rosenblat, M., Hayek, T. and Aviram, M. (2006). Anti-oxidative effects of pomegranate juice (PJ) consumption by diabetic patients on serum and on macrophages, *Atherosclerosis*, 187, pp. 363–371.

Ross, H. A., McDougall, G. J. and Stewart, D. (2007). Antiproliferative activity is predominantly associated with ellagitannins in raspberry extracts, *Phytochemistry*, 68, pp. 218–228.

Saucier, C., Jourdes, M., Glories, Y. and Quideau, S. (2006). Extraction, detection, and quantification of flavano-ellagitannins and ethylvescalagin in a Bordeaux red wine aged in oak barrels, *J. Agric. Food Chem.*, 54, pp. 7349–7354.

Scalbert, A., Manach, C., Morand, C. and Rémésy, C. (2005). Dietary polyphenols and the prevention of diseases, *Crit. Rev. Food Sci. Nutr.*, 45, pp. 287–306.

Seeram, N. P., Adams, L. S., Henning, S. M., Niu, Y., Zhang, Y, Nair, M. G. and Heber, D. (2005). In vitro antiproliferative, apoptotic and antioxidant activities of punicalagin, ellagic acid and a total pomegranate tannin extract are enhanced in combination with other polyphenols as found in pomegranate juice, *J. Nutr. Biochem.*, 15, pp. 360–367.

Skupien, K. and Oszmianski, J. (2004). Comparison of six cultivars of strawberries (*Fragaria* x *ananassa* Dutch.) grown in northwest Poland, *Eur. Food Res. Technol.*, 219, pp. 66–70.

Stoner, G. D., Chen, T., Kresty, L. A., Aziz, R. M., Reinemann, T. and Nines, R. (2006). Protection against esophageal cancer in rodents with lyophilized berries: potential mechanisms, *Nutr. Cancer*, 54, pp. 33–46.

Stoner, G. D., Wang, L.-S., Zikri, N., Chen, T., Hecht, S. S., Huang, C., Sardo, C. and Lechner, J. F. (2007). Cancer prevention with freeze-dried berries and berry components, *Semin. Cancer Biol.*, 17, pp. 403–410.

Vrhovsek, U., Palchetti, A., Reniero, F., Guillou, C., Masuero, D. and Mattivi, F. (2006). Concentration and mean degree of polymerization of *Rubus* ellagitannins evaluated by optimized acid methanolysis, *J. Agric. Food Chem.*, 54, pp. 4469–4475.

Wada, L. and Ou, B. (2002). Antioxidant activity and phenolic content of Oregon caneberries, *J. Agric. Food Chem.*, 50, 3495–3500.

Wang, S. Y., Feng, R., Lu, Y., Bowman, L. and Ding, M. (2005). Inhibitory effect on activator protein-1, nuclear factor-kappaB, and cell transformation by extracts of strawberries (*Fragaria* x *ananassa* Dutch.), *J. Agric. Food Chem.*, 53, pp. 4187–4193.

Wu, Q. K., Koponen, J. M., Mykkänen, H. M. and Törrönen, A. R. (2007). Berry phenolic extracts modulate the expression of $p21^{WAF1}$ and Bax but not Bcl-2 in HT-29 colon cancer cells, *J. Agric. Food Chem.*, 55, pp. 1156–1163.

Yi, W., Fischer, J. and Akoh, C. C. (2005). Study of anticancer activities of muscadine grape phenolics in vitro, *J. Agric. Food Chem.*, 53, pp. 8804–8812.

Zafrilla, P., Ferreres, F. and Tomás-Barberán, F. A. (2001). Effect of processing and storage on the antioxidant ellagic acid derivatives and flavonoids of red raspberry (*Rubus idaeus*) jams, *J. Agric. Food Chem.*, 49, pp. 3651–3655.

Chapter 9

C-Glycosidic Ellagitannins and Their Influence on Wine Chemistry

Michaël Jourdes, Dorothée Lefeuvre and Stéphane Quideau*

*Université de Bordeaux, Institut des Sciences Moléculaires (UMR-CNRS 5255), Institut Européen de Chimie et Biologie, 33607 Pessac, France; *corresponding author:* s.quideau@iecb.u-bordeaux.fr

9.1 *C*-Glycosidic Ellagitannins – A Special Subclass of Ellagitannins

9.1.1 Characteristic structural features and natural occurrence

The *C*-glycosidic ellagitannins constitute a subclass of hydrolyzable tannins, whose remarkable structural diversity is discussed in detail in Chapters 1 and 2 of this book. Today, over 500 members of this family of gallic acid-derived polyphenolic natural products have been isolated from various plants and fully characterized (Okuda, 2005, Okuda *et al.*, 1995, Khanbabaee and van Ree, 2001, Quideau and Feldman, 1996, Haslam and Cai, 1994, Schmidt, 1956). Among this myriad of gallic acid metabolites, the *C*-glycosidic ellagitannins present the structural particularity of having a highly characteristic C–C linkage between the carbon-1 atom of an open-chain glucose core and the carbon-2' atom of a galloyl-derived unit esterified to the 2-position of the glucose core. This

C-1-linked galloyl-derived unit is either part of a terarylic nonahydroxyterphenoyl (NHTP) unit (also known as a flavogalloyl group) that is attached *via* three ester bonds to the 2-, 3- and 5-positions of the glucose core, as exemplified in the structures of vescalagin (**1**) and castalagin (**2**), or part of a biarylic variant bridging the 2- and 3-positions of the glucose core, as exemplified in the structures of stachyurin (**3**) and casuarinin (**4**) (Fig. 9.1). The latter biarylic unit is commonly referred to as "HHDP" for hexahydroxydiphenoyl, but we prefer to use the acronym "HHBP" for hexahydroxybiphenoyl to signify more explicitly the biarylic nature of this ellagitannin unit type (Quideau *et al.*, 2004).

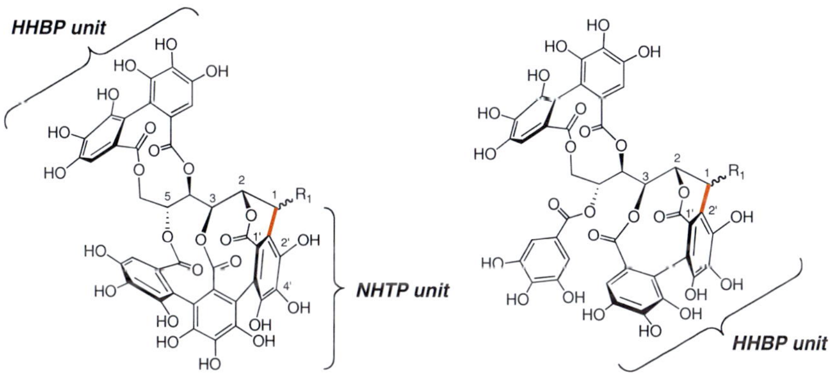

1: R_1 = β-OH, vescalagin
2: R_1 = α-OH, castalagin

3: R_1 = β-OH, stachyurin
4: R_1 = α-OH, casuarinin

Fig. 9.1 Typical examples of monomeric *C*-glycosidic ellagitannins.

Ellagitannins of the *C*-glycosidic type have been found in various plant species from the *Hamamelidaceae*, *Fagaceae*, *Betulaceae*, *Casuarinaceae*, *Juglandaceae* and *Rhoipteleaceae* families in the *Hamamelidae* subclass, from the *Rosaceae*, *Combretaceae*, *Lythraceae*, *Melastomataceae*, *Myrtaceae*, *Punicaceae*, *Trapaceae* and *Elaeagnaceae* families in the *Rosidae* subclass, and from the *Theaceae* and *Stachyuraceae* families in the *Dilleniidae* subclass (Okuda *et al.*, 1993, 2000, Okuda, 2005).

9.1.1.1 *Oak and chestnut C-glycosidic ellagitannins*

Vescalagin (**1**) and its C-1 epimer castalagin (**2**) are the first *C*-glycosidic ellagitannins that have been investigated after their isolation thirty years ago from *Castanea* (chestnut) and *Quercus* (oak) woody species of the *Fagaceae* family by Mayer and co-workers (Mayer *et al.*, 1967, 1969, 1971a). Their structures, as well as those of their 2,3-HHBP-bearing analogues, stachyurin (**3**) and casuarinin (**4**) (Okuda *et al.*, 1981, 1982, 1983), were fully determined much later after revision of their respective configuration at C-1 by the Nishioka's group (Nonaka *et al.*, 1990). The combined amount of these two epimers in the heartwood of oak species such as *Quercus petraea* and *Quercus robur* has been evaluated to vary from *ca.* 3 to 57 mg/g of dry wood, depending on the species, age and sampling position in the tree (Mosedale *et al.*, 2001, Fernández de Simón *et al.*, 1999, Vivas *et al.*, 1996, Masson *et al.*, 1995, 1996, Viriot *et al.*, 1994). The heartwood of chestnut species such as *Castanea sativa* can even contain up to 63 mg of vescalagin (**1**) and castalagin (**2**) per gram of dry wood, *i.e.*, more than 6% (!) by weight of dry material (Viriot *et al.*, 1994). Six other NHTP-containing *C*-glycosidic ellagitannins were later isolated from fagaceous *Quercus* and *Castanea* hardwood species, *i.e.*, the dimers roburins A (**5**) and D (**6**), and the lyxose/xylose-bearing monomers grandinin (**7**) and roburin E (**8**) and dimers roburins B (**9**) and C (**10**) (Fig. 9.2) (Nonaka *et al.*, 1989, Hervé du Penhoat *et al.*, 1991a/b). Among these eight typical NHTP-containing *C*-glycosidic ellagitannins, vescalagin (**1**) and castalagin (**2**) largely predominate in the fagaceous woody species containing them, representing for example between 40% and about 60% by weight of this group of ellagitannins in *Quercus petraea* and *robur* heartwoods (Fernández de Simón *et al.*, 1999, Masson *et al.*, 1994, 1995, 1996).

Many additional *C*-glycosidic ellagitannins, including monomers, oligomers and complex tannins (*vide infra*), have been identified over the years from species belonging to the selection of plant families mentioned above (for a compendium of identified structures, see Section 1.3.3 in Chapter 1 and Sections 2.1.4, 2.1.5 and 2.2.4 in Chapter 2, see also Haslam and Cai, 1994). A large majority of these structurally complex natural products (including all types of ellagitannins) has been identified thanks to the outstanding efforts of Japanese researchers, in particular from the Okayama, Kyushu and Nagazaki schools, during their more than thirty-year (still going on) systematic search of active principles in plant species used in oriental medicines.

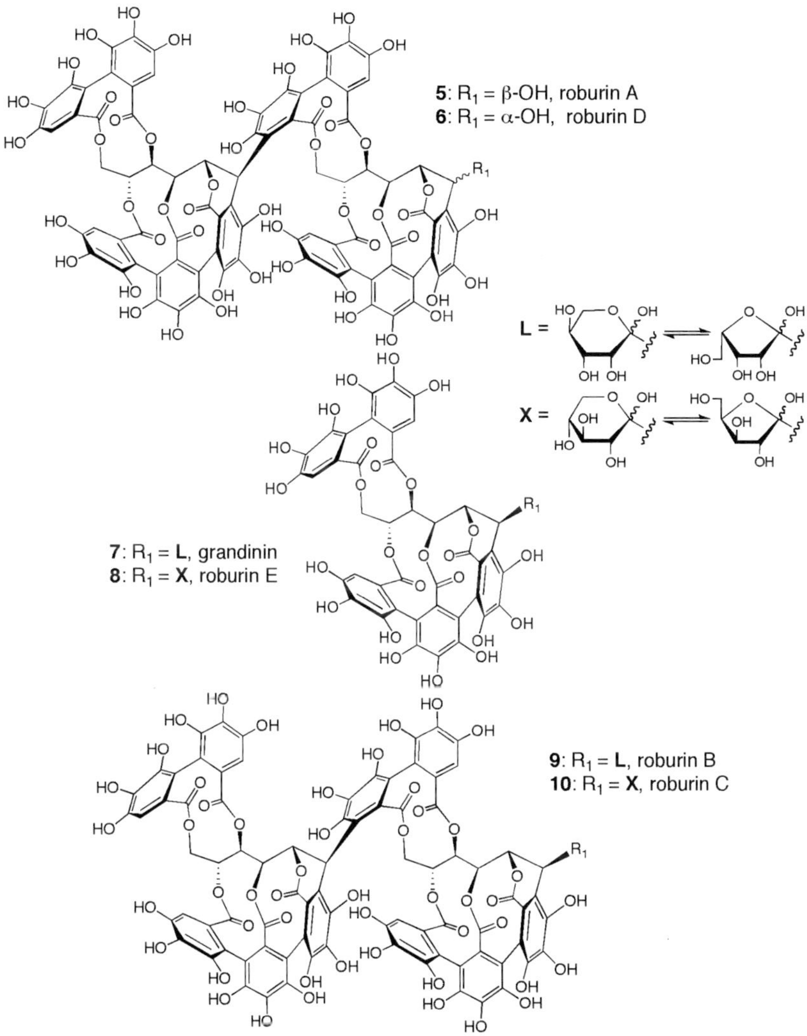

Fig. 9.2 Together with vescalagin and castalagin (see Fig. 9.1), these compounds are the major NHTP-bearing *C*-glycosidic ellagitannins found in fagaceous hardwood species.

To date the largest identified oligomeric *C*-glycosidic ellagitannin is a vescalagin/castalagin-based pentamer, called castaneanin D (**11**, Fig. 9.3), which has been isolated (and characterized) by Kouno, Tanaka and co-workers (Tanaka *et al.*, 1996) from the heartwood of the Japanese chestnut tree (*i.e.*, *Castanea crenata* SIEB. *et* ZUCC., Japanese name: Kuri).

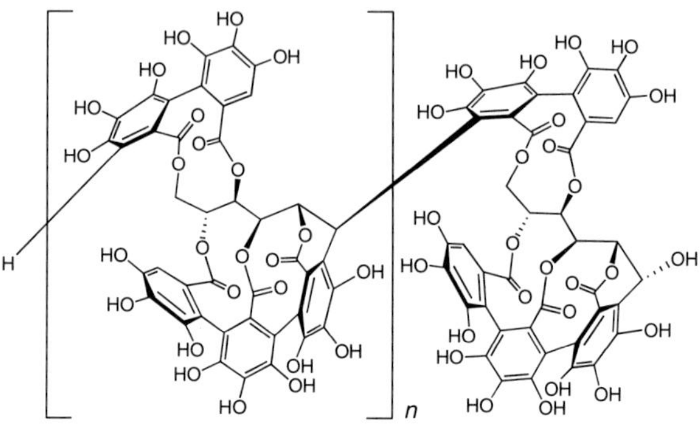

11: castaneanin D (*n* = 4)

Fig. 9.3 Castaneanin D, the largest oligomeric *C*-glycosidic ellagitannin known to date.

9.1.1.2 *Complex C-glycosidic ellagitannins*

The *C*-glycosidic ellagitannin subclass also encompasses so-called complex tannins, which are structural hybrids composed, in their simplest variations, of a *C*-glycosidic ellagitannin moiety derived for example from the monomers vescalagin (**1**) or stachyurin (**3**) and a flavan-3-ol unit such as catechin or epicatechin. In these complex tannins, both parts are connected via a C–C linkage between the carbon-1 center of the open-chain glucose core of the ellagitannin moiety and either the carbon-8 or the carbon-6 center of the ring-A of the flavan-3-ol unit. Depending on the nature of each moiety, the regiochemistry of attachment to each other, and the type of bond connectivities through which each moiety can lead to oligomeric variants, complex tannins further contribute to the ellagitannin structural diversity. Their natural occurrence appears to be limited to plant species of the families *Fagaceae*, *Combretaceae*, *Myrtaceae*, *Theaceae* and *Melastomataceae* (Yoshida *et al.*, 1992).

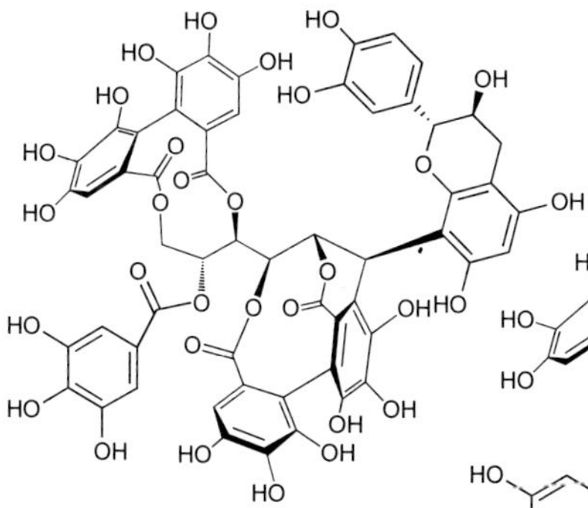

12: stenophyllanin A

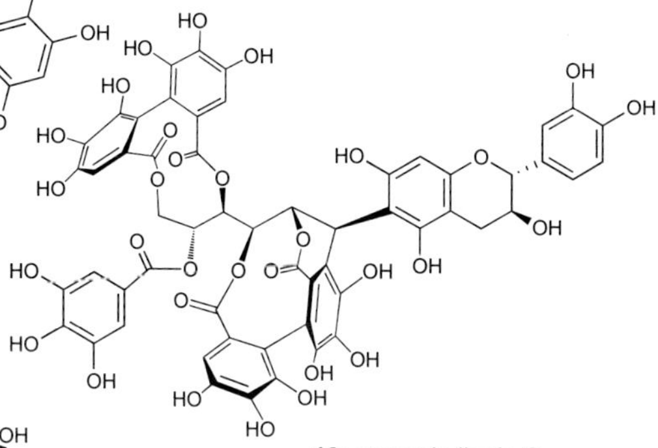

13: stenophyllanin B

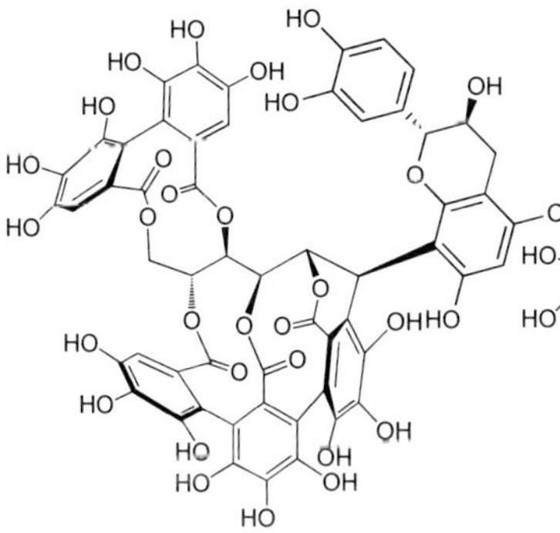

14: acutissimin A

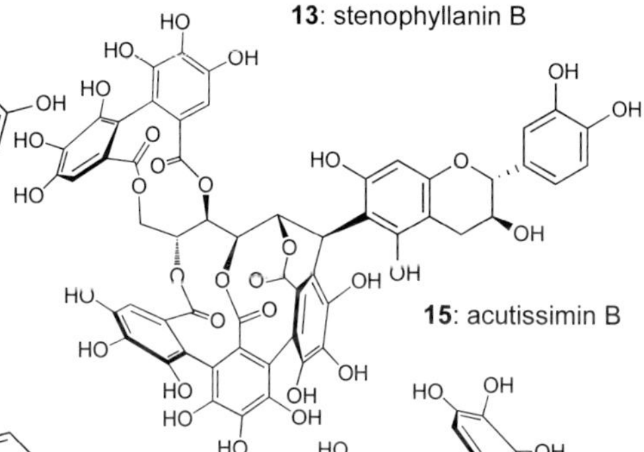

15: acutissimin B

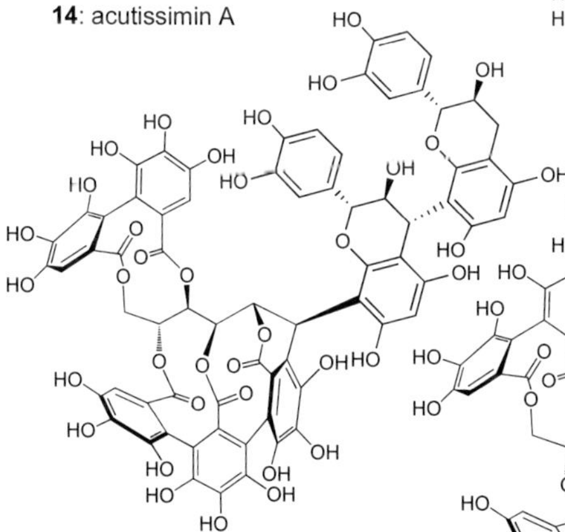

16: mongolicanin

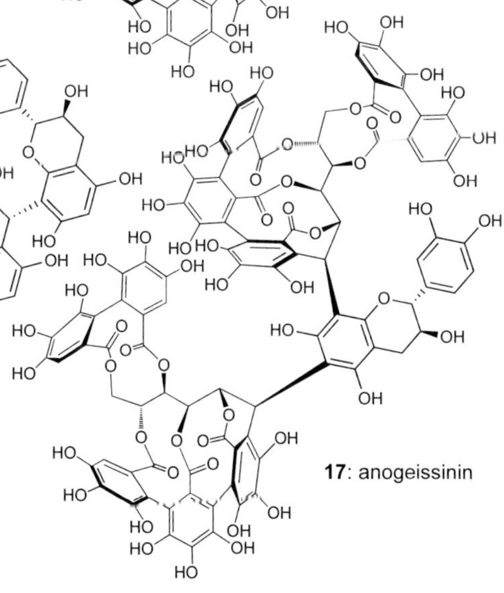

17: anogeissinin

Fig. 9.4 Typical examples of flavano-ellagitannins (complex tannins).

Typical examples of these flavanoid/ellagitannin hybrids (Fig. 9.4) are the catechin/stachyurin-based regioisomeric stenophyllanins A (**12**) and B (**13**) isolated from *Quercus stenophylla* MAKINO (Japanese name: Urajirogashi) (Nonaka *et al.*, 1985, 1990), the catechin/vescalagin-based regioisomeric acutissimins A (**14**) and B (**15**) isolated from *Quercus* and *Castanea* species, including the Japanese "sawtooth" oak tree *Quercus acutissima* CARRUTH (Japanese name: Kunugi) (Ishimaru *et al.*, 1987, Nonaka *et al.*, 1990), the catechin/vescalagin-based procyanidino-ellagitannin mongolicanin (**16**), which features the catechin dimer procyanidin B-3 as its flavan-3-ol-derived unit and which was isolated from the bark of *Quercus mongolica* var. *grosseserrata* (Japanese name: mizunara) (Ishimaru *et al.*, 1988), and the catechin/vescalagin-based anogeissinin (**17**), which features two vescalagin-derived moieties connected to the C-6 and C-8 centers of catechin and which was isolated from the bark of *Anogeissus acuminata* var. *lanceolata*, a combretaceous plant species largely distributed in South East Asia (Lin *et al.*, 1991). The camelliatannins A and B and malabathrin A (see Section 1.3.3 in Chapter 1) are examples of *C*-glycosidic flavano-ellagitannins featuring epicatechin-type flavanol moieties connected to the C-1 center of the open-chain glucose core of either stachyurin (**3**) or its 5-*O*-desgalloylated variant, desgalloylstachyurin (see Section 2.1.4 in Chapter 2).

The inherent chemical reactivity of flavano-ellagitannins renders them sensitive to additional structural transformations upon oxidation. Mongolicains A (**18**) and B (**19**) (Fig. 9.5) are two examples illustrating the consequences of such a rather unique propensity of flavanol-bearing ellagitannins to participate in further oxidation-driven reactions. These two complex tannins, first isolated (and characterized) from five *Quercus* and one *Castanopsis* species by the Nishioka's group (Nonaka *et al.*, 1988), feature a *spiro*-linked dihydrofuran-cyclopentenone motif. The formation of this motif can conceivably derive from a ring contraction of the C-1-linked O-2-galloyl part of a former NHTP unit with concomitant decarboxylation and nucleophilic attack of one of the phenolic hydroxyl groups of the A-ring of a C-1-linked catechin moiety. Hence, mongolicains A (**18**) and B (**19**) can be considered as direct metabolites of acutissimins A (**14**) and B (**15**), respectively, as first suggested by Nishioka and co-workers (Nonaka *et al.*, 1988). This proposal is further supported by the fact that these four substances invariably co-exist in their fagaceous plant sources. The initiation step of the conversion of acutissimins into mongolicains would be the dehydrogenation of the pyrogallol motif of the NHTP C-1-linked galloyl unit into an α-hydroxy-*ortho*-quinone/cyclohexenetrione system (see Section 4.3.5 in Chapter 4 and Section 9.3.4 below).

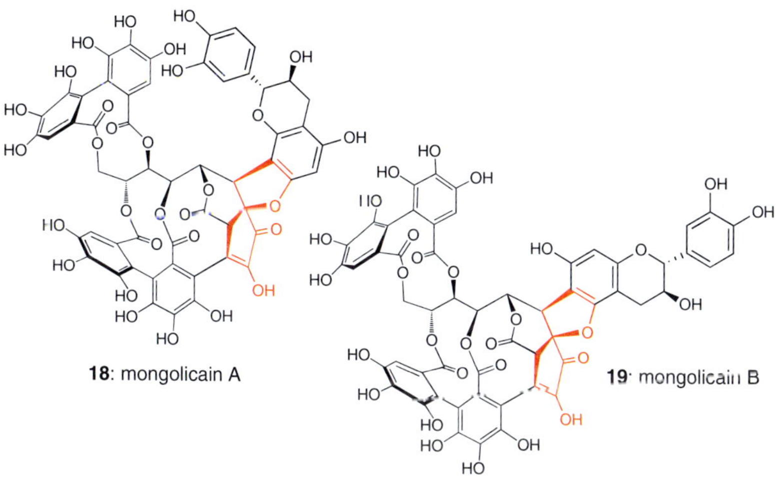

18: mongolicain A

19: mongolicain B

Fig. 9.5 Examples of flavano-ellagitannins featuring a *spiro*-linked dihydrofuran-cyclopentenone motif.

The structures of other examples of analogous flavano-ellagitannins putatively also derived from 2,3,5-NHTP- as well as 2,3-HHBP-bearing precursors are displayed in Section 1.3.3 of Chapter 1, Section 2.1.5 of Chapter 2 and Section 4.3.5 of Chapter 4, *e.g.*, psidinins A and B, guavins A and C, camelliatannin F, malabathrin E, and psiguavin.

9.1.2 Biogenetic construction – What's known !?

The biosynthetic steps leading to glucopyranosic ellagitannins from their gallotannin precursor, β-pentagalloylglucopyranose (β-PGG), are just today starting to get elucidated, essentially thanks to the work initiated by Gross and co-workers several years ago (see Chapter 3). A β-pentagalloylglucopyranose oxidizing enzyme has been found to promote the formation of the 4,6-HHBP-containing tellimagrandin II and a different laccase-type phenolase has been found to dimerize tellimagrandin II into cornussiin E by promoting the formation of a diaryl ether valoneoyl bridge between the two monomers (Niemetz and Gross, 2003a/b). This is all we know to date about enzymatic implications in the construction of ellagitannins. One have to admit that the task is enormous, considering the formidable structural diversity expressed by ellagitannins simply derived for the most part from glucose and gallic acid. No enzyme study has yet been undertaken on the steps further downhill the ellagitannin biosynthetic pathway and leading to the *C*-glycosidic ellagitannins. However, the identification of several structurally connected members of this ellagitannin subclass certainly allows for sound working hypotheses. Whether or not the elaboration of the *C*-glycosidic ellagitannin diversity is under a strict enzymatic control or is attributable to a combinatorial-type synthesis solely relying on the inherent chemical reactivity of these natural products, one have to keep in mind that the occurrence of ellagitannins harboring the characteristic *C*-glycosidic bond appears to be limited to species from a rather short selection of plant families (*vide supra*) borne out by a long evolutionary process since the colonization of land by plants (around 400 million years ago). So the truth probably lies in between the two aforementioned extremes; some plant species might have learnt to develop enzymes to ensure the formation of key bonds leading to specific structural subtypes, while still letting chemistry play its role in expanding diversity further within a given structural subtype.

The biochemical event that mediates the passage from the glucopyranosic ellagitannin class to the open-chain *C*-glycosidic ellagitannin class admitedly remains a matter of speculation, but interesting observations have nevertheless been made from the study of a plant species in which members of the two classes co-exist (Hatano *et al.*, 1986). Early spring leaves of *Liquidambar formosana* (*Hamamelidaceae*), a Chinese medicinal plant, contain high amounts of tellimagrandin II (**20**). Analysis of the same leaf material in summer indicated negligible amounts of **20**, whereas two other ellagitannins, pedunculagin (**23**) and the *C*-glycosidic casuarinin (**4**) became predominant throughout summer until autumn. Three other related compounds, tellimagrandin I (**21**), casuarictin (**22**) and the *C*-glycosidic casuariin (**24**) (Okuda *et al.*, 1983), were also isolated from the leaves collected in autumn (see photo insert, *Liquidambar formosana* in November, Japan). This seasonal variation of the ellagitannin content in *Liquidambar formosana* was proposed to reflect the following biogenetic filiation (Fig. 9.6): from tellimagrandin II (**20**) to pedunculagin (**23**) either via 1-*O*-degalloylation into tellimagrandin I (**21**) followed by oxidative C–C coupling between the O-2 and O-3 galloyl groups, or via a similar coupling to give first rise to casuarictin (**22**) followed by 1-*O*-degalloylation, and hence, from pedunculagin (**23**) to casuarinin (**4**) via casuariin (**24**) (Hatano *et al.*, 1986). This filiation would imply that the formation of the *C*-glycosidic bond precedes galloylation at the O-5 center, which is unveiled as a consequence of the glucopyranose ring opening.

Another question still remained concerning the timing of this ring-opening event from pedunculagin (**23**). Does this event precede or concomitantly follow the formation of the *C*-glycosidic bond. Further investigations on *Liquidambar formosana* enabled the Okuda's group to find one important piece missing in this puzzle. They isolated (and characterized) the open-chain aldehyde liquidambin (**25**), together with its hydrated form (Okuda *et al.*, 1987). One can thus assumed that an opening of the pedunculagin (**23**) glucopyranose ring, driven by a (fast) 5-*O*-galloylation, is the first key event that opens up the door to an access to *C*-glycosidic ellagitannins (Fig. 9.6). Then follows the intramolecular aldol-type nucleophilic addition of the 2,3-HHBP unit to the aldehyde function of the resulting liquidambin (**25**), which gives rise to the formation of the *C*-glycosidic bond such as in stachyurin (**3**) and its C-1 epimer casuarinin (**4**) (Okuda *et al.*, 1981, 1982, 1983, see also Nonaka *et al.*, 1990). Their 5-*O*-degalloylation could then give rise to the formation of 5-desgalloylstachyurin (**26**) (Lee *et al.*, 1990) and casuariin (**24**), respectively, and oxidative coupling between their 5-galloyl and 2,3-HHBP groups would in parallel furnish vescalagin (**1**) and castalagin (**2**) (Fig. 9.6) (Quideau *et al.*, 2004).

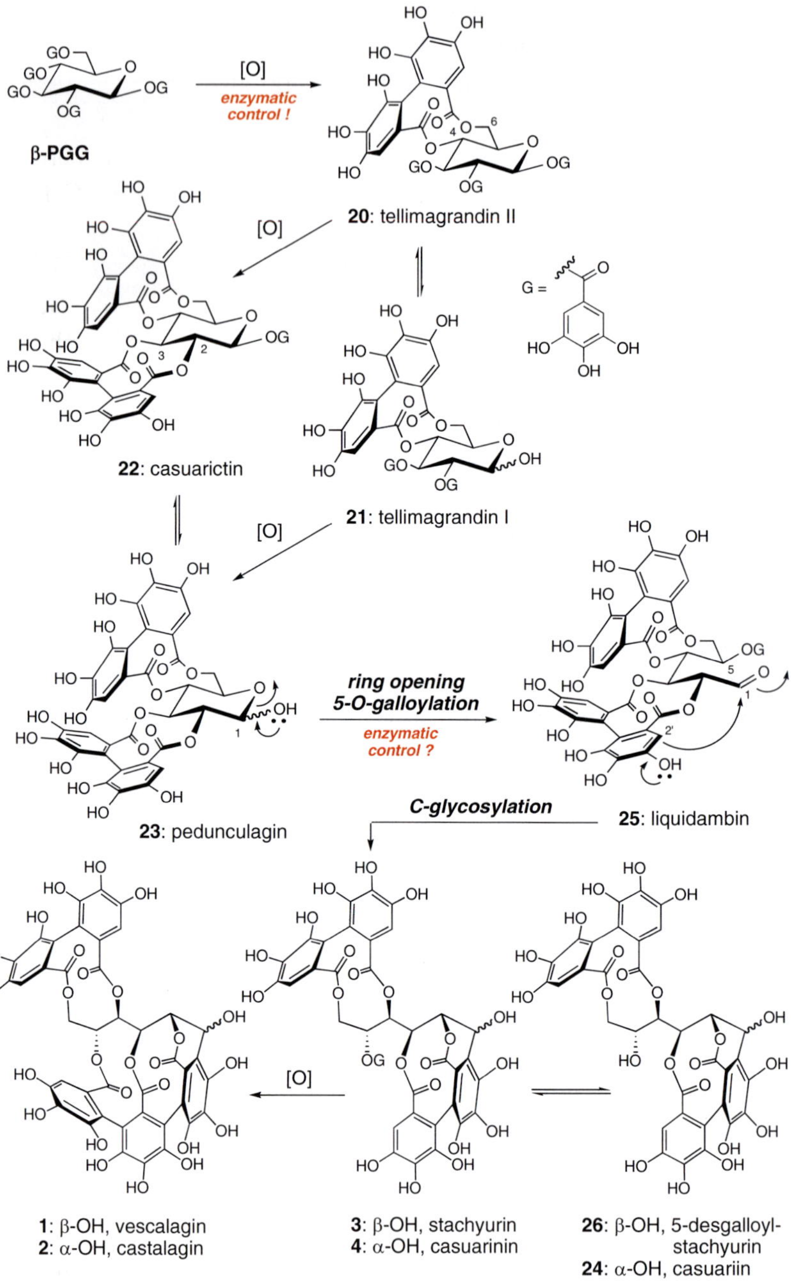

Fig. 9.6 Putative biosynthetic pathway to *C*-glycosidic ellagitannins.

Liquidambin (**25**) would thus constitute a molecular quoin for the passage from glucopyranosic to *C*-glycosidic ellagitannins. As alluded to above, a fast, probably enzymatically-controlled, galloylation at the O-5 center of the opened form of pedunculagin (**23**) would be the driving force towards this passage by maintaining open the glucose core while reinforcing the electrophilic character and accessibility of its C-1 center under the form of an sp^2-aldehydic type carbon.

Interestingly, Tanaka and colleagues managed to chemically convert pedunculagin (**23**) into casuariin (**24**, 6%) and 5-desgalloylstachyurin (**26**, 34%) by simply heating a solution of **23** in a pH 7.5 phosphate buffer (Tanaka *et al.*, 1993). This non-enzymatic transformation may also take place in some plants, but others, like *Liquidambar formosana*, seem to have, for some reasons, evolved to better handle access to *C*-glycosidic ellagitannins. Such plants might be equipped with a "trained" 5-*O*-galloyltransferase in order to biochemically forge the open-chain aldehydic liquidambin (**25**), whose enhanced electrophilicity at C-1 could then facilitate the chemical establishment of the *C*-glycosidic bond via nucleophilic addition. Of course, these thoughts are only speculative, but they could constitute the basis of working hypotheses for future studies aimed at tracking down the enzymes that might control the genesis of *C*-glycosidic ellagitannins.

In any event, once monomeric *C*-glycosidic ellagitannins such as the two epimeric pairs vescalagin/castalagin (**1**/**2**) and stachyurin/casuarinin (**3**/**4**) are available, they serve as precursors for numerous other *C*-glycosidic species. For example, ellagitannins such as the roburins A-E and grandinin (**5-10**, see Fig. 9.2), the castaneanins A-D oligomeric series (Tanaka *et al.*, 1996) exemplified in Fig. 9.3 by the pentamer castaneanin D (**11**) and the flavano-ellagitannins **12-17** (Fig. 9.4) all derive from **1** and/or **2** or **3** and/or **4**. Furthermore, all of these compounds derive from a single chemical reaction type, that is the substitution of the OH group at C-1 of their precursor by another (nucleophilic) entity, which can be their precursor itself as in the case of an oligomerization [*e.g.*, roburins A/D (**5**/**6**) and castaneanins A-D] or a sugar unit like D-lyxose or D-xylose as in the case of the formation of grandinin (**7**) and roburin E (**8**) or a flavanol moiety like catechin or procyanidin B-3 as in the case of the formation of complex tannins such as **12-17**. Here, one can ask again the question of whether the generation of each of these compounds is under enzymatic control. Probably not, for the departure of the OH group from the C-1 position of the monomers **1/2** or **3/4**, which could be chemically promoted under acid catalysis, would result in the formation of a stable benzylic cation intermediate **27** or **28** that could then be quenched by various nucleophilic species such as the ones mentioned above and conceivably others (Fig. 9.7).

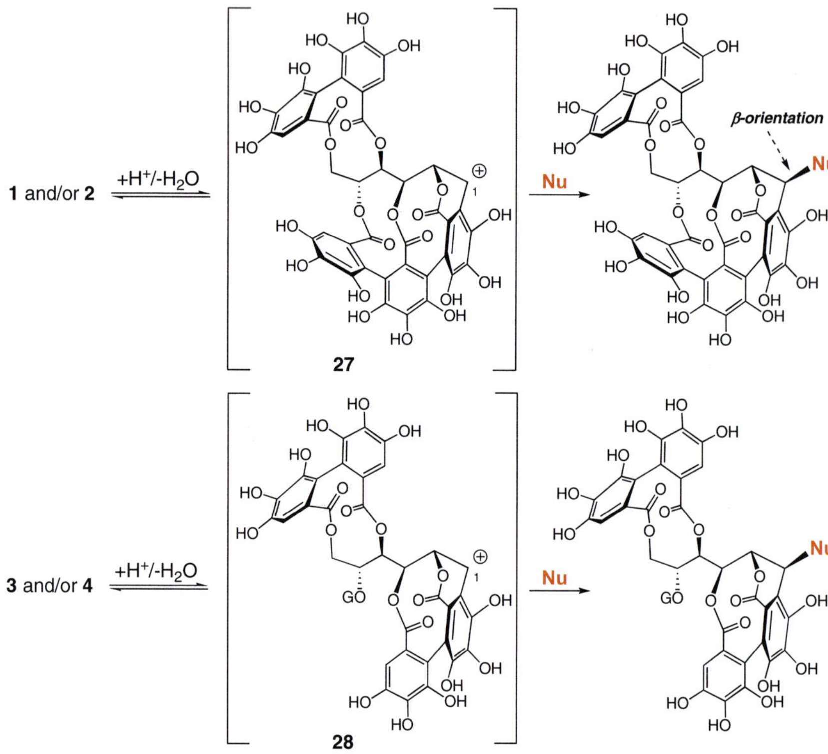

Fig. 9.7 Acid-catalyzed condensation reactions of *C*-glycosidic ellagitannins.

This chemistry taking place at the C-1 position of *C*-glycosidic ellagitannins featuring a free benzylic alcohol function at that position, *i.e.*, not only monomers such as **1/2** or **3/4** but also oligomers like the roburins A/D (**5/6**), is characterized by another intriguing apparent specificity. Nucleophilic substitution reactions proceed in a strict stereoselective manner, with the newly formed C–C bond at C-1 being always β-oriented. Indeed, this stereochemical preference is observed in all known related compounds thus derived from vescalagin/castalagin (**1**/**2**) or stachyurin/casuarinin (**3**/**4**), including all of the flavano-ellagitannins identified to date (Fig. 9.7). Such a selectivity can be argued to result from an enzymatic control, but computer-aided molecular modeling indicates that it can also strictly rely on the stereoelectronic properties of a benzylic cation intermediate such as **27** (Quideau *et al.*, 2003, 2004, 2005). This rationale is discussed in greater detail in Section 9.2.1.2.

9.2 Vescalagin *versus* Castalagin

9.2.1 Chemical reactivity dichotomy

Vescalagin (**1**) and castalagin (**2**), despite their quasi-identical structure, express drastic differences in chemical reactivity, as well as in biological activity (Quideau *et al.*, 2004, 2005). These differences are really striking when one considers that the only structural difference between these two epimers of relatively high molecular mass (*i.e.*, 934 Da) is the orientation of their small OH group at C-1. When we started to study their chemistry, we soon realized that condensation reactions under mild acid-catalyzed nucleophilic substitution conditions were possible only with vescalagin (**1**) but not with castalagin (**2**). Such a refractory behavior of **2** had been previously documented but not fully rationalized (Yoshida *et al.*, 1991, Viriot *et al.*, 1994, Vivas *et al.*, 1995, 2004). Moreover, the more reactive vescalagin (**1**) participates in nucleophilic substitution reactions with full retention of configuration at C-1. If the formation of a stable benzylic cation intermediate such as **27** is involved in such condensation reactions as suggested above (see Fig. 9.7), one might have expected mixtures of diastereomeric products via an S_N1-type mechanism. But no! Starting from vescalagin (**1**), the condensation products we obtained exclusively display a β-orientation of the newly formed bond at C-1, like in all of the related *C*-glycosidic ellagitannins isolated to date (Quideau *et al.*, 2003, 2005). Intrigued by the reactivity dichotomy of **1** *versus* **2** and the diastereoselectivity expressed from **1**, we decided to examine by molecular modeling the fine structural characterictics of each epimer and the mechanistic implications of the benzylic cation intermediate that could derive from them.

9.2.1.1 *Refractory chemical behavior of castalagin*

The first chemical reaction we investigated from vescalagin/castalagin (**1**/**2**) was their condensation with (+)-catechin in order to hemisynthesize the acutissimins A/B (**14**/**15**, see Fig. 9.4). All attempts to form these flavano-ellagitannins using **2** instead of **1** under similar conditions (*i.e.*, 1.5% (v/v) TFA/THF, 60 °C, 5 h, see Section 9.2.2) were unsuccessful. To the best of our knowledge, the only chemical hemisynthesis of an acutissimin from castalagin (**2**) was achieved using (+)-catechin under more drastic conditions in refluxing anhydrous dioxane in the presence of *p*-toluenesulfonic acid for 24 h. Only acutissimin A (**14**) could be isolated in less than 4% yield (Ishimaru *et al.*, 1987).

Molecular-mechanics calculations performed using Macromodel (MM3* force field) indicated that the minimum-energy conformer of castalagin (**2**, 501.6 kJ/mol) was slightly more stable than that of **1** (504.5 kJ/mol) (Quideau *et al.*, 2004). Admitedly, this energy difference of 2.9 kJ (*ca.* 0.7 kcal) is rather slim and the reactivity differences observed between **1** and **2** cannot solely rely upon this figure. A closer examination of the minimum-energy conformations reveals that the β-oriented OH-1 group of **1** is *exo*-located relatively to the most crowded face of the molecule, whereas the α-oriented OH-1 group in **2** is *endo*-positioned (Fig. 9.8). As first observed by Okuda and co-workers (Yoshida *et al.*, 1991), this *endo*-positioning of the α-oriented OH-1 group in **2** renders it more available to participate in an intramolecular (stabilizing) hydrogen-bond between its oxygen atom and the hydrogen atom of the phenolic OH-3' group of the galloyl-derived I-ring of the NHTP unit (Fig. 9.8). This H-bond of 2.21 Å with an O-1···H–O-3' angle of 146° may be invoked to suggest that the basicity of the O-1 atom in **2** is consequently lower than that of the same oxygen atom in **1**, hence rendering it less prone to protonation under mild acidic conditions. Furthermore, departure of a protonated OH-1 group may be energetically favored from **1**, since this OH group is rather axially oriented on the six-membered ring lactone defined by the double connection of the galloyl-derived NHTP I-ring to the O-2 and C-1 positions of the open-chain glucose core. In **2**, this OH-1 group adopts a less energetically demanding equatorial orientation (Quideau *et al.*, 2005, Haslam, 1998), thus also contributing to the lower overall strain energy of castalagin (**2**) *versus* vescalagin (**1**).

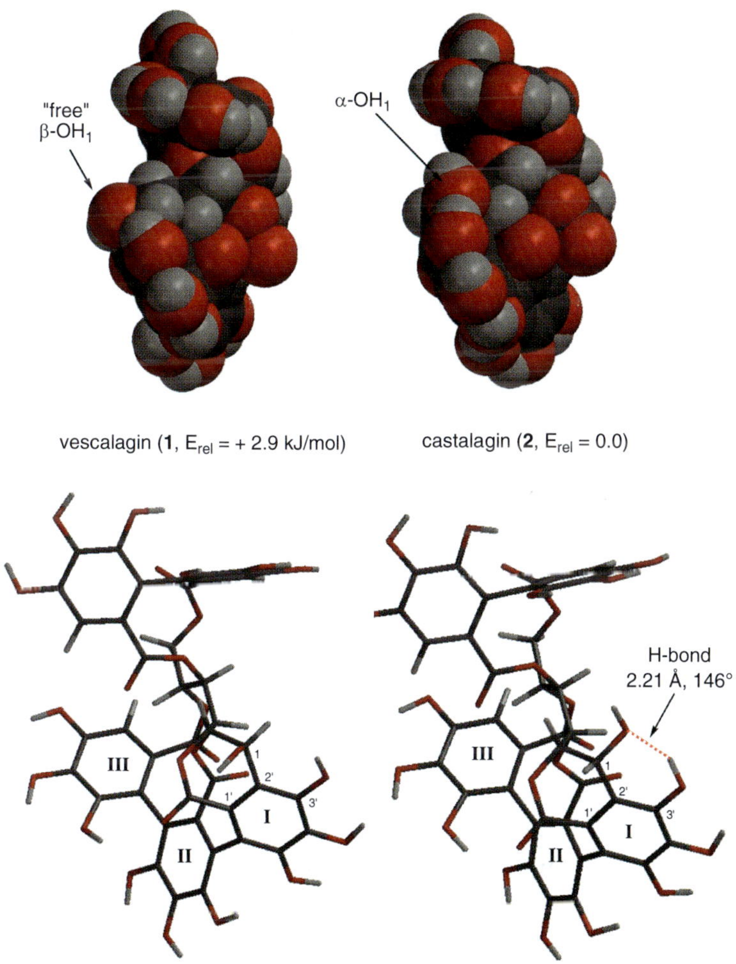

Fig. 9.8 MM3* minimum-energy conformations of vescalagin (**1**) and castalagin (**2**).

Interestingly, vescalagin (**1**) is almost systematically found in lower amounts than castalagin (**2**) in plant extracts. This intriguing observation could be a consequence of the lack of chemical reactivity at C-1 of castalagin (**2**). Thus, on the sole basis of this chemical reactivity difference between **1** and **2**, one can argue that vescalagin (**1**) is the preferred precursor of the *in vivo* formation of *C*-glycosidic ellagitannin oligomers and C-1 conjugates such as the flavano-ellagitannins and the lyxose/xylose-bearing conjugates. For example, considering the structures of the *C*-glycosidic ellagitannins typically found in significant amounts in fagaceous wood species (Fig. 9.2), grandinin and the roburins A-E would be, all six of them, derived from a nucleophilic substitution at C-1 of vescalagin (**1**). Overall, the construction of these six compounds would require nine equivalents of **1** for only one equivalent of castalagin (**2**), which would be involved in the formation of roburin D (**6**) via a nucleophilic attack of its 4,6-HHBP group onto the C-1 position of a vescalagin unit.

9.2.1.2 *Diastereofacial differentiation of the vescalagin-derived benzylic cation*

Starting from vescalagin (**1**), all of the condensation reactions we performed using various carbon-, oxygen- and sulfur-based nucleophiles occurred with retention of configuration at C-1, *i.e.*, the newly formed bond is still β-oriented (see Sections 9.2.2 and 9.3). This β-orientation is exclusively observed in all of the *C*-glycosidic ellagitannin-derived oligomers and conjugates, including flavano-ellagitannins, which have been so far isolated from plants. Since chemical reactions thus appear to follow the same stereochemistry path than biochemical transformations, the rationale for such a high diastereoselectivity cannot be invoked to rely upon an enzymatic control. Moreover, as stated above, this diastereoselectivity can at first appear surprising if one claims a passage via a benzylic cation intermediate such as **27** in the context of an S_N1-type mechanism (Fig. 9.7). However, an examination of **27** by computational means has revealed stereoelectronic effects that can explain the diastereofacial selection imposed on its electrophilic C-1 cationic center.

Calculations of the lowest-energy unoccupied molecular orbital (LUMO) of **27** showed a large and symmetrical sp^2-hybridized atomic p-orbital at C-1. Access of a nucleophile to C-1 *via* the α-face of this vacant cationic p-orbital is probably hindered by the steric encumbrance of the *endo*-face of the molecule. Most significant was the observation of a larger electron-deficiency on the "reacting" *exo* β-face of the orbital by mapping the LUMO onto the electron density surface of **27** (Fig. 9.9).

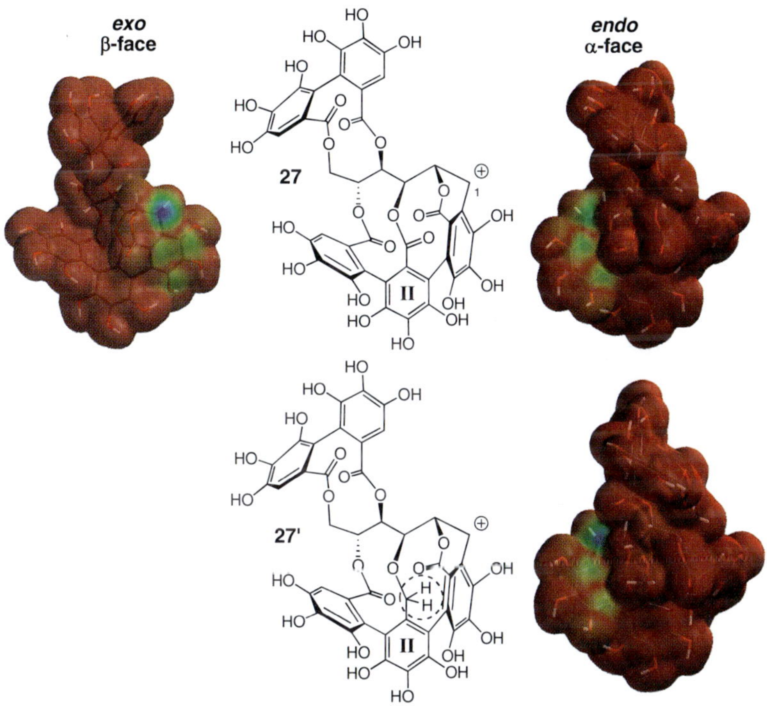

Fig. 9.9 Facial views of mapping of Spartan-generated Hartree–Fock models of the LUMO of vescalagin-derived benzylic cations **27** and **27'** onto their corresponding 0.002 electron au^{-3} electron density isosurfaces. Replacement of the carbonyl group of the NHTP galloyl-derived II-ring of **27** by a methylene unit (see dotted circle on structure **27'**) results in a recovery of the elecron-deficiency of the *endo* α-face of the vacant cationic p-orbital at C-1. According to the color code used, the bluer the color, the more electron-deficient the orbital is.

This observation suggested that the cationic p-orbital must be under the influence of an adequately α-positioned electron-donating atom. A closer examination of the structure of **27** revealed that the carbonyl oxygen atom of the galloyl-derived II-group of its NHTP unit seems properly located to get involved through one of its lone pairs in such electronic interactions. The possibility of such an intramolecular electronic "quench" of the α-face of the cationic p-orbital was further supported by the observation of a significant recovery of its electron-deficiency after having replaced *in silico* the supposedly influential carbonyl group of **27** by a methylene (–CH_2–) unit to give the cation **27'** (see Fig. 9.9) (Quideau *et al.*, 2005).

Thus, in **27**, the *exo* β-face of the cationic C-1 center is not only more accessible but also exhibits a larger electron-deficiency, hence the diastereoselectivity. Even though this proposal is only supported by the results of computer-aided molecular modeling, it nevertheless provides us with an illustrating, if not compelling, understanding of stereoelectronic factors that can control the remarkable and non-enzymatic diastereofacial selection observed during nucleophilic substitution reactions at the C-1 position of vescalagin (**1**).

9.2.2 Hemisynthesis of flavano-ellagitannins – The acutissimins story

The acutissimins A (**14**) and B (**15**) (Fig. 9.4), two regioisomeric flavano-ellagitannins that have been isolated from the bark and/or leaves of various oak and chestnut species (*Fagaceae*) such as Japanese *Quercus acutissima* (Kunugi), *Quercus miyagii* (Okinawa-urajirogashi), *Quercus stenophylla* (Urajirogashi), *Quercus mongolica* var. *grosseserrata* (Mizunara) and *Castanea crenata* (Kuri), European *Quercus petraea* (*Q. sessiliflora*) and *Castanea sativa*, and Algerian *Quercus suber* and *Quercus coccifera* (Ishimaru *et al.*, 1987, Nonaka *et al.*, 1990, König *et al.*, 1994, Lampire *et al.*, 1998, Ito *et al.*, 2002), are examples of complex tannins that can derive from an acid-catalyzed nucleophilic substitution, as described above, between vescalagin (**1**) and (+)-catechin. The flavanol unit, (+)-catechin, ends up connected to the C-

1 center of **1** via either its nucleophilic C-8 center, leading to acutissimin A (**14**), or its C-6 center, leading to acutissimin B (**15**).

Fagaceous woody species are generally very resistant to pathogens and herbivores. In particular, *Quercus acutissima*, from which the acutissimins A and B were first isolated, is usually pest-free and disease-free. Of particular curiosity is the fact that this robust oak species, native to Japan, has been selected as an ornamental tree in North America urban areas where air pollution, poor drainage and/or drought are common. Plant extracts containing acutissimins are used worldwide in folk medicines to treat diseases including gastritis and gastric ulcer (Khennouf *et al.*, 2003), diarrhea, various inflammations (*e.g.*, oral, genital and anal mucosa, skin) (König *et al.*, 1994), and are used as tonic and antitussive medicines (see Section 1.5 in Chapter 1). Kashiwada and co-workers were first to report the extremely potent *in vitro* activity of acutissimin A against human DNA topoisomerase II (Kashiwada *et al.*, 1993). We later confirmed their results in the course of our investigations on bioactive wine-related polyphenols (Quideau *et al.*, 2005).

Indeed, our first incentive to study these complex tannins raised from the fact that they are present in the bark of oak species commonly used in France to make barrels, such as *Quercus robur* and *Quercus petraea* (König *et al.*, 1994), and in which fine wines are aged. However, the bark is removed from oak tree logs, and the heartwood part from which the staves are cut to manufacture barrels (Mosedale and Puech, 1998) does not contain any acutissimin. Nevertheless, the heartwood of the aforementioned oak species used for cooperage is a rich source of *C*-glycosidic ellagitannins (*i.e.*, up to 10% by weight of dry wood, Peng *et al.*, 1991), and notably of vescalagin (**1**, see Section 9.1.1). Even though the long-term seasoning, or drying, of wood and the various pyrolytic heating stages involved in the construction of the cask (Chatonnet *et al.*, 1989, Mosedale and Puech, 1998) considerably diminish the quantity of **1** and its congeners, a non-negligible portion resists these drastic conditions (Doussot *et al.*, 2002, Cadahía *et al.*, 2001, Moutounet *et al.*, 1992). During aging in oak barrels, the hydroalcoholic and slightly acidic (*i.e.*, pH ~ 3-4) wine solution enables the solid-liquid extraction of **1** and its congeners. Once in the wine solution, these ellagitannins are slowly but continuously transformed through possible physical complexations with polysaccharides and proteins and through various chemical reactions such as oxidation, hydrolysis, polymerization (Puech *et al.*, 1999a, Viriot *et al.*, 1993) and… condensation events (Quideau *et al.*, 2005). Analyses of wines aged for 12 to 18 months in oak barrels have indicated amounts of vescalagin (**1**) comprised between 0 and 7 mg/L (Saucier *et al.*, 2006, Quideau *et al.*, 2005, Moutounet et al., 1989, 1992) (see also Table 9.1). So, vescalagin (**1**), the *C*-glycosidic ellagitannin part of the acutissimins, gets extracted from the oak into the wine solution, which does contain significant amounts of the other part of the hybrid structure of these complex tannins, *i.e.*, the grape-derived flavan-3-ol, (+)-catechin. Hence, we surmised that the formation of acutissimins could occur during wine aging in oak barrels (Section 9.3).

First, we achieved the hemisynthesis of acutissimins A/B (**14/15**) in high yield (87%) from **1** and (+)-catechin in an acidic organic medium (*i.e.*, 1.5% (v/v) TFA/THF) at 60 °C over a period of 7 h (Quideau *et al.*, 2003, 2005) (Fig. 9.10). Interestingly, the ratio of acutissimins A and B thus obtained (*i.e.*, 75:25) was similar to that observed from the isolation of these two regioisomers from *Quercus acutissima* (*i.e.*, 81:19, see Ishimaru *et al.*, 1987). Thus, the predominant formation of acutissimin A (**14**) *in vitro* as well as *in vivo* may be strictly a consequence of the better accessibility and higher nucleophilicity of the catechin C-8 center over those of the C-6 center (Delcour *et al.*, 1983, Okajima, 2001). The hemisynthesis of the as yet non-isolated epiacutissimins A (**29**) and B (**30**) was also achieved using vescalagin (**1**) and (–)-epicatechin under the same reaction conditions. Both epiacutissimins were obtained in a combined yield of 78% and in a regioisomeric ratio of 67:33 in favor of the epiacutissimin A (Fig. 9.10). As mentioned above, all of our attempts to perform the same reactions from castalagin (**2**) were to no avail. The mechanistic description of these hemisyntheses follows a classical S_N1-type nucleophilic substitution pathway, as first suggested by Haslam (Haslam and Cai, 1994). The protonation of the OH-1 group of vescalagin (**1**), under the acid-catalyzed conditions used, leads to the formation of the stable benzylic cation **27**, which is then trapped by the nucleophilic flavan-3-ol counterpart (Fig. 9.7). Starting from **1**, these nucleophilic substitutions proceed with full retention of configuration at C-1, as rationalized in Section 9.2.1.2. These hemisyntheses constitute an *in vitro* mimicry of the non-enzymatic yet diastereoselective formation of acutissimin flavano-ellagitannins.

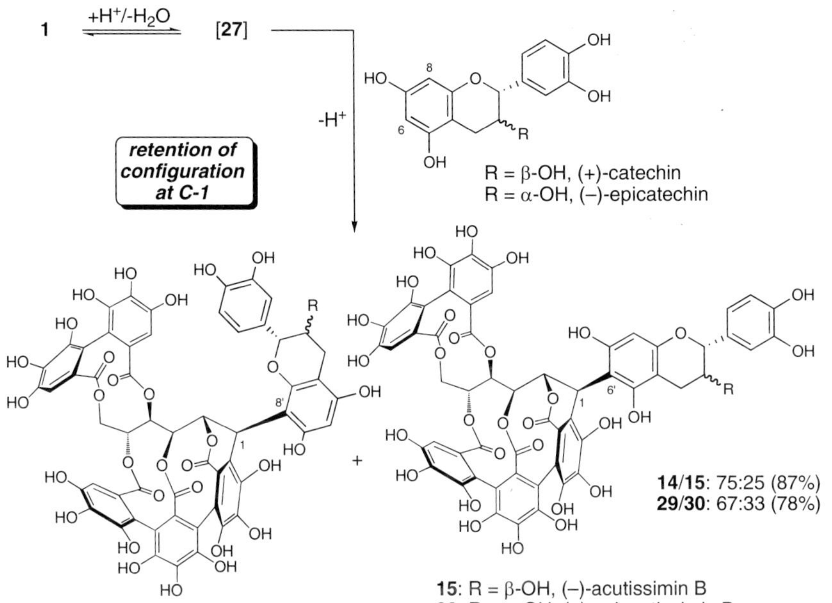

15: R = β-OH, (–)-acutissimin B
30: R = α-OH, (–)-epiacutissimin B

14: R = β-OH, (–)-acutissimin A
29: R = α-OH, (–)-epiacutissimin A

Fig. 9.10 Hemisynthesis of acutissimins (**14**/**15**) and epiacutissimins (**29**/**30**) from vescalagin (**1**) and either (+)-catechin or (–)-epicatechin, respectively, in acidic organic media (isolated yields, see text).

9.3 Impact of Vescalagin on the Chemical Profile of Wine

9.3.1 Occurrence of flavano-ellagitannins in red wine

Several hundred compounds present in red wine have been characterized to date, so the chances of finding the acutissimins in such a chemically complex medium would have been small without the compounds already to hand, thanks to our hemisynthesis work. On the one hand, we first verified that the oak heartwood used to make barrels and from which we isolated vescalagin (**1**) (Quideau *et al.*, 2004, Scalbert *et al.*, 1990), did not contain any detectable amounts of acutissimins. On the other hand, red wines typically contain (+)-catechin and (–)-epicatechin at concentrations ranging from 115 mg/L to 190 mg/L for the former, and around 80 mg/L for the latter (Carando and Teissedre, 1999, Cabanis *et al.*, 1998). As a mildly acidic hydroalcoholic solution (*i.e.*, ~ 12% aqueous ethanol at pH ~ 3-4), red wine gathers all of the prerequiste physico-chemical factors to potentially mediate the formation of acutissimins. To a wine model solution, consisting of a 12% (v/v) ethanolic aqueous solution with 5 g/L of tartaric acid at pH 3.2, was added vescalagin (**1**) and (+)-catechin in order to confirm this potentiality. The mixture was allowed to react at room temperature for 25 days, after which time both acutissimins A/B (**14**/**15**) were UV-detected as major products, together with a third major component resulting from the nucleophilic substitution of the OH-1 group of **1** by ethanol. This compound was also generated in a high isolated yield of 94% from **1** and ethanol in THF containing 1.5% (v/v) of TFA after 5 h at 60 °C, and characterized by mass spectrometry and NMR spectroscopy (Quideau *et al.*, 2005). Its structure was thus unambiguously determined as corresponding to β-1-*O*-ethylvescalagin (**31**) (Fig. 9.11). Once again, this ethyl ether of vescalagin was generated with retention of configuration at C-1. Of particular note is the isolation by Okuda's group of the methyl analog of **31** from the leaves of *Tibouchina semidecandra* (*Melastomataceae*). This compound (see **32** in Fig. 9.11) was however probably produced as an artefact during the liquid chromatographic separation procedure relying on the use of methanol (Yoshida *et al.*, 1991).

Another interesting example of the implication of such simple alcohols in the chemistry of *C*-glycosidic ellagitannins has been described by Puech and associates in the course of their investigations on the aging of spirits in oak barrels. Both vescalagin (**1**) and castalagin (**2**), and conceivably other oak-extracted ellagitannins, are transformed into ethyl-based hemiketal and ketal derivatives after prolonged exposure to 40 and 70% (v/v) ethanolic aqueous solutions (Fig. 9.11). These transformations imply an oxidation of the NHTP pyrogallol I-ring of these ellagitannins prior to the nucleophilic addition of ethanol onto the resulting α-hydroxy-*ortho*-quinone (Puech *et al.*, 1999b).

Evidence of the presence of the acutissimins in wine was obtained by analyzing a sample of red wine that had been aged for 18 months in oak barrels. Not only were the two acutissimins A and B (**14**/**15**) detected, but also the two previously unknown "epiacutissimins" A and B (**29**/**30**). A high-performance liquid chromatography-electrospray ionization mass spectrometry (HPLC/ESIMS) method was developed to quantify the presence of these four acutissimins, as well as that of β-1-*O*-ethylvescalagin (**31**) in red wines (Table 9.1) (Saucier *et al.*, 2006).

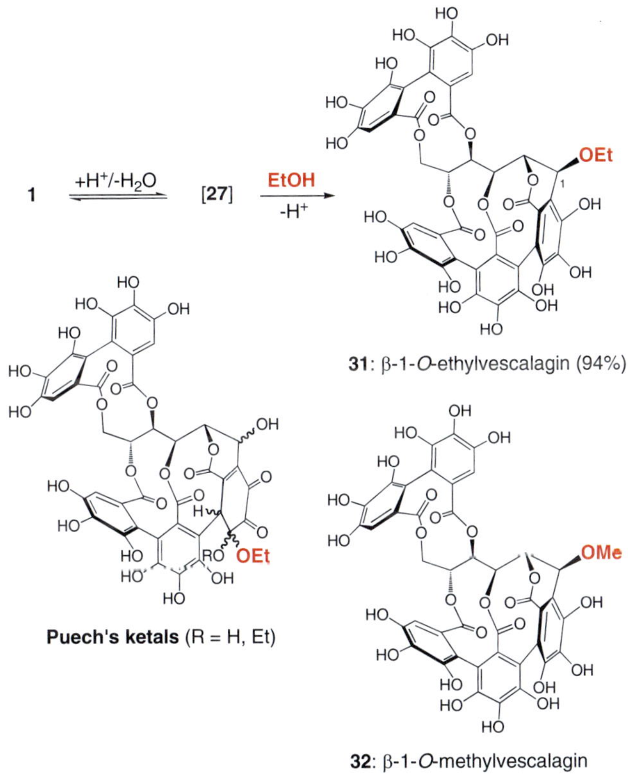

31: β-1-*O*-ethylvescalagin (94%)

32: β-1-*O*-methylvescalagin

Fig. 9.11 Hemisynthesis of β-1-*O*-ethylvescalagin (**31**) from vescalagin (**1**) and ethanol in an acidic organic medium (isolated yield, see text). Analogous compounds derived from other implications of methanol or ethanol in the chemistry of vescalagin/castalagin.

Even if these quantitative analyses indicate low concentrations of the compounds of interest, their occurrence constitutes compelling evidences that oak-derived *C*-glycosidic ellagitannins do react, via condensation at their C-1 position, with nucleophilic species in wine. Of course, this type of chemical reaction is just one among several other processes leading to the transformation of native *C*-glycosidic ellagitannins extracted from oak-made casks by wine solutions.

Table 9.1 Levels of vescalagin, castalagin, ethylvescalagin and acutissimins in a sample of red wine aged in oak barrels.

Compound	Concn (mg/L)
vescalagin (**1**)	2.20 ± 0.15
castalagin (**2**)	8.10 ± 0.31
ethylvescalagin (**31**)	0.85 ± 0.05
acutissimin A (**14**)	0.40 ± 0.03
acutissimin B (**15**)	0.28 ± 0.02
epiacutissimin A (**29**)	0.30 ± 0.02
epiacutissimin B (**30**)	0.35 ± 0.02

Moreover, it is important to keep in mind that wine is a complex multi-component reaction system, which slowly but continuously evolves under mildly acidic and oxidative conditions. As far as the acutissimins are concerned, they are further transformed in wine, as we shall illustrate below (Section 9.3.4), but they will continue to form as long as grape flavan-3-ols and oak vescalagin (**1**) are present in the wine solution. Furthermore, red wine, more than white wine, is particularly rich in various components derived from grape (skin and seeds) and its fermented juice, and contains a broad panel of nucleophilic species, which may compete in the trapping of the benzylic cation intermediate **27** derived from **1** during its condensation reactions at C-1. Examples of such nucleophiles are numerous other flavonoids (*e.g.*, from procyanidins to proanthocyanidins, and anthocyanins) and, more generally, a large library of all kinds of phenols, alcohols, amines, carboxylic acids, enolizable carbonyl compounds, and thiols (Cabanis *et al.*, 1998), including macromolecules such as proteins and polysaccharides that feature some of these functional groups. The quantitative search by any given analytical method of each of the compounds possibly resulting from a condensation reaction between one of these nucleophiles and vescalagin at any given time during the aging process of a wine can certainly fuel research for many years to come.

Prior to and beyond such considerations of strict analytical work concerns, the most important issue that should be kept in mind at the scientific level is the unveiling of a given type of chemistry at play in a given system. So, not only vescalagin (**1**), but any *C*-glycosidic ellagitannins displaying a free hydroxyl group at C-1, such as roburin A (**5**), stachyurin (**3**), 5-desgalloylstachyurin (**26**) and also vescalin (**33**) (Mayer *et al.*, 1967, 1971b) can thus act like "nucleophile sponges" if present, introduced or generated in wine. Vescalin (**33**) naturally results from the hydrolysis of the 4,6-HHBP group of vescalagin (**1**) (Fig. 9.12). It is present in the heartwood of oak species used for cooperage (Viriot *et al.*, 1994), albeit in rather small amounts estimated to range between 0.11 and 1.48 mg/g of dry wood (Masson *et al.*, 1995, 1996, Mosedale *et al.*, 2001), but can thus end up in the wine solution (Moutounet *et al.*, 1989, Puech *et al.*, 1996), in which hydrolysis of **1** into **33** can also occur. Additional examples of the expression of this reactivity between ellagitannin C-1 benzylic alcohols, again exemplified by vescalagin (**1**), and other wine nucleophiles are highlighted in the following sections.

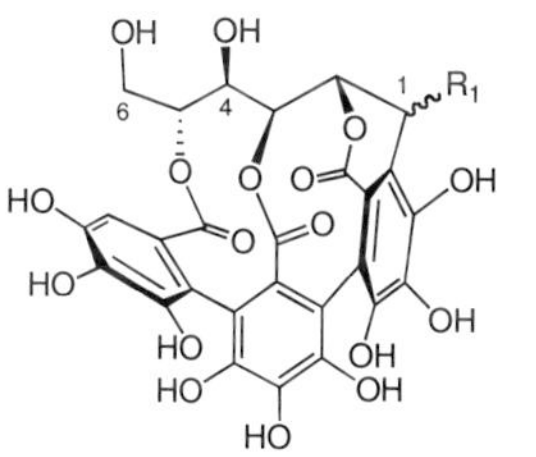

33: R_1 = β-OH, vescalin
34: R_1 = α-OH, castalin

Fig. 9.12 Stuctures of vescalin (**33**) and castalin (**34**) derived from the hydrolytic removal of the 4,6-HHBP group in vescalagin (**1**) and castalagin (**2**), respectively.

9.3.2 An example of a condensation reaction with a non-phenolic wine nucleophile

The condensation reaction of vescalagin (**1**) with the cysteine-containing tripeptide glutathione (**35**) was also investigated. This choice of nucleophile originates from observations made by enologists on the decrease of the level of **35** in wines, especially white wines, aged in new oak barrels (Lavigne-Cruege *et al.*, 2003). These observations were particularly relevant to the quality control of white wines, for glutathione

(**35**) is thought to act as an antioxidant protecting some volatile thiols that contribute to the fruity aroma of these wines. Glutathione (**35**) is known to engage in nucleophilic addition reactions with orthoquinones derived from the oxidation of caffeoylated tartaric acids in wine musts (Cheynier *et al.*, 1986, Cheynier *et al.*, 1990). This process, which is implicated in the oxidative browning of musts, can certainly contribute to the variations of concentrations of **35** in wine musts (*ca.* from 24 to 3 mg/L), but the apparent influence of aging in oak barrels on the disappearance of **35** from the wine solution led us to contemplate its participation in nucleophilic substitution reactions with oak *C*-glycosidic ellagitannins.

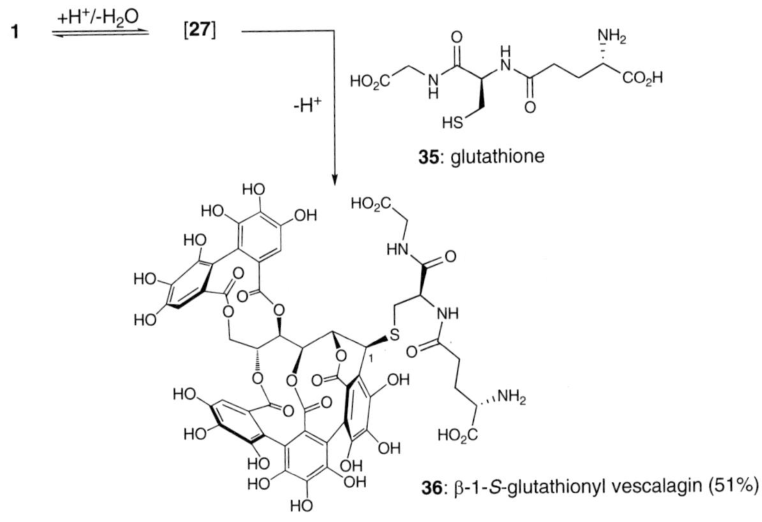

Fig. 9.13 Hemisynthesis of β-1-*S*-glutathionyl vescalagin (**36**) from vescalagin (**1**) and glutathione (**35**) in an acidic organic medium (isolated yield, see text).

On the basis of previous observations made on the condensation of vescalagin (**1**) with thiols (Jourdes, 2003), we surmised that the glutathione (**35**) could react in a similar fashion to furnish β-1-*S*-glutathionyl vescalagin (**36**). This was indeed verified under reaction conditions similar to those used for the hemisynthesis of the acutissimins (see Section 9.2.2) and the expected product **36** was obtained in an isolated yield of 51% (Fig. 9.13). The formation of **36** was also observed in a wine model solution, together with those of its sulfoxide variant and of β-1-*O*-ethylvescalagin (**31**) (see Fig. 9.11). Most importantly, its presence was also detected by HPLC/ESIMS in a sample of red wine aged during 18 months in oak barrels (Lefeuvre, 2006).

9.3.3 Influence on wine color modulation

9.3.3.1 *Anthocyanins and wine coloration – A few facts*

Together with astringency and bitterness, color is the main organoleptic criteria in red wine quality description and characterization. The maceration and alcoholic fermentation step during the wine-making process allows the extraction of anthocyanins from red grape skins and results in the intense color of young red wines. However, the native grape pigments progressively disappear during wine maturation, especially during the aging in oak barrels, to be replaced by more stable wine-specific pigments (Somers, 1971, Liao *et al.*, 1992). Numerous investigations have been conducted over the years in order to understand, characterize and perhaps control the physicochemical mechanisms underlying the color modulation of red wines. Most of these studies established and confirmed the fact that several chemical reactions take place between anthocyanins and nucleophilic or electrophilic wine components such as *inter alia* flavanols, vinylphenol, ethanal, glyoxylic and pyruvic acids to generate new pigments displaying different coloring properties (*i.e.*, either bluer or more orange tints) (Brouillard *et al.*, 2003, De Freitas and Mateus, 2006, Timberlake and Bridle, 1976, Escribano-Bailón *et al.*, 1996, Dallas *et al.*, 1996, Bakker and Timberlake, 1997, Francia-Aricha *et al.*, 1997, Fulcrand *et al.*, 1998, Remy *et al.*, 2000, Es-Safi *et al.*, 1999, 2000, Lee *et al.*, 2004, Pissarra *et al.*, 2004, Salas *et al.*, 2004). Moreover, physical co-pigmentation phenomena resulting from stacking of the colored anthocyanin flavylium cations with other wine phenolic species (Brouillard *et al.*, 2003, Boulton, 2001, Liao *et al.*, 1992, Brouillard and Dangles, 1994, Escribano-Bailón *et al.*, 1996, Gómez-Míguez *et al.*, 2006), as well as complexation with metallic cations (Ribéreau-Gayon, 1973, Dangles *et al.*, 1994), have also been postulated to contribute to the color modulation of red wines during aging and conservation. Surprisingly, covalent interactions between grape anthocyanins and oak *C*-glycosidic ellagitannins had not been considered as a possible factor contributing to this phenomenon in the case of red wines aged in oak barrels.

9.3.3.2 *Hemisynthesis of anthocyano-ellagitannin hybrids*

To address the aforementioned possibility, vescalagin (**1**) was allowed to react with the anthocyanidin malvidin (**37**) and the major *Vitis vinifera* grape 3-*O*-glucosidic anthocyanin oenin (**38**) both in acidic organic TFA/THF and wine model solutions (Quideau *et al.*, 2005). Among several other products, the anthocyano-ellagitannin hybrids **39** and **40**, in which the C-8 center of their flavylium ion A-ring is connected to the C-1 center of vescalagin, were isolated from the acidic organic reaction mixtures as expected from condensation reaction *via* acid-catalyzed nucleophilic substitution (Fig. 9.14). With these standard compounds to hand, their formation was then clearly detected in the acidic aqueous wine model solution at pH 3.2. At such a pH value, anthocyanidins and anthocyanins mainly exist under their C-2 hemiketal forms, as depicted in Fig. 9.14. In fact, flavylium ionic forms express a poor nucleophilic character at their C-8 locus relatively to their hemiketal equilibrium counterparts (Remy *et al.*, 2000), which would then constitute the preferred nucleophilic species reacting with the vescalagin-derived benzylic cation **27**. Even when performing the hemisynthesis of **39** and **40** in dry THF, the water released from the TFA-mediated dehydroxylation of **1** into **27** is available to generate hemiketalic species *via* addition to the flavylium C-2 position of **37** or **38** (Fig. 9.14).

9.3.3.3 *Anthocyano-ellagitannins and wine coloration*

The most remarkable observation made during the hemisynthesis of the anthocyano-ellagitannins **39** and **40** was the change of color of the reaction mixtures, turning from the bright red color brought about the starting malvidin (**37**) or oenin (**38**) into a deeper reddish purple color.

The visible absorption band of both anthocyano-ellagitannins **39** and **40** recorded in aqueous solutions at pH 1 and 3.2 (*i.e.*, wine pH) indeed revealed an important bathochromic shift of about 20 nm with respect to that of malvidin (**37**) or oenin (**38**) (Fig. 9.15). The respective bright red and deep purple colors of acidic (pH 1) aqueous solutions of pure **38** and **40** are clearly seen on the pictures inserted in Fig. 9.14.

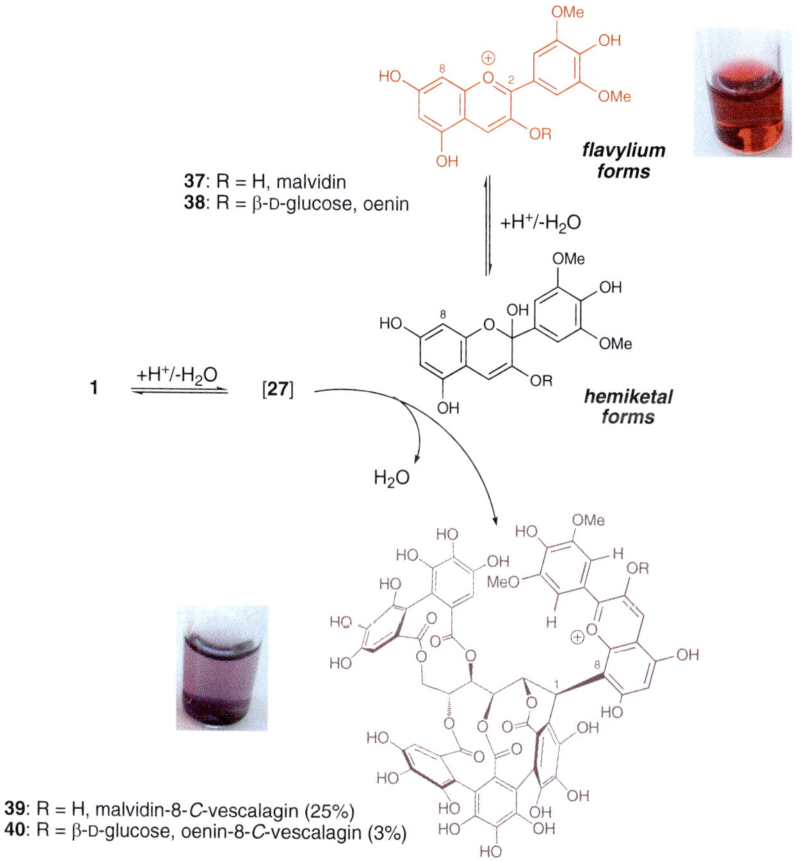

Fig. 9.14 Hemisynthesis of malvidin-8-*C*-vescalagin (**39**) and oenin-8-*C*-vescalagin (**40**) from vescalagin (**1**) and malvidin (**37**) or oenin (**38**), respectively, in an acidic organic medium (isolated yield, see text).

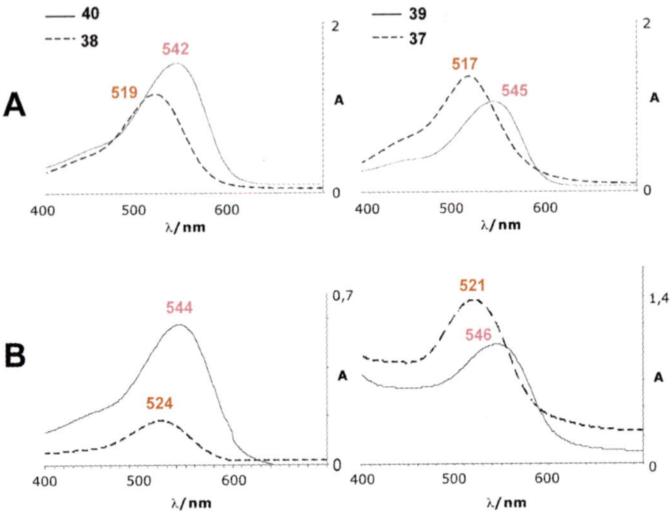

Fig. 9.15 Visible spectra of malvidin (**37**), oenin (**38**), malvidin-8-*C*-vescalagin (**39**) and oenin-8-*C*-vescalagin (**40**) in aqueous solutions at pH 1 (**A**) and 3.2 (**B**, wine model solution).

It would be obviously quite inappropriate to infer from these results that these anthocyano-ellagitannin pigments are major players in wine coloration, but they can nevertheless contribute to its modulation during wine aging in oak barrels. Their color is in agreement with the purple tints observed in young red wines, and more specifically, with the enhancement of their crimson color, which is observed in the presence of oak wood ellagitannins (Vivas and Glories, 1996). This ellagitannin-mediated color modulation had previously been attributed to the participation, upon wine oxygenation, of ellagitannins in the production of hydrogen peroxide, which can in turn oxidize ethanol into ethanal, being itself involved in the formation of ethyl-bridged anthocyanin-flavanol purple pigments (Timberlake and Bridle, 1976, Vivas and Glories, 1996, Escribano-Bailón *et al.*, 1996, Dallas *et al.*, 1996, Lee *et al.*, 2004). This process remains plausible, but the direct formation of purple-colored anthocyano-ellagitannin covalent adducts constitutes another chemically sound alternative that must be kept in mind when evaluating the evolution of anthocyanin content and color in wine during its aging relying on contact with oak wood not only from traditional barrels but also from chips, staves and even crude extracts (Barrera-García *et al.*, 2007, Del Álamo Sanza *et al.*, 2004, Del Álamo Sanza and Domínguez, 2006, Frangipane *et al.*, 2007).

9.3.4 Effect of oxygenation on flavano-ellagitannins – The mongolicain story

9.3.4.1 *Oxygen in wines – A few facts*

Oxygenation is one of the crucial factors underlying the expression and evolution of the organoleptic properties of wine such as astringency, bitterness and color. Besides the aforementioned production of ethanal from ethanol, numerous oxidations and subsequent transformations of wine components take place upon aeration. The question of whether such transformations improve or alter the quality of wine remains a matter of much concern for enologists and wine scientists, but it is generally accepted that a fast and excessive oxygenation has rather deleterious effects, whereas a slow and continuous aeration can have a positive impact on the organoleptic profile of wine (Ribéreau-Gayon *et al.*, 1983, Pontallier, 1992, Moutounet and Mazauric, 2001). In this context, the traditional aging of wine in oak barrels offers an adequate means to temper its aeration by allowing a slow penetration of oxygen through the wood (*ca.* 30-40 mg per year when using new barrels, Pontallier, 1992). Although the quantity of detectable oxygen dissolved in wines at rest during the one- to two-year period of this maturation process is low (*i.e.*, 20 to 50 μg/L, Moutounet and Mazauric, 2001), which is much lower than the maximum solubility of oxygen in air-saturated wines at ambient temperature and atmospheric pressure (*ca.* 8-9 mg/L), the capacity of wines, especially heavy red wines, to absorb oxygen is very high, and can reach up to 800 mg/L (Singleton, 1987, Moutounet and Mazauric, 2001). Once dissolved in wine, oxygen is thus progressively and rather rapidly consumed by various substrates. Phenols and polyphenols are unarguably among the first substrates in line to suffer directly or indirectly from oxygen-mediated oxidation reactions leading to coupling, condensation, polymerization and various other types of chemical (and biochemical) transformation events that are far from having been all mechanistically elucidated (Ribéreau-Gayon, 1973, Ribéreau-Gayon *et al.*, 1983, Singleton, 1987, Cheynier *et al.*, 1986, 1990, Pontallier, 1992, Moutounet and Mazauric, 2001).

9.3.4.2 *Oxidative conversion of acutissimin A into mongolicain A*

Most investigations on the fate of wine phenolics upon oxygenation at the various stages of the wine making process have concerned phenolic acids (*e.g.*, gallic acid, caftaric acid), anthocyanins, flavanols and their proanthocyanidic oligomers. Very few studies have addressed what happens to oak ellagitannins in this context at the molecular level, and the information glaned from the literature is rather contradictory. Some authors concluded that oak ellagitannins play a major role as oxidation regulators in wine, quickly absorbing dissolved oxygen and facilitating the hydroperoxidation of some wine components (*e.g.*, ethanol into ethanal, *vide supra*) (Vivas and Glories, 1993, 1996), whereas others concluded that the oxidation of ellagitannins, like vescalagin (**1**), is a very slow process (Moutounet *et al.*, 1992). One possible explanation to this apparent contradiction is that the galloyl-derived units of oak *C*-glycosidic ellagitannins are engaged in fast inter- and/or intramolecular oxido-reductive processes during which their pyrogallol moieties are reversibly converted into semiquinone free radicals and/or *ortho*-quinones through one- and/or two-electron transfers.

Support for this speculative interpretation can be drawn from a molecular-level observation made in our laboratory. In an aqueous solution left under air at 60 °C, pure acutissimin A (**14**) was converted into mongolicain A (**18**), which could be isolated by semi-preparative HPLC in a yield of 22% (Fig. 9.16). This same conversion was also observed in a wine model solution (*unpublished results*, see Lefeuvre, 2006). Interestingly, it is worth recalling here that mongolicain A is thought to be naturally derived from the oxidation of acutissimin A, as both molecules usually co-exist in their plant sources (Nonaka *et al.*, 1988, see Section 9.1.1).

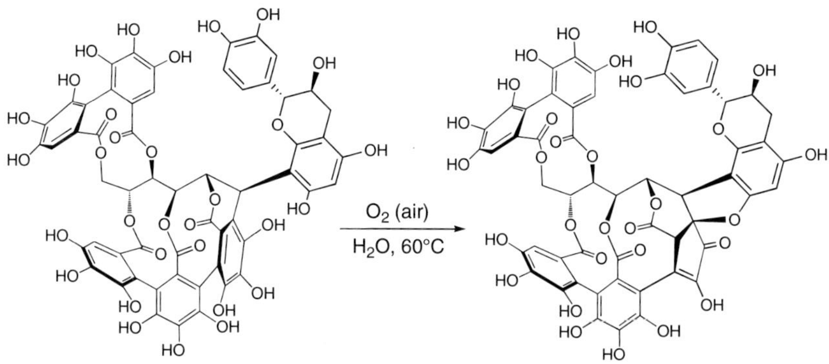

14: acutissimin A

18: mongolicain A (22%)

Fig. 9.16 Oxidative hemisynthesis of mongolicain A (**18**) form acutissimin A (**14**) in an aqueous solution (isolated yield).

The construction of the characteristic *spiro*-linked dihydrofuran-cyclopentenone motif of mongolicain A (**18**) is the result of a multi-step oxidative process starting with dehydrogenation of the NHTP galloyl-derived I-ring of the vescalagin part of acutissimin A (**14**), as depicted by Tanaka in Fig. 4.19 of Chapter 4. In the presence of oxygen, autoxidation can mediate this dehydrogenation into the α-hydroxy-*ortho*-quinone **41**, with concomitant formation of hydrogen peroxide (Fig. 9.17). The phenolic 7-OH group of the A-ring of the catechin-derived part of the molecule would then attack this *ortho*-quinone in a 1,6-addition manner to furnish **42**. Addition of water to its diketone tautomer **43** could give rise to the hydrate **44** that can then undergo a ring contraction via a benzylic acid-type rearrangement to furnish the carboxylic acid **45**. Decarboxylation of its ketone tautomer **46** can lead to the enediol **47**, which would require a final dehydrogenative oxidation into the cyclopentane-1,2-dione **48** to lead to the thermodynamically more stable cyclopentenone unit of mongolicain A (**18**).

This complex succession of events constitutes a plausible but admittedly putative mechanistic description of the generation of **18** from **14** under autoxidation conditions, and one might wonder why the galloyl-derived I-ring of acutissimin A (**14**) is the only pyrogallol unit thus sucumbing to dehydrogenative oxidation.

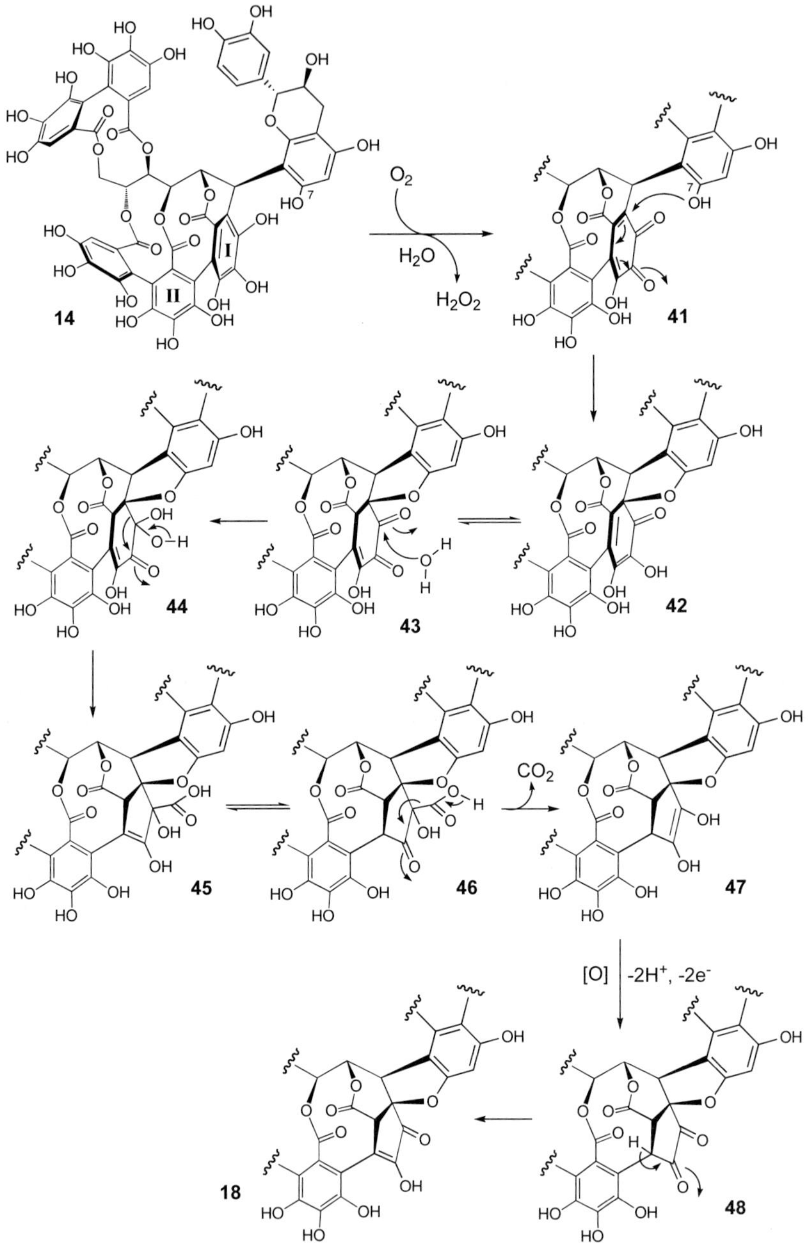

Fig. 9.17 Proposed mechanistic description of the conversion of acutissimin A (**14**) into mongolicain A (**18**) under autoxidation conditions.

In fact, the other four pyrogallol units are probably also to some extent, but reversibly, converted into *ortho*-quinones. As alluded to above, all of the pyrogallol rings of a *C*-glycosidic ellagitannin such as vescalagin (**1**) would follow the same behavior under wine oxygenation conditions in oak barrels. However, in the case of a flavano-ellagitannin such as acutissimin A (**14**, also present in wine, see Section 9.3.1), the *ortho*-quinone derived from the pyrogallol I-ring is the only one that can be irreversibly trapped by the proximal flavanol unit with its 7-OH group well-positioned to engage in such an intramolecular reaction.

9.4 Conclusion and Perspectives

The ellagitannin chemistry discussed in this chapter is yet another example of the remarkable impact of natural products on human activities. Isn't it fascinating to realize that natural products derived from the metabolism of plants for their protection against pathogens and herbivores (Hart and Willis, 1972, Scalbert and Haslam, 1987, Scalbert, 1991), then identified as active principles of plant extracts used in folk medicines and later as promising drug leads, end up being investigated for their role in the elaboration of the biophysico-chemical and organoleptic profile of a beverage such as wine. Herein, the most illustrating and fascinating example of such a natural product connection with human practices is the formation of the potent human DNA topoisomerase II inhibitor, acutissimin A (**14**), in wine as a consequence of the mixing in solution of grape catechin and oak vescalagin during aging in barrels.

The chemistry that our laboratory has unveiled at the molecular level on the physico-chemical behavior of oak *C*-glycosidic ellagitannins should help wine scientists and enologists in their long-standing and continuous endeavor toward a better understanding of wine chemistry. The various complex and hybrid ellagitannins described herein are only a few examples of a large panel of compounds that can be derived from vescalagin (**1**) and its analogous oak congeners in combination with a multitude of nucleophilic species available in wine. Other yet unpublished examples involving *inter alia* a phenolic stilbene such as resveratrol and a flavanol dimer such as procyanidin B-2 are discussed in Lefeuvre's doctoral thesis (Lefeuvre, 2006). The isolation of the vescalagin/procyanidin B-3 hybrid mongolicanin (**16**, see Section 9.1.1.2) from the bark of *Quercus mongolica* var. *grosseserrata* (Ishimaru *et al.*, 1988) supports the feasibility of the formation of such procyanidino-ellagitannins in wine.

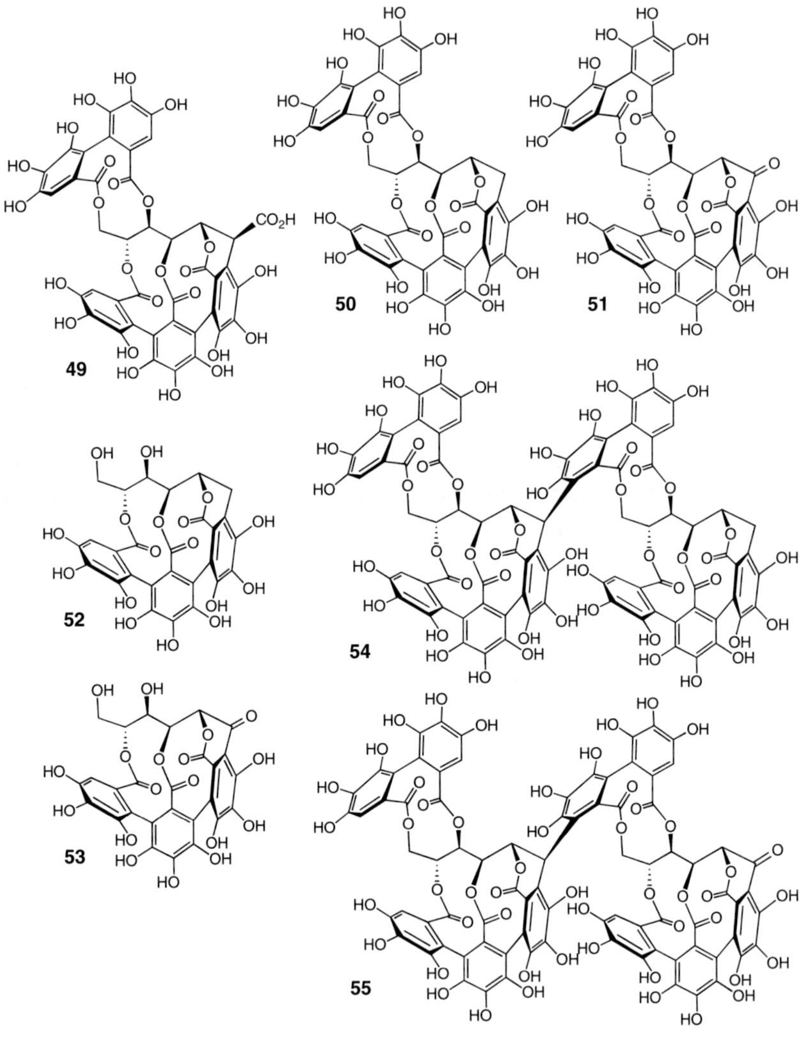

Fig. 9.18 Structures of new "taste-active" deoxycarboxyvescalagin (**49**) and ellagitannin thermal transformation products (**50-55**).

Whether or not all of these oak-derived *C*-glycosidic ellagitannins significantly contribute to the organoleptic qualities of wine still remains to be determined. The question of whether oak ellagitannins have an effect on taste characteristics has been and still is subject to debate in the literature (Vivas and Glories, 1996). Some authors suggest that the presence of oak ellagitannins can account for much of the astringency and bitterness of wine aged in barrels, whereas others believe that their concentration is too low to contribute to these taste characteristics (Quinn and Singleton, 1985, Somers, 1990, Hervé du Penhoat *et al.*, 1991b, Pocock *et al.*, 1994, Puech *et al.*, 1999a). Besides our own contributions and earlier ones notably by Puech and associates toward a better molecular-level understanding of the chemical reactivity of oak ellagitannins during cooperage and wine oak-maturing processes, Hofmann and Glabasnia have recently invested efforts in this field, and identified new "taste-active" vescalagin and castalagin deoxy- and dehydroderivatives (Fig. 9.18).

Some of these derivatives (*i.e.*, **49-50**), as well as their vescalagin (**1**) and castalagin (**2**) parents, were found to impart astringent mouth-coating sensation at remarkably low threshold concentrations comprised between 1 and 4 mg/L (Glabasnia and Hofmann, 2006, 2007). So, the debate on the contribution of oak- *versus* grape-derived polyphenols to the organoleptic properties of wine will certainly go on, but it may turn in favor of a reappraisal of oak polyphenols in view of the growing development of alternative wine-aging practices based on the use of (toasted) oak chips and micro-oxygenation treatments (Garde Cerdán and Ancín-Azpilicueta, 2006).

Acknowledgements

We gratefully acknowledge funding from the Conseil Interprofessionnel du Vin de Bordeaux, the Conseil Régional d'Aquitaine and the Agence Nationale de la Recherche (ANR-06-BLAN-0139). We deeply thank our colleagues from the Faculté d'Œnologie at the University of Bordeaux, Prof. Yves Glories, Dr. Cédric Saucier and Prof. Pierre-Louis Teissedre, for their input and advice.

9.5 Bibliography

Bakker, J. and Timberlake, C. F. (1997). Isolation, identification, and characterisation of new color-stable anthocyanins occurring in some red wines, *J. Agric. Food Chem.*, 45, pp. 35–43.

Barrera-García, V. D., Gougeon, R. D., Di Majo, D., De Aguirre, C., Voilley, A. and Chassagne, D. (2007). Different sorption behaviors for wine polyphenols in contact with oak wood, *J. Agric. Food Chem.*, 55, pp. 7021–7027.

Boulton, R. (2001). The copigmentation of anthocyanins and its role in the color of red wine: a critical review, *Am. J. Enol. Vitic.*, 52, pp. 67–87.

Brouillard, R. and Dangles, O. (1994). Anthocyanin molecular interactions: the first step in the formation of new pigments during wine aging? *Food Chem.*, 51, pp. 365–371.

Brouillard, R., Chassaing, S. and Fougerousse, A. (2003). Why are grape/fresh wine anthocyanins so simple and why is it that red wine color lasts so long? *Phytochemistry*, 64, pp. 1179–1186, and references cited therein.

Cabanis, J. C., Cabanis, M. T., Cheynier, V. and Teissedre, P. L. (1998). Tables de composition. In *Œnologie – Fondements Scientifiques et Technologiques*, ed. by Flanzy, C., TEC & DOC Lavoisier: Paris, pp. 315–336.

Cadahía, E., Varea, S., Muñoz, L., Fernández de Simón, B. and García-Vallejo, M. C. (2001). Evolution of ellagitannins in Spanish, French and American oak woods during natural seasoning and toasting, *J. Agric. Food Chem.*, 49, pp. 3677–3684.

Carando, S. and Teissedre, P.-L. (1999). Catechin and procyanidin levels in French wines: contribution to dietary intake. In *Plant Polyphenols 2: Chemistry, Biology, Pharmacology, Ecology*, ed. by Gross, G. G., Hemingway, R. W., and Yoshida, T., Kluwer Academic/Plenum Publishers: New York, pp. 725–737.

Chatonnet, P., Boidron, J. N. and Pons, M. (1989). Incidence du traitement thermique du bois de chêne sur sa composition chimique. Deuxième partie: évolution de certains composés en fonction de l'intensité de brûlage, *J. Int. Sci. Vigne Vin*, 23, pp. 223–250.

Cheynier, V., Trousdale, E. K., Singleton, V. L., Salgues, M. J. and Wylde, R. (1986). Characterization of 2-S-glutathionylcaftaric acid and its hydrolysis in relation to grape wines, *J. Agric. Food Chem.*, 34, pp. 217–221.

Cheynier, V., Rigaud, J., Souquet, J.-M., Duprat, F. and Moutounet, M. (1990). Must browning in relation to the behavior of phenolic compounds during oxidation, *Am. J. Enol. Vitic.*, 41, pp. 346–349.

Clifford, M. N. (2000). Anthocyanins – Nature, occurrence and dietary burden, *J. Sci. Food Agric.*, 80, pp. 1063–1072.

Dallas, C., Ricaro-da-Silva, J. M. and Laureano, O. (1996). Products formed in model wine solutions involving anthocyanins, procyanidin B_2, and acetaldehyde, *J. Agric. Food Chem.*, 44, pp. 2402–2407.

Dangles, O., Elhabiri, M. and Brouillard, R. (1994). Kinetic and thermodynamic investigation of the aluminium-anthocyanin complexation in aqueous solution, *J. Chem. Soc. Perkin Trans. 2*, pp. 2587–2596.

De Freitas, V. and Mateus, N. (2006). Chemical transformations of anthocyanins yielding a variety of colours (Review), *Environ. Chem. Lett.*, 4, pp. 175–183.

Del Álamo Sanza, M., Domínguez, I. N. and García Merino, S. (2004). Influence of different aging systems and oak woods on aged wine color and anthocyanin composition, *Eur. Food Res. Technol.*, 219, pp. 124–132.

Del Álamo Sanza, M. and Domínguez, I. N. (2006). Wine aging in bottle from artificial systems (staves and chips) and oak woods. Anthocyanin composition, *Anal. Chim. Acta*, 563, pp. 255–263.

Delcour, J. A., Ferreira, D. and Roux, D. G. (1983). Synthesis of condensed tannins. Part 9. The condensation sequence of leucocyanidin with (+)-catechin and with the resultant procyanidins, *J. Chem. Soc. Perkin Trans.*, pp. 1711–1717.

Doussot, F., De Jeso, B., Quideau, S. and Pardon, P. (2002). Extractives content in cooperage oak wood during natural seasoning and toasting; influence of tree species, geographic, location, and single-tree effects, *J. Agri. Food Chem.*, 50, pp. 5955–5961.

Escribano-Bailón, T., Dangles, O. and Brouillard, R. (1996). Coupling reaction between flavylium ions and catechin, *Phytochemistry*, 41, pp. 1583–1592.

Es-Safi, N.-E., Fulcrand, H., Cheynier, V. and Moutounet, M. (1999). Studies on the acetaldehyde-induced condensation of (–)-epicatechin and malvidin 3-*O*-glucoside in a model solution system, *J. Agric. Food Chem.*, 47, pp. 2096–2102.

Es-Safi, N.-E., Le Guernevé, C., Fulcrand, H., Cheynier, V. and Moutounet, M. (2000). Xanthylium salts formation involved in wine colour changes, *Int. J. Food Sci. Technol.*, 35, pp. 63–74.

Fernández de Simón, B., Cadahía, E., Conde, E. and García-Vallejo, M. C. (1999). Ellagitannins in woods of Spanish, French and American oaks, *Holzforschung*, 53, pp. 147-150.

Francia-Aricha, E. M., Guerra, M. T., Rivas-Gonzalo, J. C. and Santos-Buelga, C. (1997). New anthocyanin pigments formed after condensation with flavanols, *J. Agric. Food Chem.*, 45, pp. 2262–2266.

Frangipane, M. T., De Santis, D. and Ceccarelli, A. (2007). Influence of oak woods of different geographical origins on quality of wines aged in barriques and using oak chips, *Food Chem.*, 103, pp. 46–54.

Fulcrand, H., Benadbeljalil, C., Rigaud, J., Cheynier, V. and Moutounet, M. (1998). A new class of wine pigments generated by reaction between pyruvic acid and grape anthocyanins, *Phytochemistry*, 47, pp. 1401–1407.

Garde-Cerdán, T. and Ancín-Azpilicueta, C. (2006). Review of quality factors on wine ageing in oak barrels, *Trends Food Sci. Technol.*, 17, pp. 438–447.

Glabasnia, A. and Hofmann, T. (2006). Sensory-directed identification of taste-active ellagitannins in American (*Quercus alba* L.) and European oak wood (*Quercus robur* L.) and quantitative analysis in Bourbon whiskey and oak-matured red wines, *J. Agric. Food Chem.*, 54, pp. 3380–3390.

Glabasnia, A. and Hofmann, T. (2007). Identification and sensory evaluation of dehydro- and deoxy-ellagitannins formed upon toasting of oak wood (*Quercus alba* L.), *J. Agric. Food Chem.*, 55, pp. 4109–4118.

Gómez-Míguez, M., González-Manzano, S., Escribano-Bailón, M. T., Heredia, F. J. and Santos-Buelga, C. (2006). Influence of different phenolic copigments on the color of malvidin 3-glucoside, *J. Agric. Food Chem.*, 54, pp. 5422–5429.

Hart, J. H. and Hillis, W. E. (1972). Inhibition of wood-rotting fungi by ellagitannins in the heartwood of *Quercus alba*, *Phytopathology*, 62, pp. 620–626.

Haslam, E. and Cai, Y. (1994). Plant polyphenols (vegetable tannins): gallic acid metabolism, *Nat. Prod. Rep.*, pp. 41–66.

Haslam, E. (1998). Practical Polyphenolics – From Structure to Molecular Recognition and Physiological Action, Cambridge University Press: Cambridge.

Hatano, T., Kira, R., Yoshizaki, M. and Okuda, T. (1986). Seasonal changes in the tannins of *Liquidambar formosana* reflecting their biogenesis, *Phytochemistry*, 25, pp. 2787–2789.

Hervé du Penhoat, C. L. M., Michon, V. M. F., Ohassan, A., Peng, S., Scalbert, A. and Gage, D. (1991a). Roburin A, a dimeric ellagitannin from heartwood of *Quercus robur*, *Phytochemistry*, 30, pp. 329–332.

Hervé du Penhoat, C. L. M., Michon, V. M. F., Peng, S., Viriot, C., Scalbert, A. and Gage, D. (1991b). Structural elucidation of new dimeric ellagitannins from *Quercus robur* L. Roburins A-E, *J. Chem. Soc., Perkin Trans. 1*, pp. 1653–1660.

Ishimaru, K., Nonaka, G.-I. and Nishioka, I. (1987). Tannins and related compounds. LV. Isolation and characterization of acutissimins A and B, novel tannins from *Quercus* and *Castanea* species, *Chem. Pharm. Bull.*, 35, pp. 602–610.

Ishimaru, K., Ishimatsu, M., Nonaka, G. I., Mihashi, K., Iwase, Y. and Nishioka, I. (1988). Tannins and related coupounds. LXXII. Isolation and characterisation of mongolicanin (procyanidino-ellagitannin), mongolinin A, acutissimin C and vescalagin carboxylic acid, novel tannins from *Quercus mongolica* var. *grosserrata*, *Chem. Pharm. Bull.*, 36, pp. 3319–3327.

Ito, H., Yamaguchi, K., Kim, T.-H., Khennouf, S., Gharzouli, K. and Yoshida, T. (2002). Dimeric and trimeric hydrolyzable tannins from *Quercus coccifera* and *Quercus suber*, *J. Nat. Prod.*, 65, pp. 339–345.

Jourdes, M. (2003). Réactivité, synthèse, couleur et activité biologique d'ellagitannins *C*-glycosidiques et flavano-ellagitannins, *Doctoral Thesis*, Université Bordeaux 1, Order N°2721.

Kashiwada, Y., Nonaka, G.-i., Nishioka, I., Lee, K. J.-H., Bori, I., Fukushima, Y., Bastow, K. F. and Lee, K.-H. (1993). Tannins as potent inhibitors of DNA topoisomerase II in vitro, *J. Pharm. Sci*, 82, pp. 487–492.

Khanbabaee, K. and van Ree, T. (2001). Tannins: classification and definition, *Nat. Prod. Rep.*, 18, pp. 641–649.

Khennouf, S., Benabdallah, H., Gharzouli, K., Amira, S., Ito, H., Kim, T.-H., Yoshida, T. and Gharzouli, A. (2003). Effect of tannins from *Quercus suber* and *Quercus coccifera* leaves on ethanol-induced gastric lesions in mice, *J. Agri. Food Chem.*, 51, pp. 1469–1473.

König, M., Scholz, E., Hartmann, R., Lehmann, W. and Rimpler, H. (1994). Ellagitannins and complex tannins from *Quercus petraea* bark, *J. Nat. Prod.*, 57, pp. 1411–1415.

Lampire, O., Mila, I., Raminosoa, M., Michon, V., Hervé Du Penhoat, C, Faucheur, N., Laprevote, O. and Scalbert, A. (1998). Polyphenols isolated from the bark of

Castanea sativa Mill. Chemical structure and auto-association, *Phytochemistry*, 49, pp. 623–631.

Lavigne-Cruege, V., Pons, A., Chone, X. and Dubourdieu, D. (2003). Rôle du glutathion sur l'évolution aromatique des vins blanc secs. In *Œnologie 2003*, ed. by Lonvaud-Funel, A., de Revel, G. and Darriet, P., TEC & DOC: Paris, pp. 385–388.

Lee, S.-H., Tanaka, T., Nonaka, G.-I. and Nishioka, I. (1990). Hydrolyzable tannins from *Euphorbia thymifolia*, *Phytochemistry*, 29, pp. 3621–3625.

Lee, D. F., Swinny, E. E. and Jones, G. P. (2004). NMR identification of ethyl-linked anthocyanin-flavanol pigments formed in model wine ferments, *Tetrahedron Lett.*, 45, pp. 1671–1674.

Lefeuvre, D. (2006). Réactions de substitution nucléophile et d'oxydation des ellagitannins *C*-aryl glycosides, *Doctoral Thesis*, Université Bordeaux 1, Order N°3271.

Liao, H., Cai, Y. and Haslam, E. (1992). Polyphenol interactions. Anthocyanins: co-pigmentation and colour changes in red wines, *J. Sci. Food Agric.*, 59, pp. 299–305.

Lin, T.-C., Tanaka, T., Nonaka, G.-I., Nishioka, I. and Young, T.-J. (1991). Tannins and related compounds. CVIII. Isolation and characterization of novel complex tannins (flavano-ellagitannins), anogeissinin and anogeissusins A and B, from *Anogeissus acuminata* (Roxb ex DC.) Guill. *et* Perr. var. *lanceolata* Wall ex Clarke, *Chem. Pharm. Bull.*, 39, pp. 1144–1147.

Masson, G., Puech, J.-L. and Moutounet, M. (1994). Localization of the ellagitannins in the tissues of *Quercus Robur* and *Quercus petraea* Woods, *Phytochemistry*, 37, pp. 1245–1249.

Masson, G., Moutounet, M. and Puech, J. L. (1995). Ellagitannin content of oak wood as a function of species and of sampling position in the tree, *Am. J. Enol. Vitic.*, 46, pp. 262–268.

Masson, G., Puech, J. L. and Moutounet, M. (1996). Composition chimique du bois de chêne de tonnellerie, *Bulletin de l'O. I. V.*, 785-786, pp. 634–657.

Mayer, W., Gabler, W., Riester, A. and Korger, H. (1967). Über die Gerbstoffe aus dem Holz der Edelkastanie und der Eiche, II. Die Isolierung von Castalagin, Vescalagin, Castalin und Vescalin, *Liebigs Ann. Chem.*, 707, pp. 177–181.

Mayer, W., Seitz, H. and Jochims, J. C. (1969). Über die Gerbstoffe aus dem Holz der Edelkastanie und der Eiche, IV. Die Struktur des Castalagins, *Liebigs Ann. Chem.*, 721, pp. 186–193.

Mayer, W., Seitz, H., Jochims, J. C., Schauerte, K. and Schilling, G. (1971a). Über die Gerbstoffe aus dem Holz der Edelkastanie und der Eiche, VI. Struktur des Vescalagins, *Liebigs Ann. Chem.*, 751, pp. 60–68.

Mayer, W., Kuhlmann, F. and Schilling, G. (1971b). Die Struktur des Vescalins, *Liebigs Ann. Chem.*, 747, pp. 51–59.

Mosedale, J. R. and Puech, J.-L. (1998). Wood maturation of distilled beverages, *Trends Food Sci. Technol.*, 9, pp. 95–101.

Mosedale, J. R., Feuillat, F., Baumes, R., Dupouey, J.-L., Keller, R. and Puech, J.-L. (2001). La qualité du bois des chênes de Cîteaux pour la tonnellerie. Deuxième partie: composition en extractibles du bois en liaison avec l'espèce (chêne rouvre, chêne pédonculé). Corrélations avec la morphologie foliaire et l'anatomie du bois. *Revue Française d'Œnologie*, 187, pp. 30–33.

Moutounet, M., Rabier, P., Puech, J. L. and Barillère, J. M. (1989). Analysis by HPLC of extractable substances in oak wood. Application to a Chardonnay wine, *Sciences des Aliments*, 9, pp. 35–51.

Moutounet, M., Rabier, P., Sarni, F. and Scalbert, A. (1992). Les tanins du bois de chêne. Les conditions de leur présence dans les vins, *J. Int. Sci. Vigne Vin*, Hors Série (Le Bois et la Qualité des Vins et Eaux de Vie), pp. 75–79.

Moutounet, M. and Mazauric, J.-P. (2001). L'oxygène dissous dans les vins, *Revue Française d'Œnologie*, 186, pp. 12–15.

Niemetz, R. and Gross, G. G. (2003a). Oxidation of pentagalloylglucose to the ellagitannins, tellimagrandin II, by a phenol oxidase from *Tellima grandiflora* leaves, *Phytochemistry*, 62, pp. 301–306.

Niemetz, R. and Gross, G. G. (2003b). Ellagitannin biosynthesis: Laccase-catalyzed dimerization of tellimagrandin II to cornusiin E in *Tellima grandiflora*. *Phytochemistry*, 64, pp. 1197–1201.

Nonaka, G.-I., Nishimura, H. and Nishioka, I. (1985). Tannins and related compounds. Part 26. Isolation and structures of stenophyllanins A, B, and C, novel tannins from *Quercus stenophylla*, *J. Chem. Soc., Perkin Trans. 1*, pp. 163–172.

Nonaka, G.-I., Ishimaru, K., Mihashi, K., Iwase, Y., Ageta, M. and Nishioka, I. (1988). Tannins and related compounds. LXIII. Isolation and characterization of mongolicains A and B, novel tannins from *Quercus* and *Castanopsis* species, *Chem. Pharm. Bull.*, 36, pp. 857–869.

Nonaka, G.-I., Ishimaru, K., Azuma, R., Ishimatsu, M. and Nishioka, I. (1989). Tannins and related compounds. LXXXV. Structures of novel *C*-glycosidic ellagitannins, grandinin and pterocarinins A and B, *Chem. Pharm. Bull.*, 37, pp. 2071–2077.

Nonaka, G.-I., Sakai, T., Tanaka, T., Mihashi, K. and Nishioka, I. (1990). Tannins and related compounds. XCVII. Structure revision of *C*-glycosidic ellagitannins, castalagin, vescalagin, casuarinin and stachyurin, and related hydrolyzable tannins, *Chem. Pharm. Bull.*, 38, pp. 2151–2156.

Okajima, T. (2001). Ab initio MO investigation on the reactivity for electrophilic substitution of phenolic with oxirane and aziridine, as the model compounds of binding site of mutagen, *J. Mol. Struct. (Theochem)*, 536, pp. 73–82.

Okuda, T. (2005). Systematic and health effects of chemically distinct tannins in medicinal plants, *Phytochemistry*, 66, pp. 2012–2031.

Okuda, T., Yoshida, T. and Ashida, M. (1981). Casuarictin and casuarinin, two new ellagitannins from *Casuarina stricta*, *Heterocycles*, 16, pp. 1681–1685.

Okuda, T., Yoshida, T., Yazaki, K. and Ashida, M. (1982). Casuariin, stachyurin and strictinin, new ellagitannins from *Casuarina stricta* and *Stachyurus praecox*, *Chem. Pharm. Bull.*, 30, pp. 766–769.

Okuda, T., Yoshida, T., Ashida, M. and Yazaki, K. (1983). Tannins of *Casuarina* and *Stachyurus* species. Part 1. Structures of pedunculagin, casuarictin, strictinin, casuarinin, casuariin, and stachyurin, *J. Chem. Soc., Perkin Trans. 1*, pp. 1765–1772.

Okuda, T., Hatano, T., Kaneda, T., Yoshizaki, M. and Shingu, T. (1987). Liquidambin, an ellagitannin from *Liquidambar formosana*, *Phytochemistry*, 26, pp. 2053–2055.

Okuda, T., Yoshida, T. and Hatano, T. (1993). Classification of oligomeric hydrolysable tannins and specificity of their occurrence in plants, *Phytochemistry*, 32, pp. 507-521.

Okuda, T., Yoshida, T. and Hatano, T. (1995). Hydrolyzable tannins and related polyphenols, *Prog. Chem. Org. Nat. Prod.*, 66, pp. 1–117.

Okuda, T., Yoshida, T. and Hatano, T. (2000). Correlation of oxidative transformations of hydrolyzable tannins and plant evolution, *Phytochemistry*, 55, pp. 513–529.

Peng, S., Scalbert, A. and Monties, B. (1991). Insoluble ellagitannins in *Castanea sativa* and *Quercus petraea* woods, *Phytochemistry*, 30, pp. 775–778.

Pissarra, J., Lourenco, S., Gonzales-Paramas, A. M., Mateus, N., Santos-Buelga, C., Silva, A. M. S. and de Freitas, V. (2004). Structural characterization of new malvidin 3-glucoside-catechin aryl/alkyl-linked pigments, *J. Agric. Food Chem.*, 52, pp. 5519–5526.

Pocock, K. F., Sefton, M. A. and Williams, P. J. (1994). Taste thresholds of phenolic extracts of French and American oakwood: the influence of oak phenols on wine flavor, *Am. J. Enol. Vitic.*, 45, pp. 429–434.

Pontallier, P. (1992). The intervention of oak wood in the making of great red wines, *J. Wine Res.*, 3, pp. 241–247.

Puech, J. L., Feuillat, F., Mosedale, J. R. and Puech, C. (1996). Extraction of ellagitannins from oak wood of model casks, *Vitis*, 35, pp. 211–214.

Puech, J.-L., Feuillat, F. and Mosedale, J. R. (1999a). The tannins of oak heartwood: structure, properties and their influence on wine flavor, *Am. J. Enol. Vitic.*, 50, pp. 469–478.

Puech, J. L., Mertz, C., Michon, V., Le Guernevé, C., Doco, T. and Hervé du Penhoat, C. (1999b). Evolution of castalagin and vescalagin in ethanol solution. Identification of new derivatives, *J. Agri. Food Chem.*, 47, pp. 2060–2066.

Quideau, S. and Feldman, K. S. (1996). Ellagitannin chemistry, *Chem. Rev.*, 96, pp. 475–503.

Quideau, S., Jourdes, M., Saucier, C., Glories, Y., Pardon, P. and Baudry, C. (2003). DNA Topoisomerase inhibitor acutissimin A and other flavano-ellagitannins in red wine, *Angew Chem. Int. Ed.*, 42, pp. 6012–6014.

Quideau, S., Varadinova, T., Karagiozova, D., Jourdes, M., Pardon, P., Baudry, C., Genova, P., Diakov, T. and Petrova, R. (2004). Main structural and stereochemical aspects of the antiherpetic activity of nonahydroxyterphenoyl-containing *C*-glycosidic ellagitannins, *Chem. Biodiv.*, 1, pp. 247–258.

Quideau, S., Jourdes, M., Lefeuvre, D., Montaudon, D., Saucier, C., Glories, Y., Pardon, P. and Pourquier, P. (2005). The chemistry of wine polyphenolic *C*-glycosidic ellagitannins targeting human topoisomerase II, *Chem. Eur. J.*, 11, pp. 6503–6513.

Quinn, M. K. and Singleton, V. L. (1985). Isolation and identification of ellagitannins from white oak wood and an estimation of their roles in wine, *Am. J. Enol. Vitic.*, 36, pp. 148–155.

Remy, S., Fulcrand, H., Labarbe, B., Cheynier, V., Moutounet, M. (2000). First confirmation in red wine of products resulting from direct anthocyanin-tannin reactions, *J. Sci. Food Agric.*, 80, pp. 745–751.

Ribéreau-Gayon, P. (1973). Interprétation chimique de la couleur des vins rouges, *Vitis*, 12, pp. 119–142.

Ribéreau-Gayon, P., Pontallier, P. and Glories, Y. (1983). Some interpretations of colour changes in young red wines during their conservation, *J. Sci. Food Agric.*, 34, pp. 505–516.

Salas, E., Atanasova, V., Poncet-Legrand, C., Meudec, E., Mazauric, J. P. and Cheynier, V. (2004). Demonstration of the occurence of flavanol-anthocyanin adducts in wine and wine model solutions, *Anal. Chim. Acta.*, 513, pp. 325–332.

Saucier, C., Jourdes, M., Glories, Y., Quideau, S. (2006). Extraction, detection, and quantification of flavano-ellagitannins and ethylvescalagin in a Bordeaux red wine aged in oak barrels, *J. Agric. Food Chem.*, 54, pp. 7349–7354.

Scalbert, A. and Haslam, E. (1987). Polyphenols and chemical defence of the leaves of *Quercus robur*, *Phytochemistry*, 26, pp. 3191–3195.

Scalbert, A. (1991). Antimicrobial properties of tannins, *Phytochemistry*, 30, pp. 3875–3883.

Schmidt, O. T. (1956). Gallotannine und Ellagen-gerbstoffe, *Fortschr. Chem. Org. Naturstoffe*, 13, pp. 70–136.

Singleton, V. L. (1987). Oxygen with phenols and related reactions in musts, wines, and model systems: observations and pratical implications, *Am. J. Enol. Vitic.*, 38, pp. 69–77.

Somers, T. C. (1971). The polymeric nature of wine pigments, *Phytochemistry*, 10, pp. 2175–2186.

Somers, T. C. (1990). An assesment of the "oak factor" in current winemaking practice, *The Australian Wine Research Institute Technical Review*, 67, pp. 3–10.

Tanaka, T., Kirihara, S., Nonaka, G.-i. and Nishioka, I. (1993). Tannins and related compounds. CXXIV. Five new ellagitannins, platycaryanins A, B, C, and D, and platycariin, and a new complex tannins, strobilanin, from the fruits and bark of *Platycarya strobilacea* Sieb et Zucc., and biomimetic synthesis of *C*-glycosidic ellagitannins from glucopyranose-based ellagitannins, *Chem. Pharm. Bull.*, 41, pp. 1708–1716.

Tanaka, T., Ueda, N., Shinohara, H., Nonaka, G.-I., Fujioka, T., Mihashi, K. and Kouno, I. (1996). *C*-Glycosidic ellagitannin metabolites in the heartwood of Japanese chesnut tree (*Castanea crenata* SIEB. *et* ZUCC.), *Chem. Pharm. Bull.*, 44, pp. 2236–2242.

Timberlake, C. F. and Bridle, P. (1976). Interactions between anthocyanins, phenolic compounds, and acetaldehyde and their significance in red wines, *Am. J. Enol. Vitic.*, 27, pp. 97–105.

Viriot, C., Scalbert, A., Lapierre, C. and Moutounet, M. (1993). Ellagitannins and lignins in aging of spirits in oak barrels, *J. Agric. Food Chem.*, 41, pp. 1872–1879.

Viriot, C., Scalbert, A., Hervé du Penhoat, C. L. M., Michon, V. M. F. and Moutounet, M. (1994). Ellagitannins in woods of sessile oak and sweet chestnut. Dimerization and hydrolysis during wood ageing, *Phytochemistry*, 36, pp. 1253–1260.

Vivas, N., Laguerre, M., Glories, Y., Bourgeois, G. and Vitry, C. (1995). Structure simulation of two ellagitannins from *Quercus Robur* L., *Phytochemistry*, 39, pp. 1193–1199.

Vivas, N. and Glories, Y. (1993). Les phénomènes d'oxydoréduction liés à l'élevage en barrique des vins rouges: aspects technologiques, *Revue Française d'Œnologie*, 142, pp. 33–38.

Vivas, N. and Glories, Y. (1996). Role of oak wood ellagitannins in the oxidation process of red wines during aging, *Am. J. Enol. Vitic.*, 47, pp. 103–107.

Vivas, N., Glories, Y., Bourgeois, G. and Vitry, C. (1996). Les ellagitannins de bois de coeur de différentes espèces de chênes (*Quercus* sp.) et de Châtaignier (*Castanea sativa* Mill.). Dosage dans les vins rouges élevés en barriques, *J. Sci. Tech. Tonnellerie*, 2, pp. 25–49.

Vivas, N., Laguerre, M., Pianet De Boissel, I., Vivas De Gaulejac, N. and Nonier, M.-F. (2004). Conformational interpretation of vescalagin and castalagin physicochemical properties, *J. Agric. Food Chem.*, 52, pp. 2073–2078.

Yoshida, T., Ohbayashi, H., Ishihara, K., Ohwashi, W., Haba, K., Okano, Y., Shingu, T. and Okuda, T. (1991). Tannins and related polyphenols of melastomataceous plants. I. Hydrolyzable tannins from *Tibouchina semidecandra* COGN., *Chem. Pharm. Bull.*, 39, pp. 2233–2240.

Yoshida, T., Nakata, F., Hosotani, K., Nitta, A., Okuda T. (1992). Three new complex tannins from *Melastoma malabathricum* L., *Chem. Pharm. Bull.*, 40, pp. 1727–1732.

Index